WARHOGS

WARHOGS
A History of War Profits in America

Stuart D. Brandes

THE UNIVERSITY PRESS OF KENTUCKY

Publication of this volume was made possible in part by a
grant from the National Endowment for the Humanities.

Editorial and Sales Offices: The University Press of Kentucky
663 South Limestone Street, Lexington, Kentucky 40508-4008

01   00   99   98   97   5   4   3   2   1

Library of Congress Cataloging-in-Publication Data

Brandes, Stuart D. (Stuart Dean), 1940-
    Warhogs : a history of war profits in America / Stuart D. Brandes.
        p.   cm.
    Includes bibliographical references and index.
    ISBN 0-8131-2020-9 (cloth : alk. paper)
    1. War—Economic aspects—United States—History.
2.   Profiteering—United States—History.   I. Title.
HC110.D4B716   1997
355.02—dc21   96–53139

for Polly, Carolyn, and Marcia

When the Host goeth forth against thine Enemies,
then keep thee from every wicked thing.

—Deuteronomy 23:9

Contents

Acknowledgments

During the preparation of this manuscript, I have benefited from the assistance, advice, and encouragement of many friends and colleagues. It is impossible to recognize all of them, but several deserve special mention. Scholars who generously agreed to read and correct errors in various parts of the manuscript include Fred Anderson, E. Wayne Carp, Edward M. Coffman, Robert Cuff, John E. Ferling, J. Matthew Gallman, Laurence Gelfand, Don Higginbotham, James A. Huston, Ludwell Johnson, Mark Leff, and Harold Selesky. Colleagues who offered helpful suggestions include David Huehner, James Lorence, Robert Storch, and Kerry Trask.

The manuscript could not have been completed without the generous financial support of several institutions. The Institute for Research in the Humanities, University of Wisconsin/Madison, directed by Robert Kingdon, invited me to spend a stimulating and rewarding year as a visiting fellow. The University of Wisconsin Centers permitted me to take leave and provided a sabbatical grant at a crucial stage of the project. At other stages I received helpful support from the Eleanor Roosevelt Institute and from the University of Wisconsin Graduate School.

Librarians are the unsung heroes of the scholarly world, and many have assisted me. Special recognition is due, however, to the late Gary J. Lenox, a great bibliophile, a warm friend, and the very resourceful director of the library at the University of Wisconsin Center-Rock County. At the State Historical Society of Wisconsin in Madison I received indispensable aid from Loraine Adkins, Michael Edmonds, and John Peters. Carolyn Brandes Crawford, then an undergraduate student, served efficiently as my research assistant. My wife, Polly Powrie Brandes, endured absence, tardiness, and weariness with her customary equanimity. She also read the entire manuscript, and her keen eye corrected countless errors in grammar and usage. She deserves special appreciation.

Introduction

It must give great Concern to any considerate Mind that when this whole
Continent at a vast expence of Blood & Treasure is endeavoring to establish its
Liberties, . . . there are Men among us so basely sordid as to Counteract all our
Exertions for the Sake of a little Gain.

GEORGE WASHINGTON, APPEAL TO NEW YORK (1775)

Jail? You want me to go to jail? . . . Who worked for nothin' in that war? . . . When
they work for nothin', I'll work for nothin'. Did they ship a gun or a truck outa
Detroit before they got their price? . . . Half the Goddam country is gotta go if I go!

JOE KELLER, IN ARTHUR MILLER'S *ALL MY SONS* (1947)

In the evening of what otherwise had been a bright and promising April day in
1607, a band of native warriors concealed themselves upon a spit of land
known to others as Virginia. After secretly observing the strangers who had
rowed ashore that morning, they decided, correctly as it turned out, that the
uninvited callers were not simply casual visitors but were instead the dangerous vanguard of an occupying force. The bowmen crept within close range of
the trespassers, loosed their arrows, and stained the sand with English blood.
The English detachment answered with a fusillade of musketry and then pru-
dently withdrew to the safety of their squadron. American history, defined as
commencing with the arrival of the first permanent English settlers in North
America, was only a few hours old, but it had already experienced its first military engagement. Two Englishmen had been wounded.[1]

Over the next four centuries, countless skirmishes, raids, invasions, and
similar acts of war would follow. Millions of contestants would be engaged,
and costs and casualties would rise geometrically. "War" in its many forms—
aggressive or defensive, Revolutionary or Civil, colonial or general, declared or
undeclared, low-intensity or limited, hot or cold—was fought, threatened, or, at
the very least, prepared for almost continuously as the American story un-
folded. Mars called on the North American community regularly, and each time
he exacted a terrible toll in blood and treasure.

When English settlers first fought Indian warriors along the Atlantic coast-
line, both sides maintained firmly established military customs, derived from
centuries of experience, about why and how to engage in warfare. Although
both had goals that were broadly economic, Europeans had long fought for
purposes that were, in varying degree, more explicitly monetary. Romans were

skilled in the art of plunder, and looting and booty remained common features of the Middle Ages. The capturing of prisoners for ransom was a well developed industry in Europe.

Both sides soon realized that a very different enemy who fought in different ways required a modification of strategy and tactics. An American way of warfare must necessarily develop. But although it was soon apparent that New World conditions would dictate a new way of fighting, it was less obvious (but equally inevitable) that these new circumstances would change the way that soldiers would be supplied with the sinews of war. The military customs of feudal Europe must gradually give way to more American and ultimately more democratic means of waging warfare.

While much has been written about the evolution of American military strategy and tactics, economic mobilization in support of the troops has received comparatively little attention. This book addresses three aspects of this topic: the form and appearance of American traditions of mobilization, the ethical problems associated with mobilization, and the growth and success of administrative procedures intended to ensure that mobilization is carried out with efficiency and equity.

The first issue considered is how Americans have striven to distribute the burden of mobilizing economic resources for military purposes differently from their European ancestors. Since the first colonial settlements, Americans have never fully agreed on how, if at all, vital military supplies could be procured fairly and practically under American conditions. Nevertheless, they very early acknowledged that the ways of Europe by which feudal lords richly benefited from war while the lower and middle classes sacrificed must cease. The old military system must certainly be transformed, which they agreed would be a welcome step, but a question lingered: Could the New World find a way to share war's economic burden more equitably? Could the economic cost of war be distributed among the population so that no one paid unjustly and no one acquired undue gains? In short, could war be financed in an *American* way?

There were several reasons why an American tradition of funding the cost of war must necessarily appear. In contrast to populous Europe, military participation in America generally engaged a broader spectrum of the community. America possessed no monarch, no feudal nobility, no powerful legislature, no crown official, and no standing army capable of compelling ample financial support for military service. Instead, the American population was widely dispersed, prosperous, literate, and independent. In America, more so than in Europe, wars could be neither fought nor funded without substantial public consent. If wars were to be fought by Americans, at the very least a limited consensus in support of both a war's aims and its means was required. Americans who have been critical of military purposes have often found it more convenient to censure the price of a conflict, as measured in bloodshed or in appropriations or both, than to assail its aims.

The second purpose of this book is to examine the history of the American struggle with a fundamental moral problem: Can one rightly profit during wartime? If this question could be answered affirmatively, as it eventually would be, a corollary issue appears: how much or in what degree can one rightly profit during a war disaster? These are enduring problems, which like war itself have threaded their way through American history. While they are by no means a central issue in American history, like Indian-white relations, the evolution of participatory democracy, sectionalism, or the rise of the welfare state, the story of the struggle against war profits is as revealing as it is enduring.

As the American community prospered, its people established distinctions about how wealth could be ethically acquired. There were ways of accumulating wealth that were recognized as acceptable and others that were regarded as illegitimate. Americans have always generally recognized that wealth extracted from the land through agriculture or animal husbandry or earned from fair competition in the commercial marketplace was legitimate. Conversely, wealth obtained by theft or fraud has always been stigmatized. But between these discrete categories, an island of middle ground gradually formed. War profits have normally been regarded as falling within a special category of semilegitimate wealth—although not specifically illegal, war profits are not entirely ethical, either. This separate, odd classification affords an opportunity to deepen the understanding of how Americans have measured the respectability of wealth accumulation.

Americans have generally honored wealth obtained slowly above that obtained quickly. They have also normally valued earnings derived from the manufacture of tangible articles more than invisible earnings acquired from inventory appreciation. Earnings gained in supplying civilian markets have usually commanded more respect than those obtained in defense contracting. Because war wealth has often been acquired quickly, through price inflation or through the manufacture and sales of weapons of destruction, it has often borne a streak of impropriety. Neither honest wages nor ill-gotten gains, war wealth has generally displayed a gray veneer.

Several ethical problems associated with the accumulation of wealth during war-time help to explain this ambiguity. Is it proper for an individual to gain or enlarge a private fortune while the community at large faces great peril and while some of its members are sacrificing life and limb for its protection? Is that kind of wealth legitimate? If so, is all of it legitimate or only part of it? Some wartime gains were obtained by methods that would be illegitimate in peacetime, whereas others were ethically indeterminate. What of wealth obtained by stealing from the enemy, as privateers did? Was that wealth, which would be illegitimate in peacetime, honorable in war? In short, what role did extraordinary wartime circumstances play in establishing the propriety of acquired wealth?

War was (and is) an exceptional occurrence which had (and has) a peculiar capacity to distort the wealth curve so that those whom society deems least deserving can become economically advantaged and vice versa. It is well recognized that there is a stochastic or random element that affects wealth distribution. The political philosopher Hal R. Varian has described this factor as "acts of God, depressions, [and] accumulations of fortunes that may be transferred to new generations."[2] In other words, both natural and social events produce random but unintended changes in wealth holding. A society that seeks to distribute its assets in a just and equitable fashion cannot ignore stochastic elements. (The economist Lester C. Thurow estimates that they account for 70 to 80 percent of the variance in individual earnings.)[3] War is one of these random elements, no less than drought or earthquake or epidemic. That horror, death, and destruction fall on some while others profit is a cruel injustice.

This wrong has seldom gone unnoticed. The recurring controversy over war profits serves to illustrate the perennial American concern about the fairness of the race to wealth. By skewing the peacetime distribution of wealth unfairly, wartime circumstances reveal an ability to defeat the search for economic justice. Although Americans have generally shared a fundamental, perhaps excessive, trust in the fairness of the race, there has also existed a concomitant, deep-seated suspicion of the tactics of some of the competitors. Warfare has an odious but undeniable ability to negate an equitable distribution of life's blessings, one of which is wealth.

Behind many, perhaps most, of the public controversies that serve to define the American past lurk subtle disagreements about wealth distribution. Despite its importance, Americans generally choose not to address the question directly, preferring that it creep into public discourse indirectly.[4] Following each of America's major wars, the distortion of wealth distribution introduced by the contest has become a subject of earnest discussion, varied in form but always passionate. Usually the issue has been a matter of political debate, but another avenue by which this issue has gained entry into public discourse is cultural. The legacy of wartime gains forms a minor but tenacious theme in American literature. Although not a main topic of this study, the literary dimension receives some discussion.

There has been little or no precise consensus about what constitutes a just and honorable division of wealth. This study does not presume to answer the question of what composes an ethical wealth distribution, which is a complicated, controversial, and perhaps impossible enterprise. Even determining whether wealth distribution has changed, and why it has changed, is an exceptionally difficult problem in itself, although recent works by the economist Peter McClelland and the historian James L. Huston testify to the growing interest of social scientists in this question.[5] Nevertheless, normative judgments about the distribution of wealth have been made regularly, both explicitly and implicitly.

Although the importance of the pattern of wealth distribution has waxed and waned, Americans have never been shy about government intervention in economic matters in time of war. During wartime, wealth distribution has seldom been unimportant, for that is when public issues have direct, compelling importance for the everyday life of ordinary people. Nevertheless, the experience of war has attracted scant attention from scholars interested in wealth distribution. The goal here is to trace the continuing American disapproval of war profiteering and thus to establish that Americans have consistently, although sometimes furtively, judged that an equitable distribution of wealth must take into account how the wealth was acquired.

Successive generations of Americans have grasped this reality in varying degrees. As they grappled with the problem of defining, preventing, or restricting war profits (with varying success), they devised numerous ways of implementing their ethical ideas about shared sacrifice. The third purpose of this book is to outline the development of the administrative science of managing war profits. The national will to control war profits has been constant (in varying degree), but changing circumstances have required the fabrication of new methods for expressing it. The need for these institutions was a subordinate (but not insignificant) reason for the rise of the modern state.

The problem of controlling war profits offers an excellent opportunity to sketch the evolution and maturation of the administrative procedures employed to put into effect the practical and ethical principles of war leadership. In colonial America, war profiteers were exercised by moral exhortation, price controls, and legal intervention in the form of fines and incarceration. The Revolutionaries hoped by appealing to civic virtue to win their independence without gain by either soldier or civilian, but found to their chagrin that it was necessary to add open competitive bidding as a protection against avarice. The Federalists built state-owned armories so as to obtain arms on a nonprofit basis, but many Jeffersonians thought even this modest measure unwise. The Jacksonians found that laissez-faire might have its limits if it meant large profits on supplies for the war with Mexico.

During the Civil War Lincolnian nationalists supplemented these procedures by appeals for patriotic sacrifice, by an income tax, and by financial payments to citizens who reported fraud. Late-nineteenth-century liberals fought profits on arms sales to France in 1871, on naval armorplate in the 1890s, and on supplies for the Spanish-American War. Progressives railed against profits on powder sold to Britain and France before the United States entered the Great War and against price increases in basic commodities after Americans joined the alliance. In the 1920s and 1930s the question of how to restrict war profits seemed nearly as important as how to prevent war itself, and both Hooverian associationalists and Rooseveltian New Dealers anguished over how to contain profiteering. In the twentieth century, the administrative art of controlling war profits included government-owned weapons plants, legal profit restrictions, civilian audits of military contracting, contract renegotiation, and excess profits taxes. The economic managers of World War II possessed both the finest arsenal of weapons against war profits and the greatest national resolution to use them, but they too could report only qualified success.

A legion of American luminaries have agonized over war profits, including General George Washington and Presidents Theodore Roosevelt, Woodrow Wilson, Warren Harding, Herbert Hoover, and Franklin D. Roosevelt. In this respect as in others, Washington and Franklin Roosevelt were the greatest leaders, but even the less highly regarded Dwight D. Eisenhower was impelled in part by a concern about war profiteering when he circulated his famous warning against the "military-industrial complex." (Eisenhower was himself a veteran of World War I and thus a member of the generational cohort most concerned about war profits. As a junior staff officer in 1930, Eisenhower became directly engaged in the war profiteering controversy. He was assigned to the Hoover administration's War Policies Commission as chief military aide.) In 1967 he intimated that his specific ideas about a military-industrial complex—which he defined generally as "the role of pressure groups in every area of our social and economic life"—formed during his presidency of Columbia University from 1948 to 1951. "The aggressive demands of various groups and special interests, callous or selfish, or even well-intentioned," Eisenhower contended, "contradicted the American tradition that no part of our country should prosper except as the whole of America prospered."[6]

There are two main difficulties with studying war profits. The first hazard is its scope—what can be gained in comprehensiveness can be lost in depth. To minimize this problem, this study concentrates on the United States and excludes Europe.[7] The four major American military mobilizations—the Revolution, the Civil War, and World Wars I and II—receive primary attention. The smaller conflicts—the colonial wars, the Indian campaigns, the War of 1812, the Mexican War, and the Spanish-American War—are less emphasized. The "limited" wars of the cold war period, the Korean, Vietnam, and Persian Gulf conflicts, remain open for another study.

Also omitted is the history of profiteering by enemies of the dominant Anglo-American culture: Indians, French, British Loyalists, and Mexicans. There is some reference to the profiteering problems of the Confederacy, but this subject is not fully developed. The term used in the title, warhog, applied solely to men, and the present investigation applies mostly to them. Although there was a concordant term for women, warsow,[8] no evidence has been found of how female profiteers differed from their male counterparts (if at all). The gender dimension remains unexplored, and there is no known ethnic dimension.

The second hazard is definitional. The word profiteering is disturbingly imprecise and nearly as pejorative as the term military-industrial complex.

Although it would be preferable to avoid such a slippery term, *profiteering* is in common parlance and has been so since the early twentieth century.[9] It conveys a generally recognized (albeit vague) meaning. It can be defined with reasonable precision, but nevertheless must be used with considerable care. The term selected for the title, *warhog*, originated during the Great War. It is a successor to (but less archaic than) eighteenth- and nineteenth-century expressions such as *engrosser* or *forestaller*. It is less well known, less sweeping, and therefore less limiting than *profiteer*.

Nevertheless, *profiteering* deserves a definition. It may be defined as a gain in economic well-being obtained as a result of military conflict. The gain is usually monetary, but it may also come in the form of appreciated stock prices or payment in kind, such as the acquisition of government facilities.[10] The assets are usually acquired during wartime, but the term *wartime* presents definitional problems of its own. In colonial America the delineation between war and peace was less distinct than in later periods.[11] In the twentieth century the greatest "war" profits were obtained during the preparedness period before the United States entered the Great War. The postwar reconversion immediately after World War II was also a time of substantial economic gain.

Normally an adjective is attached to the gain, as in "fair profit," "reasonable profit," or "just profit." The opposite case is referred to as an "unreasonable profit," "excessive profit," or "exorbitant profit." Thus a distinction is conventionally drawn between a remuneration that is normal, necessary, and customary and one that is unusual, unwarranted, and abnormal. This differentiation invariably conveys an ethical judgment, implying that the level of benefit obtained exceeds that which is proper. As in all ethical judgments, war profiteering is more easily defined conceptually than empirically.

War profiteering has appeared in many different forms, and for centuries Americans have struggled to find a satisfactory way of identifying them in practice. The best-known type occurs when a vendor raises his prices to take advantage of a seller's market arising from the sectoral shifts that develop as a result of military purchasing. For the purposes of this study, this type of profiteering is termed "extortionate pricing" or, more colloquially, "price gouging." It involves an increase in profit above a prewar standard; profit rates that are above a general or industrial standard; profits that are unrelated to levels of risk or investment; and profits that exhibit evidence of unusual market power. More than one of these factors may be present.[12]

After the Pequot War of 1636-37, a Puritan divine offered what was undoubtedly the first American attempt at defining this kind of price gouging. According to John Cotton, a merchant might raise his price "when there is a scarcity of the commodity," but otherwise he must maintain the "current price." In modern terms, Cotton meant that "demand-pull" inflation associated with wartime conditions must be resisted by an ethical merchant; to raise a wartime price above the going peacetime rate for personal benefit was improper. On the other hand, Cotton saw "cost-push" inflation as an irresistible Act of God. Cotton may be excused for failing to realize that the difference between these two types of inflation is often blurry.[13]

In 1776 a new attempt at precision emerged. The Secret Committee of Trade of the Continental Congress declared itself "willing to allow what might be a reasonable compensation without being willing to submit to extortion." But what was a "reasonable compensation"? The Secret Committee defined "reasonable" as a 2.5 percent markup on each sale. This may or may not have been the first attempt at a quantitative definition of reasonableness, but it established the known record for the all-time low figure.[14]

The quantitative definition varied continually, but after rising later in the Revolution it followed a variable but revealing trendline. In 1917 the Progressive-oriented American Committee on War Finance made perhaps the first distinction on wartime profits made on weaponry—limited to 3.5 percent—and on agricultural commodities, where 6 percent was acceptable. The American Legion thought that 7 percent was allowable, splitting the difference between railroads, which were making 6 percent, and public utilities, which were at 8 percent. Definitions of acceptable profits (on sales after costs) used during the Progressive era by the Congress and by the Army Ordnance Corps were 5 percent and 10 percent, respectively. Military officers generally approved of somewhat higher figures than did civilian managers; the record for the lowest known figure endorsed by a military leader was 6 percent (by General Douglas MacArthur, 1935). In 1934 a congressional compromise limited nominal profits on warships to 10 percent but set the limit on warplanes at 12 percent. The discrepancy was cleared up in 1940 by cutting both to 8 percent.[15]

In the early twentieth century, the giants of American politics took up the quest for a working definition of reasonableness. Theodore Roosevelt, who was part-militarist and part-Progressive, expressed that duality well. He wrote, "It is criminal to halt the work of building the Navy or fitting out our training camps because of refusal to allow a fair profit to the businessman who alone can do the work speedily and effectively; and it is equally mischievous not to put a stop to the making of unearned and improper fortunes out of the war." Herbert Hoover, serving as food administrator in 1917, advised his superior, Woodrow Wilson, that the law prevents "unjust, unreasonable, unfair . . . profit." Hoover was perplexed by the predicament of applying this vague rule to practical situations, which, he warned, "is arising daily." Hoover proposed as a definition of these terms "any profit in excess of the normal pre-war average profit of that business and place where free, competitive conditions existed." Wilson responded by defining a "just price" as one that would maintain efficiency, provide a living for those who conduct military business, and enable them to pay good wages and to expand their enterprises. With customary eloquence (but with misplaced confidence) Wilson declared: "No true patriot will permit himself . . . to grow rich by the shedding of [soldiers'] blood. He will give as freely and with as unstinting self-sacrifice as they. When they are giving their lives will he not give at least his money?"[16]

Less frequently, critics of munitions profits quantified what they meant by excessive or unreasonable gains. In 1778 a Revolutionary patriot denounced a vendor for charging "a most exorbitant price . . . nearly one quarter more than the market price." He advised his commander that "if you are not under a pressing necessity for provisions, prudence & policy dictate that we should discover some indifference in the purchase of this Cargo 'till they fall to a more reasonable price." Another principled compatriot denounced a "Clothier [who] has placed three hundred Per Cent upon the Stirling Cost. This is totally repugnant to the Act of Congress. And to be an Accessory to an Act deviating from the Decisions of that August Body might be Construed into high Treason."[17]

In 1918 Robert S. Brookings, a business executive serving on the War Industries Board, said that a 25-30 percent after-tax return on capital was improper. In 1935 Senator Bennett Champ Clark of Missouri affirmed that a 30-35 percent markup on the cost of constructing a warship was "excessive." In 1924 an official with the Department of Justice whose duty was to resolve disputes on military contracts left over from the late war submitted perhaps the most perfectly rounded of profiteering's many circular definitions. He defined an "unconscionable consideration" as "that which shocks the conscience of the chancellor." "But," lamented the befuddled bureaucrat, "such definition is most elusive when one seeks to apply it to a particular case."[18]

Although price gouging has been the most consistently recurrent form of profiteering, there have been many others. Throughout American history, contractors with a keen if shady view of the main chance have been engaged in fraudulent dealings with the nation's military forces. Matériel furnished to the troops has often been of degraded quality, sometimes necessarily, sometimes not. In the seventeenth and eighteenth centuries, wartime gains could also be obtained by plunder, privateering, ransom, and trading with the enemy. During the Civil War, stock speculation and the price and quality of transportation were special concerns. In the twentieth century, many of these types continued, but large corporate salaries, commodity speculation, excessive construction costs for military posts, black market trading, evasion of price controls, and gains from postwar reconversion sparked new controversies.

Despite the great longevity and numbing regularity of American military action, warfare remains fortunately only a subordinate experience in American history. Nevertheless, the story of how Americans have endeavored during their many wars to assemble resources equitably in the American way, to define the ethical questions of mobilization, and to manage economic sacrifice responsibly is a revealing aspect of the American struggle to achieve a just society.

1

A Provoking Evil

Engage in War, only in a just Cause. Not to gratify Pride, Avarice and Ambition, to increase and enlarge our Possessions.
REVEREND WILLIAM WILLIAMS, BOSTON (1737)

Elizabethan England, poised to plant English culture in the North American wilderness, presented a striking example of the interrelationship between war and wealth. England's first attempts at colonization confronted twin challenges. One was how to assemble the enormous financial resources necessary to sustain a colony. The other was how to mobilize and transport the military forces necessary to meet powerful adversaries—French, Dutch, and, most importantly, Indians. Raleigh's Roanoke failed surely because of lack of money and probably because of military weakness. The English settlements established in Virginia and Massachusetts after 1600 survived (if only narrowly) because they were able to find the means of prevailing against one military challenge after another. They did so by adapting the militia customs of England to a system of collective self-defense suited to the New World. After a century of experience in America, the military policy of collective self-defense was gradually replaced by a policy based on individual self-interest.[1]

In the early seventeenth century, England was less militaristic than most of its European rivals, yet its eventual supremacy as a colonial power rested upon triumph in a series of highly successful wars. While the Continental powers were experiencing a military revolution that would produce professional armies, England continued to rely on a centuries-old tradition that emphasized amateur soldiery.[2] Elizabethan troops, scoffed a military historian, were unprofessional: "very unsatisfactory gatherings of haphazard material," with a "terribly deficient organization," and whose training was "usually contemptible." For the student of military tactics, he maintained, "the whole period is singularly dull."[3]

For the student of the relationship between war and wealth, however, American colonial history has much to offer. The American experience was a blend of war and peace, of soldier and civilian, of licit and illicit gain. The English settlers imported from the Old Country established ways of financing war and transformed them to suit the colonial conditions. They created American versions of plunder and ransom, of fraud and smuggling, and of deep conflict between public authorities and defense contractors.

The American military tradition sprang from the English penchant for a citizen army. The English preference for amateur soldiers is so ancient that it has baffled scholars seeking to identify its origin. There is agreement, however, that the tradition was in existence long before the Norman Conquest.[4] The English dislike of mercenaries and the high taxes associated with them was one of the grievances against King John that led to the signing of the Magna Carta.[5] In place of a professional army, England developed a tradition of universal military service based on the concept of a nation-at-arms. Although this general obligation to accept military service was better observed in theory than in practice, it nevertheless became the foundation upon which the seventeenth-century American colonists organized their defense. Tracing its lineage back to the English fyrd, the colonial militia was an institution better suited to the requirements of colonial defense than it was to defending the home island.

By 1400 the original concept of the English militia had been modified. It was expensive and inconvenient to train the entire male population, there was no serious danger of invasion, and militiamen objected on constitutional grounds to service abroad. As a result only a minority of the population received military training, although this training was supposed to be more intensive. The older idea of the militia was replaced by a new institution, the trained bands or simply "trainbands," which became the immediate ancestor of the American colonial militia.[6]

In England, the trainbands were militia companies whose members had other occupations but who would form occasionally (usually four times a year) and would perform simple military maneuvers with crude weapons. They functioned primarily as a law-and-order body rather than as a true fighting force. During the eighteenth century Britain turned toward a professionalized army, but the colonies lagged. In America, where the primary opponents were Indians who might strike at any time and then melt into the forest but who seldom conducted a prolonged engagement, the trainbands were a modestly successful arrangement. England was unwilling to spend the money to protect the colonists from Indians, so the colonists had no choice but to defend themselves. Because there were so few colonists and because they often lived separately, it was vital that each of them become a citizen-soldier. Only New Netherlands and New Sweden relied primarily upon professional soldiers. Of the English colonies, only Pennsylvania, the Quaker colony, omitted a militia (until 1755).[7]

Although seventeenth-century American colonies lacked a professional military, and although many of them were settled as religious societies, they were nevertheless quite warlike (with the exception of the Quakers). In 1607, of the 104 settlers in the new colony of Virginia, six held the military rank of captain. This military background proved useful immediately, as Indian arrows struck the first English targets on the very first day in 1607 that white men arrived in Virginia. When in 1628 the General Court of the Massachusetts Bay Company laid plans for the new settlement that would harbor the Great Migration, it hired Samuel Sharpe at the rate of £10 for three years "to have oversight of the ordnance to bee planted in the ffort to bee built upon the plantation and what ells may concerne artilery bisines to geeve his advise in." The Bay Company also hired Thomas Graves, a miner, and assigned him the duty of discovering lead and saltpeter deposits. It is arguable, therefore, that a close working relationship between civil and military authorities originated in England before colonization and was transported across the ocean in the first ships.[8]

Englishmen had few qualms about fighting. They believed that most wars were fought for selfish ends or, as the Reverend Roger Williams put it, "for greater dishes and bowls of porridge." But they distinguished between wars for avaricious purposes and "just" wars. The latter included both Indian wars and defense against the king's troops. In 1634, when the Massachusetts Bay Colony was still very new, it prepared defenses against a possible invasion by government forces seeking to revoke its charter, as recommended by the Puritan archenemy, the Anglican bishop William Laud. In 1637 the first great Puritan-Indian conflict occurred (the Pequot War) and Roger Williams, who was certainly no friend of Massachusetts, deemed it a justifiable "defensive war." In 1675 the Reverend Increase Mather similarly justified King Philip's War, which he said had been provoked by the Indians' "insolencies, and outrages, Murthering many persons, and burning their Houses in Sundry Plantations."[9]

The military discipline of the trainband resembled the religious discipline that Puritans sought to inculcate into their community. In 1653 a dedicated and militant Puritan prophetically exhorted his companions to "see that with all dilligence you incourage every Souldier-like Spirit among you, for the Lord Christ intends to atchieve greater matters by this little handfull then the World is aware of." A later cleric likened soldiers to Puritans because both shunned "Idleness and Sloth, Carelessness and Inactivity." And in 1694 the Reverend Moses Fiske went so far as to claim that "'tis utterly impossible to be christians without [war]; they must fight their way through the wilderness of this world, or lose Heaven." New England clergy opened and closed militia training days with prayer, voted in militia elections, delivered prebattle sermons, and accompanied militiamen into combat.[10]

Of course, not all colonists were Puritans. But English settlers shared similar views on the morality of war. The Reverend Gilbert Tennant, Pennsylvania's famous Presbyterian leader, called for military preparedness by reminding his followers, "The Lord *is* a man of War," quoting Exodus 15:3. Only the Quakers demurred, and even they were willing to compromise their pacifist opposition to military service. Quakers consistently appropriated money for war, but they would claim that the funds were only "for the King's Use," "for the Governor's Use," "for the Use of the Crown," or some other such euphemism. On one occasion Quakers even boasted that they had contributed more money for defense than "what is given in some neighboring colonies."[11]

By modern standards, the American militia was an extremely crude system of defense. When it was established in the early seventeenth century, the long bow, which had been England's national weapon for centuries, had only recently been superseded by firearms. Guns were scarcely an improvement in killing ability over bows, but they made noise and gave off clouds of smoke that temporarily frightened Indians. The lance remained in use (due to its cheap price) until 1676, when its worthlessness in the wilderness was finally established. Firearms were approximately ten times as expensive as wooden weapons, and their introduction coincided with the first examples in American history of fraudulent practices in defense procurement.[12]

The American militia was a simple social system. Following English precedent, colonial militia laws customarily required that each man of military age report on muster day equipped with his own musket, balls, and powder. The only responsibility resting upon colonial authorities was to provide a few cannon and to ensure that each militiaman was adequately equipped. The latter was more difficult than it seemed. Militiamen frequently failed to provide themselves with firearms, and if they did so their guns were often broken or obsolescent. Their supply of powder and balls was commonly neither fresh nor ample.

Firearms were expensive. In Virginia a settler was expected to furnish himself with a musket, sword, bandoleer, twenty pounds of powder, and sixty pounds of shot. This equipment constituted about one-fourth of the total cost of the settler's tools. Fathers were expected to furnish muskets for their military-age sons, and masters were required to arm their servants. Since colonial families were large and masters often had several servants, expenses could mount quickly. Sometimes weapons were available only from the organizers of a colony. Prices in the American wilderness were normally higher than in England, but bitterness about costs was nearly unavoidable when settlers were dependent upon a single provider for a major part of their equipment. Faced with these considerable expenses, refusal to pay on grounds of poverty was common.[13] But how could colonial leaders determine whether this excuse was legitimate? And if the reason was sound, how could arms be provided, since to leave men unarmed was extremely dangerous?

Colonial authorities used a number of methods to ensure that militia requirements were met. The most popular technique, derived from English practice, was to assess a fine for noncompliance. Imposing fines was a simple matter, but collecting them in the cash-starved wilderness was difficult. Crafty militiamen sometimes passed arms back and forth so that it would appear to an unwary inspector that everyone had one. Inspectors frequently shirked their duty by failing to impose fines upon their friends and neighbors. Rhode Island's sad experience with this problem was typical. In 1658 the General Court attempted to clamp down by ordering the "Town serjant of each Towne . . . to pay all the fines hee neglects to take." This reform failed, so eight years later the Court turned over authority to magistrates, also threatening that "in case of opposition" the magistrates could "require and take sufficient ayde" to collect the fines. But by 1680 the Court was still attempting to fine officers who failed to collect fines. North Carolina tried the opposite approach, offering a 5 percent commission to collectors, but this led to corrupt sheriffs. Maryland allowed the captain of the militia to make house inspections, and if muskets were not present, he was to sell them to the soldier at not more than 100 percent profit.[14] But nothing worked well.

The earliest approach to this problem was to declare that poverty was no excuse. In case a person pleaded inability to contribute to defense on grounds of poverty, his magistrate was to provide him with a musket at public expense and then bind him out as a servant until he had repaid the cost. But this did not answer the questions of whether the man was truly indigent, what wage he should receive while working off the payment, or how the town was to obtain the money to buy him the gun in the first place. To avoid cheating on the question of indigency or wages, committees to decide the issue were established. Funds for initial purchase of weapons were usually obtained from militia fines, but Rhode Island granted an annual subsidy of nine shillings per company to provide arms for the poor or, as the law read, "for the encouragement of the meaner sort." Since many of the poor were recently released servants, by 1700 some colonies began to require that masters provide ex-servants with a musket upon their release from indenture.[15] This was an early example of the application of the principle of providing for defense according to the ability to pay.

Assurance that guns were provided solved only one aspect of the general question of defense. There were associated problems of providing powder and ammunition and ensuring that weapons were maintained in good repair. Again, fines were imposed for failure to maintain on hand the minimum required supply of powder (usually one or two pounds), and again the system was difficult to enforce as powder was extremely scarce in the colonies and was needed for hunting. Musket balls were so dear that they were used as coin. In 1635 the Court of Assistants of the Massachusetts Bay Colony declared them to be legal tender at the rate of one farthing each. Anticipating the problems of using a soft metal such as lead for money, the Court specified that each ball must be "of a full boare." Musket balls shaved to substandard diameter must have been among the first instances—there would be many more—of profiteering by degrading the quality of military goods. Keeping an adequate supply of ammunition was so important, moreover, that in 1632 two Massachusetts men were fined by the court "for their wastefull expense of powder and shot." Alexander Miller and John Wipple thus had the dubious distinction of becoming the first defense providers in American history to be prosecuted for waste.[16]

If guns, powder, and ammunition were available, there was still no certainty that militiamen were adequately equipped. The reason was that colonial firearms were chronically in poor repair. The first muskets shipped to the colonies were obsolete matchlock models. These were slow to load, heavy, and inaccurate and the match went out in wind or rain. The smoke from the match could reveal the soldier's position in daylight, and so could the glow of the match at night.[17] Matchlock weapons were soon replaced by doglocks and flintlocks, which were more suitable for Indian fighting. But repairing them in the wilderness was difficult. There were few gunsmiths present, and they charged a premium price for their work.

Colonial leaders struggled constantly to control gunsmiths. Establishing a reputation that would plague defense contractors for centuries, colonial gunsmiths were charged with being slow to complete their assignments, doing poor work, and charging high prices.[18] They were even suspected of preferring to fix muskets owned by Indians. The Indians were quick to obtain firearms (usually by illicit trade), but they were much slower in learning to repair them. The colonial leaders were well aware that the ability to repair malfunctioning weapons was (then as now) a major military advantage, and they were properly concerned that their gunsmiths might compromise it. During the Pequot War, the Court of Assistants of Massachusetts Bay felt it necessary to order "that no man within this jurisdiction shall, directly or indirectly, amend, repair, or cause to be amended or repaired, any gun, small or greate belonging to any Indian."[19]

The tone of other colonial orders was similarly accusatory. In 1650 Rhode Island ordered that the colony's six gunsmiths, "all excuses sett aparte, shall mend and make all lockes, stocks and pieces . . . presented to them, for just and suitable satisfaction in hand payed, without delay, under the penalty of ten pounds." Massachusetts similarly directed that "all smiths and other workmen are to attend the repairing of armes and other necessaries and may not refuse such pay as the country affords." But perhaps South Carolina's experience with Thomas Archcraft, a local gunsmith, is most indicative of early governmental helplessness in the face of a private technological monopoly.

In September 1671, the Grand Council of South Carolina commanded Archcraft to stop manufacturing Indian hatchets without a license, to cease all other work, and to repair all arms brought to him. A month later, the council threatened that if he continued to "disobey or unnecessarily prolong the performance of such directions" he would be "severely punished." Nine months passed before he was again ordered "imediatly to sett about the [repairs] and that he doe interpose noe other worke till such armes be compleatly fixed and finished." A month after that, the council complained that "the Smithes worke . . . is held at such unreasonable rates that many inhabitants in this Province are alltogether disapoynted in getting their Armes well fixed and in good Order to the great danger of this settlement." Finally, after five more months, the council lost its temper. In December 1672, it moved that "for as much as complaints doe come unto the Grand Councill dayly against Thomas Archcraft . . . [that] he either, alltogether neglects, the mending of them, or else returnes them as ill, sometimes worse then when he received them . . . and for as much as the said Archcraft now hath many Armes lying by him as use-less . . . it is therefore ordered the said Archcraft remaine in the Marshall cus-tody, untill he has sufficiently fixed up all the defective Armes he hath received into his care."[20] Archcraft thus became the first American defense contractor jailed for poor workmanship and nonfulfillment of a contract.

One of the most common forms of military fraud was cheating on military payrolls and provisions for the troops. This practice was pervasive in Europe, frequent in England, and not unknown in America. The earliest evidence ap-peared during the Pequot War, when some Connecticut warriors evidently began to observe a hoary old English tradition of stealing government-provided supplies at the conclusion of a campaign. It was easy to claim that a weapon or other utensil was lost in battle, when in actuality it had been sold (sometimes to the enemy), gambled away, traded to the captain in return for a discharge, or simply secreted home. The tightfisted Queen Elizabeth had gone so far as to proclaim the death penalty for such pilfering, but to no avail. The Connecticut General Court was unwilling to go that far, but in 1637 it did direct "that if there be any armor, guns, swords, belts, bandoleers, kettles, pots, tools, or anything else that belongs to the commonwealth, that were lost, landed or leafte (behind) . . . they are to be delivered into the hands of the Con-stables . . . and if there be any things found in any man's house or custody . . . they shall be subject to the censure of the Courte for their concealing."[21]

Censure evidently was not strong enough, for in 1704 Connecticut imposed a stiff fine upon soldiers who sold or embezzled ammunition, and as late as 1747 the colony was still trying to identify its embezzlers. Massachusetts sol-diers who claimed to have lost ammunition were ordered to ride the wooden horse for an hour and to have the cost of replacement deducted from their pay. Soldiers who reported that they had lost their musket in action were not to be discharged but were to remain in service at no pay until the colony was com-pensated for the loss. During Queen Anne's War, England sent arms to Virginia and Maryland for their defense, but according to the complaints of crown of-ficials, the colonists stole most of them.[22]

Military payrolls were tempting targets for officers with larcenous minds. In the first place, some people wanted to be on the muster list, whereas others wanted to avoid service, and both were willing to pay for the privilege. One lucrative practice was to sell commissions in British army units operating in the colonies, often to the dismay of better qualified Americans. Vendible commissions were customary in England, and any rank up to colonel could be bought or sold. Since holding a military appointment offered opportunity for profit beyond the base salary, purchase of a commission could be a good investment, as some Americans discovered. One such source of profit was the sale of false certificates of medical disability that disqualified the bearer from military service. Although there were hefty fines against this bribery, in 1702 the General Court of Massachusetts Bay observed that "divers persons fit and able for service, by corrupt and fallacious means do obtain such certificates."[23]

A more common device, one that was very widespread in Europe, was falsifying or padding the military payroll. This was so common in the British army that a word was even coined to describe the practice. When a captain continued to collect and pocket a soldier's wages well after his death or desertion, the soldier was called a "dead-pay." The practice was so widespread that it even became accepted; Elizabethan captains negotiated with the government to raise the limit of dead-pays from 6 to 10 percent of a company's strength. This form of fraud was less frequent in America than in England, but not absent. Americans sometimes enlisted blind, deaf, or otherwise disabled men into military service to fulfill their quota of troops.[24]

The most serious scandal involving payroll padding in the American colonies occurred over several years in New York. Four companies of regular troops were stationed on the New York frontier from 1690 to 1705. The full complement of each company was ninety-nine men, but in reality only about half of them were normally present. The remainder were either dead, deserted, or working at much higher wages in civilian jobs. As long as the absentees reported once a year and allowed their captain to pocket a percentage of their pay, they could come and go as they wished. Since half of the troops were missing, the provisions account offered inviting opportunities for sticky-fingered officers. Between the payroll and the provisions, the total stolen amounted to as much as £1,500 to £2,000 per year.[25]

This kind of fraud could succeed because of the involvement of low-paid Crown officials. In 1705 Captain John Nanfan, the lieutenant governor of New York, was jailed briefly for participating in the scheme. In 1698, the governor of New York, the Earl of Bellomont, stated that the opportunities for corrupt profit among the "factious," "impudent," "insolent," and "cocksure" New Yorkers were extremely tempting. Bellomont wrote, "If I would make New York the mart of piracy, confederate with the merchants, wink at unlawful trade, pocket the off-reckonings, make £300 a year out of the victualling of the poor soldiers, muster half-companies, pack an assembly which would give me what money I pleased, and pocket a great part of the public moneys, I could make this government more valuable than that of Ireland, which I believe is reckoned the best in the King's gift." New York may have been the best place for stealing within the British Empire, but there is evidence that payroll padding also occurred in Massachusetts, Connecticut, Rhode Island, and South Carolina.[26]

The well-established European custom of ransoming prisoners for profit crossed the Atlantic, but it did not become firmly established in the wilderness. In 1609 Pamunkey warriors were shocked when Captain John Smith held a pistol to the head of their *werowance* Opechancanough and demanded a ransom paid in maize. In 1613 the Indian princess Pocahontas was captured by Captain Samuel Argall of Virginia and held for ransom. Argall demanded the release of all men, weapons, and tools taken by the Indians. Powhatan, the Indian sachem, offered seven prisoners and seven muskets, but the offer, even at seven to one, was unacceptable to the British. By 1676, however, Massachusetts was less demanding. The New England settlers ransomed 188 prisoners for a price of £397 13s. Indians learned from this that English colonists would often pay ransom for white prisoners.[27]

The English colonists modified the Old World system of ransoming prisoners for profit to meet the peculiar defense needs of the wilderness. The whites preferred not to return Indian captives to their tribes, so some Indian prisoners were either executed or sold into slavery. These practices were eventually replaced by a more coldly effective convention that derived from wilderness practices—the policy of paying bounties for Indian scalps. As in Europe, the purpose was to give soldiers a financial incentive to fight and also partially to defray the cost of military action. The scalp bounty plan varied from colony to colony, but the earliest and simplest form was simply to pay a flat rate for each scalp delivered. North Carolina required only that a soldier display a captured warrior or his scalp for verification by two justices of the peace. This entitled the soldier to collect a bounty of £30. During King Philip's War, Massachusetts and Connecticut paid bounties for the head of an adult Indian, but by King William's War a quarter century later a scalp alone was sufficient.[28]

As the scalp bounty system matured, it became more complex. Colonial legislatures learned to discriminate among the Indians whom they considered to be dangerous. The scalp of an adult male generally commanded a greater price than that of an adult female, and the scalps of adults were more prized than those of children. In 1696 Massachusetts paid £50 for the scalp of an adult male, £25 for a female, and £10 for a child. In 1764 (over Quaker protests), Pennsylvania paid 134 Spanish pieces of eight for the scalp of a male Indian and 50 pieces of eight for a female's scalp.

As their experience with Indian warfare increased, colonial authorities reconsidered the way they paid their troops. At first, soldiers received both regular wages and financial incentives (scalp bounties). Later, colonial leaders questioned why they should pay twice. Their solution was to devise a sliding scale. Soldiers could choose whether they wanted regular wages, piecework, or a combination, but they could not receive both full wages and the maximum scalp bounty. In Queen Anne's War, Connecticut troops could collect wages plus receive £20 for each Indian scalped or captured, or they could provide their own subsistence and receive £40 per scalp. Massachusetts had a scale that ranged from £10 to £50 a scalp; a soldier generally received a higher payment only if he furnished his own provisions. Only New Hampshire remained generous; it set its rate at £100 a scalp and once paid a bounty even though only a blood trail was found.[30]

Although the scalp bounty system seems cruel by modern standards, it was not a violation of the standards of western civilization in the seventeenth century. According to the legal scholar Hugo Grotius, the killing of prisoners of war was lawful, as was assassination (as long as it was not accompanied by perfidy). Since the colonies lacked the facilities to hold prisoners, and since it was difficult to transport Indian prisoners through the forest, the scalp bounty system was a practical if gruesome response to new conditions. Of course, it too was open to fraud. It was difficult to distinguish between a friendly Indian and an unfriendly one by the scalp alone, as it was difficult to tell an adult from a child or a male from a female. The scalp bounty system also intensified the violence. It encouraged colonists to go scalp hunting—in other words, to seek out Indians and to kill them for scalp money alone. Some of this was deliberate: in South Carolina in 1712 John Barnwell led a small army made up mostly of friendly Indians against hostile Indians. Barnwell requested that the colonial legislature enact a scalp bounty, in part because he feared that the friendly Indians would refuse to kill the hostiles unless a financial reward were available. Although scalp hunting continued well into the eighteenth century, as early as 1704 a French commander insisted upon live prisoners, describing scalp bounties as "too inhuman."[31]

Plunder was another long-standing European practice that was imported to the English colonies nearly intact. As early as the fourteenth century, England had employed the practice. During the Scottish Wars (1334-36), soldiers were paid at the rate of 4 pence a day plus all they could plunder. Three centuries later, when Englishmen established colonies in America, they relied partially on plunder as a way of compensating troops. Plunder was a convenient institution for the New World because the Indians possessed valuables that were highly coveted by whites. In 1609 English settlers ravaged Indian camps along the James River in Virginia, plundering pearls, copper, and bracelets.[32]

In the Plymouth and Massachusetts Bay colonies, early settlers donned heavy corselets to deflect Indian arrows. But when they found that the weight of the armor interfered with the amount of corn they could plunder from Indian villages, they abandoned the corselets in favor of lighter leather or quilted armor. By the time of the Pequot War, they often fought without the cumbersome protection. Plymouth also used plunder against the French. In 1635 a ship captain was offered seven hundred pounds of beaver pelts plus all the plunder he could seize to attack a settlement on Penobscot Bay. In 1676, during King Philip's War, Connecticut troops were promised "all the plunder as they shall seize, both of persons and corn or other estate." Captured Indians could be sold as slaves or scalped for a bounty. The troops were cautioned that plunder was not to be taken from "innocent persons," although that term was not defined. And in the same year, a squabble over plunder formed one of the grievances of Nathaniel Bacon in his famous rebellion against Governor William Berkeley of Virginia. Bacon and his followers hoped to plunder Indian settlements for profit, but Berkeley was against the idea.[33]

The practice of plundering continued to be accepted as a legitimate means of warfare by the American colonies at least until the Revolution. In 1702 South Carolina asked young men to join an invasion of Spanish Florida, promising a share of the plunder. In 1740 Governor George Thomas of Pennsylvania recruited prospective soldiers for an expedition that he claimed would lead to "Attacking and Plundering the most valuable Part of the Spanish West-Indies." (Seven hundred men joined up.) As late as the French and Indian War, the governor of Pennsylvania proposed paying American troops in land seized from the French. This was to be divided on a sliding scale that went from two hundred acres for a private to one thousand acres for a colonel. This was a departure from usual American practice, however. By the mid-eighteenth century the egalitarian American colonies usually gave equal shares of plunder to all soldiers and sailors, regardless of rank.[34]

The most lucrative form of plunder during the colonial period was nautical. During wartime (which was much of the time), privateering was one of the colonies' leading industries. Privateering was an ancient custom, entirely permissible under international law (until 1856), and it became an ancillary reason for colonization. Nevertheless, the practice of legitimizing capture at sea by issuing letters of marque created a blurry ethical distinction between piracy and legalized theft.[35]

New York and Newport were the ports where privateering was most resolutely practiced. During King George's War (1740-48), 466 privateering voyages originated in the British colonial ports, about half from New York and Newport alone. Many more Americans served in these private navies than served in the king's ground forces, and the reasons were assuredly economic. Sailors who went privateering could earn twice as much as merchant seamen and six times as much as a sailor in the Royal Navy. American merchants could obtain a return of 130-140 percent of their investment in a single successful voyage. During King George's War alone, privateers captured 829 prizes worth about £7,561,000. Toward the end of that war, the limiting factor on profitability was the absence of French and Spanish vessels in sufficient numbers to warrant investment in a profiteering voyage.[36]

The English settlement in America spawned a legion of competitors. The adversaries included pirates, several kinds of Europeans (Spanish, French, and Dutch), and, most frequently, Indians. But paradoxically, these groups were also the favored trading partners of the English colonists. In theory, when warfare erupted, trade between the erstwhile customers and suppliers was supposed to cease. But in practice the urge to profit often overcame the sense of obligation to the community. Trade continued, but now on an illegitimate basis. No form of profiteering was more robustly condemned than trading with the enemy. A North Carolinian denounced illicit traders as "leeches who are sucking our blood to enrich themselves, and then spue it up into the possession of our enemies."[37]

Charges of trading with the enemy appeared everywhere. Only a year after settlers first arrived in Virginia, sailors from English ships were conducting "Night Marts" with the Indians, exchanging stolen weapons for Indian goods. Captain Christopher Newport defied Captain John Smith by exchanging twenty swords for twenty turkeys. In 1628 the most disliked person in Plymouth Colony was a trader named Thomas Morton, who offended his Pilgrim neighbors by dancing around a Maypole with Indian women. More dangerously, Morton sold guns to his Indian friends. Plymouth and nearby towns asked Captain Myles Standish to arrest Morton and send him back to England. Morton was captured and sent home on the next ship, but he returned a year later and resumed his trade.[38]

Trading with the enemy proved to be an exceptionally difficult form of profiteering to control. In 1629 Massachusetts Bay began to consider what could be done about "reforming so great and unsufferable abuses," but found little success. When the General Court of Connecticut first met on 26 April 1636, the very first topic taken up was a charge against Henry Stiles for trading a gun to the Indians for corn. The court ordered Stiles to retrieve the weapon "in a faire and legally waye, or els this corte will take it into further consideration." Two months later, Stiles had not complied, so they directed that he "personally . . . appear to answer his neglect." In 1640 Connecticut also fined George Abbott, a servant, five pounds "for selling a pystoll & powder to the Indeans." Half a century later Massachusetts was still seeking to punish traders who "put knives into the hands of those barbarous infidels to cut the throats of our wives and children."[39]

But trading with the enemy, like other forms of profiteering, raised nettlesome questions. There were no simple solutions to such problems as these: Which Indian (if any) was the enemy? Who was a pirate? What did "trading with the enemy" mean to a Quaker, who believed he had no enemies? Was trade with a neutral port on Spanish Hispaniola improper because the ultimate destination of the cargo was probably (but not necessarily) the enemy end of the island? And how could trade violations be discovered and punished?

Aside from the fact that trading with the enemy was an ancient English practice that even the death penalty could not stop, a major problem was that most of the colonial wars originated in Europe. An American commitment to victory existed but with less intensity than in England. Similarly, American colonists who were not directly endangered by an outbreak of violence with the Indians were often uncooperative about suspending commerce.

Albany, New York, acquired the worst reputation as a center of illegal trade with the Indians. Increase Mather claimed that during King Philip's War, Dutch traders at Albany supplied the Indians with powder. "Men that worship Mammon," declared Mather, "notwithstanding all prohibitions to the contrary, will expose their own and other men's lives unto danger, if they may but gain a little of this world's good." Even Massachusetts Bay did not wish to suspend all trade with Indians during the war. It inaugurated a system of licensing, the first such in American history. But Albany remained a thorn, even after the Dutch demise. During Queen Anne's War, an English prisoner discovered that illegal trade was "constantly carry'd on" at Albany. "It was customary for the Indians, in their return from a trading journey to Albany, to go upon some of the frontiers of Massachusetts, and do great spoil and mischief." But Albany was by no means the only place where a load of furs could be exchanged for a musket and powder.[40]

Although arms trading with the Indians was of greatest concern to the colonists, wartime trading with the French, Spanish, and pirates was more deeply troublesome to British officials. Most of the colonies were accused of wartime trading at one time or another, but New England harbored the largest concentration of smugglers, with Rhode Island being "notoriously guilty."[41]

Royal officials were disturbed by the American willingness to sell weapons to Britain's enemies, but what angered them most was the openness by which Americans flouted the law. In 1688 the lieutenant governor of Barbados charged that pirates were able to walk the streets of Newcastle, Pennsylvania, "with as much confidence and assurance as the most honest men in the world, without any molestation whatever." One British admiral told of discovering forty-two American ships at once trading illegally with French Haiti during King George's War. During the French and Indian War, the governor of Jamaica submitted affidavits to show that ten New England ships were trading with the Spanish port of Monte Christi (for transshipment to Haiti), selling lumber, fish, and barrels and buying sugar and molasses. His estimate was undoubtedly low: in 1760 400-500 ships visited Monte Christi in order to trade with French customers.[42]

The British also claimed that Americans of both high and low station were aiding the smugglers. In 1696 the governor of Pennsylvania was charged with entertaining pirates, and in 1702 the British secretary of state objected to the appointment of seven men to the Governor's Council in New York on ground they were known or suspected to have traded illegally with pirates. In 1702 th

lieutenant governor of New Hampshire was charged with illegally selling mast trees, a vital resource for the Royal Navy. Perhaps typical of these complaints was that of Governor Robert Hunter of Pennsylvania, who during Queen Anne's War notified his British superiors of "very black practices lately perpetrated in Pensilvania by one Parks, master of the ship *St. John Baptist.*" The *Baptist* was seized while "impudently loading provisions for Petit Guave [French West Indies]." The sheriff of New Castle, Pennsylvania, freed Parks and his men, and allegedly after a "sham tryall" was held, "all was smother'd in a country Court, which acquitted ye traitors."[43]

Americans responded to these charges either by denying that illegal trade existed or by minimizing its extent. In 1709 the governor of Massachusetts, Joseph Dudley, pointed out that there were only four customs agents stationed in his colony, and there were forty harbors to police. (This would not be the last occasion in American history in which a campaign against profiteering suffered from lack of enforcement officials.) The key to enforcement, however, was whether or not legal proscriptions enjoyed public support, and in the case of wartime trade embargoes it was evident that American colonists did not respect them deeply. One exasperated customs official summed up the frustrations of his colleagues when in 1702 he wrote of Americans: "The people do seldom or never pay any taxes for the support of the Church or State; they entertain and encourage pirates; they carry on all manner of illegal trade, violate all the Acts made to prevent those evils; they affront the King, his Laws, Authority and Officers, and by all these disloyal and unjust actions grow rich and get estates, and have hitherto excaped the punishment and just reward of their wickedness."[44]

Besides trading with the enemy, the most enduring and controversial form of profiteering during the colonial period concerned the pricing of war matériel. Abrupt changes in the level of military activity always bring sectoral shifts in the economy, and when prices rose there were frequent charges of extortionate pricing—in colloquial terms, price gouging. The moral problem this presented was tangled. There was no sharp distinction, but only a fine shading, between prices that reflected the natural scarcity of a wilderness setting and prices that were deliberately inflated in order to gain an abnormal profit. America's premier moralizers, the Puritans, were the first to grapple with the enigma.

Even before leaving for the New World, the General Court of the Massachusetts Bay Company discussed the question of how to control defense costs. In 1629 the court ruled that investors in the company should provide cannon, powder, and ammunition at cost. Once the colony was established, a magazine could be established by the proprietors as long as the selling price was not more than 25 percent greater than the cost of the commodity. By the time of the Pequot War of 1637, the leaders of the Bay Company were disturbed by sharply rising prices for ammunition and armour, particularly in the settle-

ments lying outside Boston. They appointed a three-man committee to "speak with" defense providers about their prices. Unfortunately, informal persuasion proved ineffective as a price control measure. In the future, many other Americans would try, but few would succeed in controlling profiteering by moral injunction.[45]

The problem of high prices led Puritans to deep soul-searching. The center of the controversy was Captain Robert Keayne, who was one of Boston's most prosperous merchants as well as the founder and commander of its trainband company. In 1637 (and periodically afterward) Keayne was charged with excessive profit seeking. He was fined £80 and told not to be so acquisitive, since (1) he was a professor of religion; (2) he was a man of eminent ability; (3) he was already wealthy and had only one child; (4) he had come to America "for conscience's sake"; and (5) he had been warned against doing so. In a tearful mea culpa, Keayne asked forgiveness.[46]

During the following week, the Reverend John Cotton delivered a lecture in which he attempted to set forth principles which ought to guide the profit-seeking Puritan businessman. Cotton argued that no one should sell above the "current price," meaning the price that was usual in that time and place. A businessman also ought not to attempt to make up in one transaction what was lost in another, and especially he should not seek to recoup losses from casualties at sea (since "it is a loss cast upon himself by Providence, and he may not ease himself casting it upon another"). But Cotton left a large loophole. "When there is a scarcity of the commodity," Cotton wrote, "there men may raise their price, for now it is a hand of God upon the commodity, and not the person." Of course, there was little reason to raise a price *other* than that a scarcity existed. Cotton's theories simply justified the working of the market and offered no practical guideline for restraining profit making. Many years later, Robert Keayne included John Cotton in his will.[47]

Despite Cotton's admonishments, the dispute over arms prices remained unresolved. The Puritans claimed not to mind the high cost of defense. Captain Edward Johnson, a trainband commander, legislator, and militant Puritan, exhorted his co-religionists to "spare not to lay out your coyne for Powder, Bullets, Match, Armes of all sorts, and all kinde of Instruments for War . . . for the Lord Christ intends to atchieve greater matters by this little handfull then the world is aware of." But enough was enough. In 1639 the General Court of Massachusetts Bay banned arms sales in which the profit margin exceeded 100 percent.[48]

By 1675 the Puritans regarded profiteering as a Provoking Evil. Shopkeepers and merchants who set "excessive prizes on their goods" tempted God's wrath. This arrived in terrible severity in the form of King Philip's War, which Puritans blamed in part on their growing greediness. Although they declared their cause to "appear to be both just and necessary, and . . . only a Defensive War," they resolved to punish profiteers severely. In 1676 Richard Scammon was convicted of a "breach of the law of oppression in taking 500 feet of boards for mending a pistol lock . . . by his own statement not worth more than 6s. 6d." Scammon was sentenced to return half the boards to his customer and to forfeit the remainder to the county treasurer as a fine.[49]

A revealing incident that illustrates the relationship between the military and the merchants took place in Boston during Queen Anne's War. In 1711 a fleet of ships carrying 12,600 men arrived in the city in preparation for an invasion of French Canada. Overnight, the population of Boston more than doubled, severely straining the inventory of provisions. British officers offered to purchase provisions at prices that were customary in England, but the Bostonians refused to sell. The British suspected that the Americans were temporarily withholding goods in hopes of obtaining an extortionate price later. The British demanded that the General Court of Massachusetts impose a schedule of fixed prices. When the court complied, some merchants became "enraged." Massachusetts authorities ordered that no one should conceal provisions, but as one British officer noted, "there was no penalty for any person endeavouring to make a monopoly," and hoarding continued. It was only after the British commanding officer was invited to assist at the June commencement at Harvard that provisions arrived "in sufficient quantitys."[50]

The heart of the controversy was not the schedule of fixed prices but rather the exchange rate between Massachusetts currency and the British pound. Boston merchants demanded that a Massachusetts pound be valued at 83 percent of a British pound, but the British commanders balked. After several days of negotiations, the merchants grudgingly agreed that a Massachusetts pound would be valued at 71 percent of the British counterpart. The merchants of Boston accepted the lower figure, according to the British general, only "after they found that no body would conive at or share with them in their exorbitant gain on the publick's necessity." The agreement stipulated that there be no fixed price on wine because soldiers consumed it in quantity, it varied considerably in quality, and it could easily be adulterated with water. Instead, it was agreed "that sworn teasters might be employed to come as near as they could to the present value of the quantity."[51]

Despite the compromise, bad feelings persisted. The British government took its commanders in America to task for paying excessive prices. Admiral Hovenden Walker, the senior naval officer present, protested that while the prices "were very exorbitant and excessive, . . . we were obliged to comply with them, they being resolved to make an advantage of our Necessities." The governor of Massachusetts defended the Americans, claiming that the Massachusetts legislature "set a rate on all victualls below the ordinary price that Her Majestie's forces might reasonablely by supplyde."[52]

Who was correct? Doubtless the dispute could never have been resolved in a way that was completely satisfactory to all concerned, but the British system of supply practically ensured disagreement. By relying on military officers as the procurement agents, the British used men who were neither interested in the problem nor qualified to resolve it. Nearly three years later, as the matter continued to simmer, Admiral Walker offered views that applied to other military men in his position in other times and other places: "I must confess, I always thought it more the business of a military Officer to furnish himself as well as he could, with such Stores and Materials as might enable him to put in Execution and Enterprize he was commanded upon, with the best Appearance and View of Success, than the nice Calculation of the Charge, which seems to me rather what belongs to another Province."[52] In other words, a military officer is a soldier, not an accountant. A soldier wants only the best equipment available, and hang the cost!

By the French and Indian War, procedures for the control of defense costs had improved, but the disagreement between British commanders and American merchants did not lessen. Americans were charged with speculating in currency, trading with the enemy, cheating on coinage, and most frequently, committing extortion. One British officer complained that Americans objected when British soldiers cut trees with which to repair a log fort and even wanted to charge him for "Ground to encamp on." Douglas Leach has reported an occasion when American colonists actually refused to replenish British warships until the navy paid for the fresh water from their wells. A British army commander, the ill-fated General Edward Braddock, employed American purchasing agents, but complained that "none . . . will serve without exorbitant pay and [I] am forc'd to make more Contracts than I otherwise should to guard against the failure of some of them, in which Contracts the people take what Advantage they can of our Necessity."[54]

Connecticut's legislature became so concerned about "exorbitant prices" charged to equip its troops that it imposed a comprehensive price control schedule. Sutlers were limited to a 50 percent markup on all liquids, cheese, tobacco, sugar, tea, raisins, coffee, candles, and soap, and a 25 percent markup on clothing and all other articles sold to soldiers. Even so, there were so many soldiers who refused to pay the charges they incurred that the army became burdened by the "sundry law-suits" initiated by the sutlers. Connecticut authorities demanded "that the sutlers and traders . . . shew reasons, if any they have, why their said accounts should not be examined in some proper manner and a just and reasonable price be fixed for each kind of article." In May 1761, a committee of seven was appointed to "set a just and reasonable price for each kind of article according to the circumstances of the time and place when and where same was delivered." This was the first known instance of the renegotiation of a defense contract in American history.[55]

Another issue that became serious was the ongoing disagreement between Britain and the American colonies in respect to sharing war costs. It certainly was of long duration. As early as the 1690s the conflicting positions were clear: Americans thought that Britain owed them defense and were consequently reluctant to pay for it themselves; Britain thought that the colonies owed loyalty to the Crown, and the colonies ought gladly to sacrifice for the common defense. In 1698 Governor Francis Nicholson complained bitterly about his Maryland neighbors: "Those who are obliged by law . . . to have arms and ammunition of their own cannot amongst them all equip a twentieth part. . . . Some were very willing that the King should deliver them from popery and slavery and protect them in time of war, but now that by the King's valour and conduct all these troubles and fears are ceased, they are not satisfied with his government because it curbs in their former atheistical, loose and vicious way of living and debars them of their darling illegal trade." A royal customs official notified London that "when [the Marylanders] found that all their threats could not frighten me from my duty and therefore their beloved profitable darling, illegal trade, must be ruined, they resorted, having no other game to play, to open disobedience and contempt. . . . [The Americans] are a perverse, obstinate and turbulent people who will submit to no laws but their own, and have a notion that no Acts of Parliament are of force among them except such as particularly mention them."[56]

The war profits dissension was a raw old wound. In 1742, during the War of Jenkins' Ear, both New York and Massachusetts petitioned the Privy Council to appropriate funds for defense. The British considered New York's request "a matter of Surprize." Noting that the New Yorkers had failed to tax themselves to provide powder, the British denied the request, commenting that "from the Scituation, Soil, and Extent of this Province [we] cannot conceive them unable to Furnish themselves with Powder. . . . The Inhabitants of this Province may . . . very well postpone the restoring of their Publick Buildings and lend an helping hand at least for their own Security without throwing the whole Burthen upon Your Majesty." Massachusetts got the same answer: the Privy Council refused to send any ammunition "till it appears that this Province shall have paid for the Small Arms sent thither in 1704."[57] Britain was still angry about a bill that was nearly forty years old!

In 1711 Colonel Richard King, the quartermaster general who had sought to purchase provisions for the British attack upon Canada, echoed this theme. King complained of "the interestedness, ill nature and sowerness of these people, whose government, doctrine, and manners, whose hypocrisy and canting are insupportable. . . . With what great truth one may affirm, that till all their charters are resum'd by the Crown, or taken away by act of parliament, till they are all settled under one government . . . , they will grow every day more stiff and disobedient, more burthensome than advantageous to Great Britain."[58]

Thus by the close of the colonial period, the war profits problem had fully emerged. The American military system had been organized as an amateur, paramilitary structure whose basic concept was collective self-defense. Originally financed by individual soldier-settlers and by English investors, it relied for its support upon weapons, ammunition, and supplies from home, supplemented by plunder and by local contributions. The system gradually converted from collective self-defense to one based on individual self-interest. This self-seeking basis led to abuses such as cheating on supplies, scalp bounties, trading with the enemy, and price extortion. In the early eighteenth century, a more modern, semiprofessional army appeared, but by the eve of the Revolution this change was incomplete.[59] The distinction between individual interest and collective military obligation remained unresolved during the colonial period. Its unsettled, ill-defined nature led to a quarrel over war profits between Americans and British. This fundamental disagreement over the price of Anglo-American defense helped to point the way toward the American Revolution.

2

Virtue Tested

There is such a thirst for gain, and such infamous advantage taken to forestall, and engross the articles the Army cannot do without, thereby enhancing the cost of them to the public fifty or a hundred pr. Ct., that it is enough to make one curse their own Species, for possessing so little virtue and patriotism.

GEORGE WASHINGTON

British military policy was a key ingredient in the coming of the American Revolution. Although perhaps secondary to the army's perceived threat to political liberty, the army's role as a major economic institution was profoundly repugnant to Americans. Although there were constitutional proscriptions against a standing army in peacetime, the American colonists had virtually nothing to say about the army's composition, location, activities, or cost. The army's economic role formed an important part of the American critique of Britain, and it contributed mightily to the development of a complex and sophisticated Revolutionary ideology.[1]

The development of American military practices was an embodiment of that ideology. The American attitude toward the relationship between war and wealth coalesced during the War of Independence and was deeply affected by past experience and by Revolutionary ideology. American revolutionaries viewed the War of Independence as a struggle between a virtuous society and a corrupt society. The mere existence of the British army as a peacetime institution, regardless of whether it was stationed at home or abroad, was clear evidence to some that Britain was a society rotten with corruption. This belief was shared not only by American revolutionaries but also (and not coincidentally) by British writers of the "Country" persuasion. According to this thinking, the best of all possible societies contained a tripartite balance between monarchy, which provided power, aristocracy, which offered wisdom, and the people at large, which imparted national spirit. It was vital that all three elements be reasonably independent, else their contribution would be negated. The British army had the insidious effect (from the American point of view) of co-opting the independence of Parliament and thus tipping the balance toward autocracy. The way it did this was by monetary corruption. A substantial cluster of members of Parliament were also army officers, thereby dependent upon the king for promotion and pay. They were no longer independent freeholders, able to weigh issues carefully and to arrive at the public good objectively. They were now "placemen" or (to use the Americans' scornful term) "interested," meaning that the king had placed them in Parliament and that they served their own personal interest as opposed to the larger public interest.[2]

Americans were aware that in twisting the Lion's tail they were dangerously arousing the ire of the world's leading military power. In part, American hopes of victory depended on certain physical advantages, such as distance and vastness. In the heady days of 1775, the crucial advantage that Americans believed they possessed was not natural but moral. Thus an American captain could write that his men fought with the "confidence of the martial superiority of free men to slaves."[3] British soldiers were professionals, mercenaries; they fought for pay rather than for patriotism. This was doubly true of Britain's hired allies, the Hessians, who seemed to have no patriotic motivations whatsoever. Mercenary soldiers, so the argument went, lacked valor; when the ultimate trial by fire came, they would not risk their lives for money. European armies would lack the staying power of the American fighters, who served a higher ideal.[4]

The word Americans most frequently used to describe their supposed moral advantage was virtue. Stated forthrightly, Americans were virtuous whereas their enemies were corrupt. Although superficially a simple word, in the context of the American Revolution (and particularly in its early phase), *virtue* was saturated with meaning. On one level, it meant that America's traditionally amateur warriors lacked the vices that were associated with their increasingly professional cousins. The licentiousness of professional soldiers, according to American theorists, included not only such obvious forms of debauchery as gambling, profanity, drunkenness, and prostitution, but also milder varieties such as fickleness, infidelity, and even levity. George Washington vaguely feared that if given free rein, soldiers would "run mad with pleasure" and would engage in "diversions of all kinds." American citizen-soldiers, conversely, were expected to display a degree of asceticism that was religious and utopian. Massachusetts clergymen were among the staunchest critics of the debased morality of the British army, and it was partially to satisfy them that Revolutionary theory placed such emphasis on an amateur military.[5]

In a broader sense, the Revolutionary concept of American virtue carried an economic meaning. Regardless of their military or civilian standing, all patriotic Americans were expected to display their virtuousness by sacrificing economic self-interest to the good of the cause. The paragon of virtue was

George Washington, who rejected the salary of $500 per month offered by Congress and who served for eight years without pay (although he was compensated for his expenses).[6] Another celebrated example was that of the three Americans who apprehended the famous British spy Major John André. André offered his captors a thousand guineas for his freedom, but they delivered the prisoner to American authorities. This exposed Benedict Arnold's treason and perhaps saved the Revolution. In gratitude for their "virtuous and patriotic conduct," Congress voted each of the men an annual payment of $200 and presented each with a silver medal.[7]

The leaders of the Revolution used the concept of "virtue" in exactly this economic sense. George Washington, who yielded to no man his dislike of war profiteering, delivered a jeremiad against the practice soon after he assumed command of the Continental army in 1775. "It is a matter of great Concern," Washington wrote, "to find, that at a time when the united efforts of America are exerting in defence of the common Right and Liberties of mankind, that there should be in an Army constituted for so noble a purpose, such repeated Instances of Officers, who lost to every sense of honour and *virtue* [emphasis added], are seeking by dirty and base means, the promotion of their own dishonest Gain, to the eternal Disgrace of themselves, and Dishonour of their country—practices of this sort will never be overlooked."[8] In 1778 Washington became so angry at defense contractors for charging extortionate prices that he declared that it was "enough to make one curse their own Species, for possessing so little virtue and patriotism."[9]

Congressmen employed similar usage. John Adams, in appointing James Warren to the Navy Board in 1777, remarked, "The profit to you will be nothing, But the Honour and the Virtue the greater. I almost envy you this Employment." Richard Henry Lee complained that "the inundation of money appears to have overflowed virtue. . . . Look around you, do you anywhere see wisdom and integrity and industry prevail? . . . The demon of avarice, extortion, and fortune-making seizes all ranks." In 1779 William Whipple decried "the failure of public virtue." "Speculating miscreants . . . have been sucking the Blood of their country."[10]

The existence of virtue among the American people implied a willingness to sacrifice that had both military and economic dimensions. Theoretically, all members of a virtuous society ought willingly to set aside personal interests and to take up arms in the common defense. The concept of the Minuteman, drawn from the ancient idea of a nation-at-arms, symbolized the willingness of the virtuous American to sacrifice at a moment's notice. George Washington personified this ideal. The British, on the other hand, were supposedly sated by luxurious living, and large numbers of them allegedly would not risk their lives. Although the British were more numerous, a smaller proportion would be available for combat, thereby partially overcoming the American numerical disadvantage.[11]

The willingness to make an economic sacrifice took several forms. Soldiers were expected to serve for little or no pay. General Charles Lee exhorted innkeepers to "give testimony to your virtue and patriotism" by refusing to serve persons who were disloyal to the cause. But most important it meant a refusal to profit from the war. No virtuous American should plunder civilians, steal from the government, or raise prices to extortionate levels. Colonel John Laurens, aide-de-camp to General Washington, perhaps expressed this concept best in a letter to his father, Henry Laurens, who was president of the Continental Congress. "If we are as virtuous as we ought to be," wrote young Laurens, "we should have those who are enriching themselves by commerce, privateering, and farming, supplying the army with every necessary convenience at a moderate rate." Although Laurens doubted that this was entirely possible, he asked his father as the leader of Congress simply for "such prices as bear some proportion to our pay."[12]

Although Revolutionary ideology provided the concept of a virtuous people, its application to the matter of war profits originated in other experiences. These included recollections of the Great War for the Empire, which lived in the social memory as a time of rampant war profits (anticipating the remembrance of the Civil War and World War I). In debate in Congress in September 1775, three congressmen remembered how the troops had been "imposed upon" by contractors in the "last War." They declared their determination not to allow it to happen in the Revolution. Washington himself may have profited from his service in the earlier conflict. As colonel of the Virginia militia, he received a salary, expenses, and a 2 percent commission on all supplies purchased for his troops. After the war he was able to buy 500 acres for £350 and was also able to pay £300 for slaves to cultivate the land.[13]

Contracting policies of the British army before and during the Revolution were a second irritant. A commission in the British army was a valuable prize, for which the applicant paid handsomely. A lieutenant-colonelcy cost £4,500, but it conferred numerous prerogatives. A colonel handled the money designated for soldiers' wages, could sell luxuries to the rank and file, and received a very large share of spoils won in battle. In 1780 Britain had fraud in payments to troops estimated at £30,000. One American captain expressed his contempt for the nepotism of the British officer corps: "It was no unusual thing in the [British] army, for a Colonel to make drummers and fifers of his sons, thereby, not only being enabled to form a very snug, economical mess, but to aid also considerably the revenue of the family chest. . . . In short, it appeared that the sordid spirit of gain was the vital principle of this greater part of the army." In 1779 a German immigrant serving as a captain in the Continental army, following European practice, attempted to extract a fee for licensing sutlers. A court-martial excused him with a reprimand on the grounds that his "conduct might possibly have arisen from a misconception of the nature of his office." General Washington "acquiesced" in the sentence, which he found "lenient."[14]

A third source of the antiprofiteering ethic were the views of many American clergymen. Men of the cloth were among the most outspoken critics of the British army, which they considered a fountain of moral degradation, a serpent in the American garden. They found the greatest source of sin to be everyday business transactions, and when these took place in a military context, the temptation was deepened. The favorite scriptural reference of American ministers who decried war profits was Deuteronomy 23:9, which advises, "When the Host goeth forth against thine Enemies, then keep thee from every wicked thing." As one American private recorded during the peak of Revolutionary fervor in 1775: "Our minister preached 2 sermons. He preached from Dutrinomy 23th and 9th verse in the forenoon and afternoon, and he preached 2 very good sermons to the soldiers how it was best for us to do and what not to do."[15]

The republican ideal that pervaded the Revolutionary era included several implications for the way an army should be organized. The basic theory was that virtuous citizens would volunteer for service with no expectation of pecuniary reward. In short, duty and money would be completely divorced. There would be no enlistment bonuses, no handsome wages, and no severance pay. Every soldier would provide his own weapon and would train himself to fire it. Because an army needed organization, soldiers would accept military discipline graciously. Civilians would sacrifice to support the troops without complaint. Enlistments would be short because to allow permanent service would invite the corrupting influences of a professional army. The brash young Americans planned to take on the British regulars with what was essentially a national militia.[16]

These ideas were put into effect in the early days of the Revolution. On 30 November 1775, congressional leaders reached a decision not to pay enlistment bonuses but rather to rely on patriotic exhortation to raise the army.[17] A companion policy adopted by the various states established enlistments on a short-term basis. These steps guided the army's development during the first year of the war.

The militia system failed. In 1775, during the initial excitement, sufficient numbers of men volunteered to take up arms. As a long-term solution, more tangible incentives proved necessary. The idea of short-term enlistments was found wanting in the American invasion of Canada. The final and unsuccessful assault on the city of Quebec had to be scheduled for 30 December 1775, because the soldiers' enlistments were due to expire on the following day. During the summer of 1776, the concept of valorous citizen-soldiers was again tested when British regular forces invaded New York. Washington's men were soundly whipped and forced to flee into New Jersey.

American leaders faced a painful choice: reorganize or surrender. Their answer was obvious, and in September 1776, Congress approved a plan to pay enlistment bounties. This reversal was an admission that Americans were not so virtuous that they would risk their lives for ideals alone. But if the flame of republicanism flickered, it did not die out entirely. "I do not mean to exclude altogether the Idea of Patriotism," wrote Washington. "I know it exists, and I know it has done much in the present Contest. But I will venture to assert, that a great and lasting War can never be supported on this principle alone. It must be aided by a prospect of Interest or some reward. For a time, it may of itself push Men to Action; to bear much, to encounter difficulties; but it will not endure unassisted by Interest."[18]

The new army fielded after the experiments of the first year took a new approach to the question of soldiers' pay. A disappointed George Washington admitted that few men "act upon Principles of disinterestness," so few, indeed, that they were only "a drop in the Ocean." The solution was to be found in financial incentives: handsome bounties and better pay. Enlistment bounties were set at $20 and 100 to 200 acres of land in 1776. By 1780 recruits were receiving $150 in specie for five months' service. Wage rates were comparable to the earnings of civilian laborers in 1776, except that because soldiers were guaranteed continuous employment they were theoretically better compensated. Inflation quickly destroyed this wage advantage, but bounties remained attractive. As one disgruntled congressman remarked, "Bounties before unheard of have been given to but little better purpose than to hire the populace to visit the army; this instead of checking the growing avarice of the country (which ought to have been their object) has cherished it, till they pay their devotion to no other shrine than mammon's."[19]

The establishment of bounty payments for enlistment opened a new field for military fraud. New recruits or veteran soldiers occasionally deserted their regiments and reenlisted in others (usually under a different name) for the purpose of obtaining multiple bounties. This was a perennial problem throughout the war, and it was particularly shocking to George Washington. When in 1776 he first learned of double enlistments, he reacted "with great astonishment and surprise," and he warned grimly that "so glaring a fraud . . . will be punished in the most exemplary manner." This was the first of a dozen warnings issued by the commander, and he backed up his words with action. The original punishment for bounty fraud was thirty-nine lashes, which was more severe even than the penalty for desertion, which was thirty lashes. They were later made crimes of equal severity, but the punishment was increased, first to 100 lashes "well laid on" and later to death. The actual execution of a soldier for bounty fraud was extremely rare (there is only one known case), but sentences of 100 lashes, while infrequent, were common enough to be a frightening deterrent.[20]

In practice, the wage policy of the Continental army was ambivalent. While in theory wages were comparable to those of civilians, in practice they lagged far behind. Rampant inflation eroded or destroyed soldiers' buying power, but even worse was the irregularity of payment. One intrepid warrior who served from 1776 to 1783 recalled that he was paid regularly for only the first five months of his service. In the following six years of service, he had only one real payday. Continental soldiers did not fight for monetary reasons, and their officers were in even worse shape. Inflation struck them harder, as one surgeon suffering at Valley Forge explained: "The present Circomstances of the Soldier is better by far than the Officer—for the family of the Soldier is provided for at the public expence if the articles they want are above the common price— but the Officer's family are obliged not only to beg in the most humble manner for the necessaries of Life,—but also to pay for the afterwards at the most exorbitant rates—and even in this manner, many of them who depend entirely on their Money, cannot procure half the material comforts that are wanted in a family."[21]

The claim that officers were sometimes forced to beg to avoid starvation was not far-fetched. In 1779 Connecticut's congressional delegation pleaded for public assistance for the troops, which it described as being as poor as orphans and widows. During the "hard winter" of 1779-80, conditions equaled the squalor of Valley Forge, two years earlier. Soldiers ate birch bark and roasted old shoes, and officers slaughtered their pet dog for food. Even a major general, John Sullivan, was obliged to borrow money to support his family, and he pleaded with Washington for relief (unsuccessfully). William Heath, another general whose officers resorted to borrowing from local citizens for their necessities, informed Washington of their dismay. "I need not paint to your Excellency their feelings when they are dunned by the peasants for small sums, and [I am] at a loss what answer to give them." The officers' families had become indebted to their neighbors and were unable to repay them.[22]

Washington hoped to alleviate his officers' plight by endorsing a plan to give them half-pay for life—a form of deferred compensation. This scheme went awry on ideological grounds: it was too British and too unrepublican. The opposition in Congress argued that it created a public-supported aristocracy that had profited from the war. Another British-oriented plan advocated by Colonel Theodore Bland would grant officers knighthood, vendible commissions, half-pay during peacetime, and pensions for widows. Washington adamantly opposed this idea as a violation of Revolutionary principle, and Congress agreed. A compromise eventually settled the half-pay controversy. Officers were granted half-pay for seven years. Throughout their travail, the men of the Continental line displayed a kind of deep stoicism that was the ultimate measure of their commitment to the cause. As General William Heath wrote at the coldest moment of the "hard winter," "Notwithstanding all their tryals and sufferings they exhibit a patience, health, good temper and spirit not to be paralleled."[23]

The remarkable resiliency of the Continental army in the face of extreme adversity has been often noted but seldom explained. American morale weakened under the weight of British attack, but ultimately did not break, according to John Shy, because the British war effort greatly deepened the American dislike of everything British. Yet another reason that Americans supported their army longer than the British supported theirs was the differing attitude of the two armies toward graft. Most Americans, following George Washington's determined and unflagging leadership, supported aggressive measures to ferret out every form of fraud. The British, following the lead of every commander except Charles Cornwallis, winked at graft. The Royal Army dripped with corruption.[24]

The ingenuity displayed by Royal officers in milking military appropriations for their personal profit during the American Revolution was remarkable. One device was simply an old custom: padding military payrolls and keeping the overcharge or "off-reckoning." By 1780 the fraudulent off-reckonings amounted to £30,000, and the paymaster had absorbed another £100,000. A very lucrative new trick was to charge the government for one measure of supplies but to substitute a smaller quantity while pocketing the difference in price. This could be accomplished by charging for a "country cord" of firewood, which was 19 percent larger than the standard cord actually delivered, or for a "Winchester" bushel of provisions, which was 25 percent larger than the bushel in actual use. There were also extra rations for officers' nonexistent families, currency manipulation, overcharges for the construction of defensive installations, fraudulent receipts for forage, and overcharges in renting horses and wagons to the army. These practices varied from mildly questionable to outright larceny, but the result was similar in that some officers amassed extensive fortunes in the process. Three deputy quartermasters retired shortly after the war and settled on their newly purchased country estates, one of which totaled 20,000 acres. The chief engineer of the Royal Army purchased two estates and a colonel's commission, and sent his sons to expensive military schools. In 1778 Horace Walpole acidly remarked, "General Howe is returned, richer in money than laurels."[25]

In contrast, the American army took vigorous steps from the beginning to protect its reputation from charges of corruption. The first officer charged with fraud, a quartermaster, was court-martialed on 5 July 1775. Others followed soon thereafter. "As nothing can be more fatal to an Army, than Crimes of this kind," General Washington wrote in August 1775 to John Hancock, "I am determined by every Motive of Reward & Punishment to prevent them in the future." In a fine expression of republican rhetoric addressed to Richard Henry Lee, Washington wrote: "In short I spare none & yet fear it will not all do, as these People seem to be too inattentive to every thing but their Interest." Washington even despaired that his vigor in prosecuting fraud would ruin his reputation: "My life has been nothing else since I came here but one continued round of annoyance and fatigue; in short no pecuniary recompense could induce me to undergo what I have, especially as I expect, by shewing so little

countenance to irregularities and public abuses to render myself very obnox-
ious to a greater part of these people."[26]

Washington proved to be as good as his word. While British officers were
making their fortunes on defense fraud, American officers were being prosecuted and dismissed by the dozen. At least thirty-two officers of the Continental Line were punished during the war for a wide variety of fraudulent
practices (see appendix A). Most were dismissed from the service. (In November 1775, Congress established dismissal as the punishment for officers; enlisted men were to receive from fifteen to thirty-nine lashes.) In addition,
officers were humiliated and ostracized. Congress specified that upon conviction for fraud, an officer's name was to be published in newspapers, "after
which it shall be deemed scandalous in any officer to associate with him."[27]

Captain-Lieutenant Theophilus Parke, an artillery officer, was chosen as an
example to others. Convicted in March 1780 of fraud and forgery, he was sentenced to be dismissed from the army "with infamy." This meant that his sword
was broken over his head before an assembly of his troops, he was excluded
from holding civil or military office, and the charge was published in the press.
Yet even this humiliation was far preferable to the fate of some enlisted men
convicted of fraud. In May 1780, a military surgeon witnessed a frightening
scene: "The criminals were placed side by side, on the scaffold, with halters
round their necks, their coffins before their eyes, their graves open to view, and
thousands of spectators bemoaning their awful doom. At this awful moment,
while their fervent prayers are ascending to Heaven, an officer comes forward
and reads a reprieve for seven of them, by the commander in chief. Only one
man was actually hanged, for the crime of forging discharges that allowed a
hundred men to leave the army.[28]

Nothing better illustrates the developing American attitude toward war
profits than does the hoary old military custom of plunder. To some extent,
both the British and the Continental armies were guilty of plundering.
Scholars agree, however, that the weight of the evidence shows that the British
and their Hessian allies plundered somewhat more freely than did the American troops.[29]

From the outset of hostilities, official American policy opposed plunder.
European troops brought with them a centuries-old plundering tradition that
was difficult to control and impossible to eradicate. American commanders
considered the limitation of plunder an important advantage in their campaign to achieve broad support. With certain notable exceptions, and despite
many violations, the army attempted to observe the policy throughout the
conflict. The exceptions were, first, during the American invasion of Canada in
1775, when undisciplined American troops plundered their northern neighbors; second, during the "hard winter" of 1779-80, when farmers were unwilling to sell food to the troops; third, when American commanders consistently

allowed and even encouraged the seizure and sale of property owned by British soldiers stationed in America; and finally, when Americans seized the property of "Tories," a category which, unfortunately, was defined with a considerable degree of elasticity.[30]

George Washington's dislike of war profits was never displayed more graphically than in his hatred of illegal plunder. Shortly after assuming command in the summer of 1775, Washington established procedures to encourage his troops to seize property belonging to British soldiers. As befitted the egalitarian warriors of the infant Republic, Washington ordered that this legal booty was "to be equally divided between the Officers and men, that took it." (The last phrase led to jurisdictional squabbles over who had actually captured the property.) Although Washington continued to encourage the plundering of British forces (as an incentive to motivate his troops to fight) until at least the end of 1776, he also issued orders against "unlawfull and irregular plundering." This meant the seizure of property not belonging to enemy troops. In practice, distinguishing between the two categories proved quite difficult. After the Battle of Monmouth Court House in 1778, Washington checked his soldiers' packs for possession of illegal plunder.[31]

The Continental Army had good reason to restrict plundering. The army, and the Revolution it advanced, depended upon the goodwill of the civilian population for support. Plundering would win few friends for the cause. No one understood this fact better than Washington. He summoned his considerable powers of persuasion and all his eloquence to exhort his men to stop the practice. In 1777, upon receiving reports of illegal plundering—a practice he called "wicked, infamous, and cruel"—Washington wrote: "We complain of the cruelty and barbarity of our enemies; but does it equal ours? They sometimes spare the property of their *friends:* but some amongst us, beyond expression barbarous, rob even *them!* Why did we assemble in arms? Was it not, in one capital point, to protect the property of our countrymen? And shall we to our eternal reproach, be the first to pillage and destroy? Will no motives of humanity, of zeal, interest and of honor, restrain the violence of the soldiers?"[32]

Washington had decided that unless plundering ceased, "the people will throw themselves, of Choice, into the Hands of the British Troops." This was a threat to the Revolution itself, and it justified, in the commander's eyes, Draconian measures. In July 1777, Washington directed that a court-martial sentence of a soldier convicted of plundering be carried out. To set an example for others, the man was to be executed, Washington ordered, "in the most public manner your situation will admit."[33]

On other occasions Washington covered his iron fist with a velvet glove. He tried to shame his troops into halting the plundering. "How disgraceful to the army is it," he remarked, "that the peaceable inhabitants, our countrymen and fellow citizens, dread our halting among them, even for a night and are happy when they get rid of us?" But when words alone were not enough, Washington issued regulations intended to curtail plundering. Only officers of the rank of colonel or above could issue passes to leave camp; no soldier could leave quarters with his firearm; rolls were called between 8 and 10 P.M. to isolate plunderers; and wagonmasters were directed to remain close to their horses to prevent unauthorized use. In 1777 Washington granted to dragoon commanders authority to impress horses to replenish lost animals, but when he found that the cavalrymen "go about the country, plundering whomsoever they are pleased to denominate Tories, and converting what they get to their own private profit and emolument," he withdrew the authority. In fact, Washington withdrew all authority to plunder Tories. A General Order of 21 January 1777 prohibited "the infamous practice of plundering the Inhabitants, under the specious pretence of their being Tories—Let the persons who are known to be enemies to their Contry, be seized and confin'd, and their Property disposed of, as the law of the State directs—It is our business to give protection and support, to the poor, distressed Inhabitants; not multiply and increase their calamities." Washington's harsh restriction of plundering and plunderers bestowed a tangible moral advantage on the American cause.[34]

If British commanders had been equally determined, they might have triumphed in the contest for American loyalty. But reports of plundering by Redcoats and Hessians circulated continually. Fences and buildings were burned for firewood; household furniture was stolen or destroyed; cattle were slaughtered with much of the carcass wasted; mechanics' tools were taken. There was even a report that cherry trees were cut down when Royal troops were unable to reach the fruit in the high branches. A. R. Bowler has suggested that the miserable quality of rations shipped to America from England impelled the British and Hessian troops to plunder. This is probably true (and may explain their commanders' permissiveness), but it is unlikely that such an excuse would have persuaded many American farmers that British actions were justified.[35]

If George Washington was angered by graft and plundering, he was equally outraged by a third source of war profits: trading with the enemy. This practice was a direct and obvious violation of Revolutionary principles. A virtuous republican who possessed more goods than he could consume himself ought properly to donate them (or sell them for a modest price) to the good of the cause. Only a grossly corrupt countryman or an ardent royalist would sell them to the British to obtain a greedy profit. In August 1775, at the same time he was punishing graft and exhorting his troops against plundering, Washington issued the first of many lamentations on trading with the enemy. In a letter to the New York Provincial Congress, he wrote: "It must give great Concern to any considerate Mind that when this whole Continent at a vast expence of Blood & Treasure is endeavouring to establish its Liberties . . . there are Men among us so basely sordid as to Counteract all our Exertions for the Sake of a little Gain."[36]

Washington's moral outrage was just one reason to oppose trading with the enemy. Goods shipped to the enemy not only strengthened them (and undercut the American logistical advantage) but also weakened the American effort in that there were fewer goods available to the patriots. Hard money transferred to the British stimulated inflation and made it more difficult to collect taxes. Trading with the enemy bred distrust among compatriots, and after 1778, it threatened the alliance with France. The French justifiably felt that, in return for aid rendered to the American cause, they could expect to obtain a larger share of the American market. When Americans returned to pre-Revolutionary trade channels, even as the war raged on, the French were rightfully annoyed. Trading with the enemy had only one redeeming virtue: it provided a source of intelligence to the American leadership. In George Washington's opinion, this value was not worth the cost. "Those people who undertake to procure intelligence under cover of carrying produce into New York," Washington complained, " . . . attend more to their own emoluments than to the business they have charged, and we have found their information so vague and trifling, that there is no placing dependance upon it."[37]

As with any illicit activity, it is impossible to determine exactly how much trading with the enemy occurred. Contemporary reports were so extensive that it is obvious that the amount of proscribed trade was substantial. During the war the value of British goods exported to Nova Scotia increased to a figure that was ten times as large as that which the colony could possibly consume. Consignments from Scottish ports to North American destinations from which cargoes could be transhipped to American customers tripled over prewar levels.[38]

Anticipating later attempts to exclude British goods (the War of 1812 embargo, Prohibition), the embargo leaked badly, and the most capacious leak was through Canada. It is also clear that trading with the enemy increased as the war dragged on and the idealism of the first year waned. In March 1776, the British evacuated Boston, substantially because of the difficulty of obtaining supplies there. But when they occupied New York four months later, a steady trickle of supplies commenced. These supplies were not sufficient to meet the needs of the British forces, which depended primarily on supplies transported from home, but the trade became considerable. Although the much-maligned American militia enjoyed some success at interdicting this trade, New York's commerce with outlying regions may have returned to prewar levels by 1780. Connecticut became a center of illegal trade, and ship captains hired to stop it began to engage in it themselves. When the British occupied Philadelphia in 1777, dozens of new stores opened, some of which were specialized firms catering specifically to the needs of British officers. Washington found it necessary to dismantle some flour mills near Wilmington, Delaware, in order to keep them from supplying enemy troops in Philadelphia.[39]

Trading with the enemy during the American Revolution was among the most ethically ambiguous matters in the entire history of the war profits issue. The vexing question was how to tell just when normal commerce shaded into trading that was morally offensive. Unless a person believed implicitly that British control of America was wholly illegitimate, entering into a transaction which might help to sustain that control was not reprehensible, except in the eyes of dedicated rebels. The British did not consider their occupation of Boston or New York to be an invasion. They believed they were simply reoccupying soil that was rightfully theirs. A merchant who traded with the British could persuade himself that he was acting ethically if he were in sympathy with the British, neutral toward them, or even lukewarm on the war. There was also a matter of timing. The fighting began more than a year before independence was declared. When did an individual's moral obligation to suspend trading commence? With the first battles, with the call for an embargo, or with independence?

British traders complained that "villainous" American customers commonly made purchases and then reneged later on their obligations. From the American point of view, "trading with the enemy" was not unethical as long as nothing of value was given. Finally, there was the question of price. American traders, in the tradition of their fathers and grandfathers, greeted the arrival of the Royal Army by happily jacking up their prices.[40] If items of little military value were sold at sufficiently high prices, it might be reasonably argued that the British cause was harmed rather than helped.

Another problem was the utter impracticality of completely severing trade with Britain. Indeed, this was not even desirable. Besides gunpowder, the item in shortest supply in America in the early years of the war was salt. Since salt was used for preserving meat, it had a vital military value as well as a civilian use. Salt shortages were so severe that riots broke out, ration cards were issued (in New York), and salt was even used as a medium of exchange. To alleviate the shortage, the Virginia legislature made an exception to the restrictions on trade with Britain, allowing ships to clear for Bermuda, as long as they engaged in that trade alone. Still, scarcity had driven up the price of salt, so that an unusual profit might be obtained in reselling it. A North Carolina county attempted to limit the gain by restricting the markup on salt to 200 percent, but dissatisfaction remained.[41]

Considering that Revolutionary leaders temporarily allowed trading with the enemy for salt, was profiting from that trade unethical? An answer might be found in the magnitude of the profits obtained. As one example, Nicholas Brown, a merchant of Providence, Rhode Island, was able to triple his investment in the salt trade in 1777. If this had become public knowledge, he certainly would have been denounced as undevoted. In 1776 the Pennsylvania Council of Safety learned that certain importers had marked up a shipment of salt by 87.5 percent. The Council resolved to express its disapproval of "the

Conduct of those mercenary men, who, regardless of the public Good and the Interest of their Country, in this Unreasonable and avaritious manner, monopolize those articles so necessary to the Community and lay on such enormous Profits as puts it out of the power of the Industrious Poor to procure a necessary supply for their Families." The Council confiscated the salt and sold it at cost. In 1781 Congress voted to prohibit the salt trade with Bermuda for the duration of the Revolution.[42]

In another transaction, the ethical dimension was affected by the ultimate use of the British product. In 1778 Henry Laurens purchased some hard-to-find buff cloth that was manufactured in Britain. Normally, such a purchase by the president of Congress would be highly questionable. But Laurens sent the cloth to his son John, a lieutenant colonel in the Continental Army, for the purpose of making a new uniform.[43] Henry Laurens might well have preferred to purchase French cloth, but since the final use of the product was military, there was little to criticize.

A final ambiguity involved the complicity of customers in the illegal trade. When the war broke out, there was a rough consensus in America that English goods should be shunned. American merchants turned away from their customary sources of supply in the British empire in favor of opening up new ones elsewhere. Opportunism and patriotism were impelling forces: now that the Navigation Acts were inoperative, good profit might await the merchant who was the first to tap the new trade. In particular, the large and potentially lucrative market for American tobacco in France beckoned to opportunistic American traders. If French products were suited to the taste of American consumers, Britain could be neatly excluded. Exchanging French finished goods for American tobacco (with Britain kept out) might serve as a basis for a permanent trading relationship.

Although it attained some success, this plan was better in conception than in execution. American consumers were used to purchasing British products, and this preference did not disappear with the outbreak of hostilities. The experience of Nicholas Brown of Rhode Island is illustrative. Early in the war, Brown patriotically refused to purchase British products. But he later reversed his position, specifying that he would buy *only* British goods. He found that French goods did not sell well in America, and New England goods did not sell well in France. French traders were slow to fulfill contracts, and they did not wish to extend credit. Brown's dilemma was whether to continue with his French venture, while his customers turned either to less-principled vendors or to those with Tory inclinations, or else to engage in trading with the enemy. Brown and other merchants preferred the latter alternative, or, as Brown described it, "the Clandestine Way." There was no question in George Washington's mind about the ethical level of this trade. In 1781 he wrote, "Men of all descriptions are now indiscriminately engaging in it, Whig, Tory, Speculator. By its being practiced by those of the latter class, in a manner with impunity,

Men who, two or three years ago, would have shuddered at the idea of such connexions now pursue it with avidity and reconcile it to themselves (in which their profits plead powerfully) upon a principle of equality with the Tory."[44] The heart of the issue was whether a businessman who was sympathetic to the Revolution was thereby obliged to allow his customers to be absorbed by his Loyalist competitors. Loss of livelihood was a greater sacrifice than most men were willing to make.

Trading with the enemy involved primarily civilian goods. Nevertheless, its existence expressed a lack of support for the American war effort. The system of procuring vital military supplies also exhibited some lack of concern for the army's needs. Domestic suppliers and manufacturers sometimes sought to obtain unusual profits out of the irregular conditions created by the war. This occasionally materialized in the fashion of degraded quality of military supplies, one of the most persevering forms of profiteering.

Food and other provisions supplied to soldiers have been a notorious source of complaint since soldiers first marched, but the Continental army had genuine reason to feel aggrieved. Quartermasters often ordered beef but received horsemeat, forcing the army to establish a board of butchers to examine the meat supply. Salt pork was frequently ruined when teamsters drained the brine from the barrels in order to lighten their loads. Imported blankets were so small that two and sometimes even four were required to cover a man. Shoes were legendary for their dilapidated quality. A pair often lasted for only a few days. In 1779 General Washington complained of "great abuses" whereby shoes were made "by putting in small scraps and parings of leather and giving the Shoes the appearance of strength and substance, while the Soals were worth nothing, would not last more than a day or two's march."[45]

The amount of profits received by American military suppliers during the Revolution was controversial at the time, and their degree of excessiveness will never be completely resolved. The recent work of E. Wayne Carp argues persuasively that charges of corruption in the supply offices were exaggerated.[46] Nevertheless, it is an undeniable fact that there was a widespread public perception that speculative buying, hoarding, price gouging, degraded quality, and various other fraudulent procedures were very common. There were several sources of this belief.

In the first place, the administrative procedures for supplying the military forces were poorly developed or nonexistent. The fledgling nation fought its first war according to the methods that had served the various colonies in the skirmishes with the Indians and the French. When such a war broke out, it was customary for the colonial legislature to appoint a war committee. This body held the authority to award contracts for military supplies to local providers. When the Revolution commenced, the Continental Congress assumed the responsibility for procurement and appointed committees to carry it out. The responsibilities of these committees overlapped, and members could not devote full attention to any single assignment. When the day-to-day work of purchasing military supplies was carried out by harried congressmen, public suspicions about the efficiency of their supervision was easily aroused.

A second controversial feature was the reliance upon a contracting system that gave the defense contractor an incentive to raise prices. The Continental army employed purchasing agents who were compensated by being allowed to keep a percentage of the cost of each contract. This percentage was normally 2.5 percent, but sometimes higher. Therefore, the larger the contract, the larger the supplier's commission. This system had its origins in peacetime business practices. American merchants frequently employed purchasing agents at a customary rate of 5 percent. This rate rose considerably during wartime. In June 1777, Congress partially replaced the discredited commission system with a system of fixed pay and rations for the Commissary Department. The quartermaster general continued to be remunerated by commission.[47]

A third circumstance leading to charges of profiteering was the method of financing the war by inflation. A characteristic common to most inflationary periods is the general belief that others are gaining economically faster than oneself. Since others seem to be advancing, it follows that they must be doing so as the result of active encouragement, that is, by aggressively raising their prices. In other words, to the people of the Revolutionary era, a general price increase was necessarily the result of a conspiracy to raise prices. Along with poor congressional oversight and faulty contracting systems, inflationary psychology contributed greatly to charges of price gouging.

Although these factors led to a somewhat exaggerated perception of the amount of profits obtained during the war, there is ample evidence that some Americans did profit handsomely from the conflict. The largest profits, however, did not go to those whom the public believed to be gaining. They went instead to persons who were on the periphery of the supply system and who capitalized on unusual opportunities that cropped up because of the war.

A prime example of this phenomenon was the case of Robert Morris, the man whose name is most closely associated with Revolutionary War profiteering. As a congressman, until 1778 Morris chaired the Secret Committee, which gave him broad authority over procurement. He then left government and remained in private business until 1781, when he returned as superintendent of finance, charged with the responsibility of overhauling the creaky supply system. Morris successfully advocated that the government switch from using its own procurement officers to advertised contracting with private vendors, on the grounds that open contracting would be cheaper and much less arbitrary than impressment. When the war ended, Morris received much public criticism for enlarging his own fortune while serving in public office. Morris did earn considerable sums during the war, and he did mix public business with private. During the eighteenth century the distinction between private interest and public interest was much less clear than it was to become, and the greatest gain Morris made during the war came while he was out of public service from 1778 to 1781. Undoubtedly the knowledge he acquired as head of the Secret Committee served him well later, but this certainly did not violate any contemporary ethical standard.[48]

Perhaps the single most lucrative enterprise brought about by the Revolutionary War was raiding British sea lanes. At least 800 British ships were captured, 600 by privateers and 200 by the Continental navy. These ships had a total value of more than £5 million and were sold by their captors as quickly as possible. There was certainly nothing unethical about this established enterprise: Congress had approved it, all had a chance at it, and there was substantial risk involved. Even George Washington himself invested in a privateering enterprise, as did other Revolutionary leaders.[49]

Yet there was room for some question. The substantial value of the ships and cargoes seized often led to considerable squabbling about the distribution of the prize shares. American prize agents in Paris resold captured property, realizing a usual markup of twenty-five times but perhaps as high as eightyfold. Congressman John Langdon of New Hampshire resigned his seat in order to serve as U.S. "Agent of Prizes" in that state. Langdon received a commission (usually 5 percent) on the sale of every prize captured by a holder of a congressional commission. He amassed a great fortune during the war, and within six months of its conclusion he began work on a new home that was as opulent as any mansion in New England. Langdon exemplified the opportunity privateering afforded for aggressive, opportunistic men to rise in economic status during the Revolution.[50]

Although most equipment for the Continental army came from abroad, Congress and the state governments gave great encouragement to domestic production. Cannon and gunpowder were the supplies needed most. The total number of cannon forged in the United States is unknown, but it was certainly in the thousands, and most states had foundries.[51] The best documented enterprise was that of Nicholas Brown and his brother Moses. The Brown venture experienced difficulties that would become commonplace in the defense industry two centuries later: unpredictable markets, changing military requirements, inflationary increases in materials and labor, cost overruns, and controversy over prices.

By January 1776, the Browns had begun to expand their foundry in order to produce cannon. Their offer to sell included a requirement that payment be made even if the war ended before production was complete. This transferred the risk of peace (and loss of market) to the government. The Navy Board changed the patterns for the cannon after signing a contract to commence production. During the delay, inflation took a hefty bite out of the profits, but the government was reluctant to cover this cost. It was also slow to pay its bill. Instead, the Browns were labeled "extortioners" by Congressman John Langdon, who himself emerged from the war as a wealthy man! As an antidote to inflationary increases in costs, the Browns refused to quote firm prices on future orders until the cannon were completed. (This anticipated the flexible pricing of World War II.) Despite all the problems, however, it is clear that the Browns profited handsomely, although the amount cannot be gauged precisely. When the war ended, the Browns had paid for a modern foundry that could be converted to civilian production.[52]

Not all American defense industries were so profitable. Powder mills and lead mines failed to reward their owners, largely because of production impediments. Lead mines were either unproductive (as in Massachusetts) or were taken over by the government to prevent extortionate prices on a critical commodity (as in Virginia). When the mine operators of Virginia refused to accept a fixed price of £33 per ton of lead, the Virginia convention took over the mines, offering only to pay "a reasonable annual rent." In the case of gunpowder, there was a national shortage of its vital component, saltpeter (potassium nitrate). No powder mill could operate without it, and Congress controlled the supply. Early in 1776 Congress agreed to send fifty tons to Pennsylvania, ten tons to Massachusetts, and ten tons to New York. In the eighteenth century, as in the twentieth, large states received most of the defense contracts. Actual losses were sustained by men who used their personal wealth to subsidize production or purchases in the expectation that Congress would reimburse them. The director of a state-owned gun factory in Virginia died deeply in debt for that reason, and General Nathanael Greene received his compensation posthumously.[53]

Impelled by deep ideological convictions, the Revolutionary generation took sweeping and innovative steps to restrain war profits. Although by modern standards their governing mechanisms were rudimentary, the methods they devised in hopes of controlling war profits anticipated nearly all of the programs utilized by the federal government during the great wars of the twentieth century.

In the first place, Revolutionary leaders exhorted the public to restrain their appetites. As noted, George Washington was the most frequent and most forceful opponent of war profits in whatever guise. He was by no means alone. In 1780 Major General William Heath issued a public proclamation pleading that "no person or persons will be so ungenerous to their friends and allies . . . as to enhance the prices of the necessaries of life above the current prices." Just in case anyone could not resist the urge to overcharge, Heath threatened to post the names of the gougers conspicuously in public places. Of twentieth-century American leaders, only Franklin D. Roosevelt approached the frequency and eloquence of Washington's moral exhortations.[54]

A second approach was to make the tax structure more progressive. In the maritime towns of New England, flush with profits from privateering, there was a strong demand for luxury goods. In 1777, anticipating the Civil War income tax and modern excess profits taxes, Connecticut began to assess personal property, especially coaches, clocks, and silver plate, at substantially higher rates. From 1779 to 1781 the Maryland legislature imposed a progressive property tax on luxuries, such as silver plate. In 1780 New York approved legislation designed to keep manufacturers' profits, merchants' prices, and the wages of mechanics and laborers at the levels of 1774, the last prewar year. New Jersey established a tax directly pointed at deputy quartermasters general on the assumption that these officials were profiting unfairly from the war. (The Continental Congress resolved that New Jersey ought to reverse its decision.)[55]

Rationing was a third device that presaged twentieth-century techniques. To deal with a critical shortage of salt in several localities, in 1775-76 Revolutionary leaders commandeered private supplies in order to maintain even distribution and prewar prices. In December 1775, the Virginia Convention seized 3,600 bushels, which it alloted to thirty County Committees of Safety to be sold at five shillings per bushel to persons selected by the committees. To counter charges of favoritism, the justices of the peace replaced the Committees of Safety as the distribution agent. Each family received one peck of salt for each family member. In 1776 Pennsylvania allowed the price of salt to triple to 15 shillings per bushel. No family was to receive more than one-half bushel, but only the honor system enforced that edict. Each family was required to make a public declaration "of what Quantity they are possessed of, more than their Just proportion of this necessary Article, at a time of such great Scarcity of it."[56]

Related to rationing as a means of spreading the burden of the war was price control. Most of the states participated in price control efforts, and six conventions were held to establish ceilings. North Carolina was typical. In 1780 its legislature resolved that prices had reached "extravagant height" because of the "wicked arts of a set of men called speculators." The Carolinians attempted futilely to license commerce and limit profits to 25 percent on each transaction. But not all the states joined the effort, and those that did lacked the power to enforce their programs. Prices increased to more than two hundred times greater than prewar levels. In a feeble effort to fight the increase, some authorities resorted to impressment. Since the use of force to bring about public compliance was foreign to republican theory, the authority to impress military supplies was circumscribed by a degree of due process. Before impressment could take place, a justice of the peace had to issue a warrant. Two or three "indifferent" private citizens were to adjudicate disputes as to the reasonableness of the prices offered.[57]

Rampant inflation exposed a conflict in the theory of republicanism, which protected private property and which was averse both to excessive war profits and to using force to control prices. In the case of hoarding to obtain a windfall profit, was the government justified in using impressment to force sales? George Washington was normally and wisely reluctant to undertake impressment, because he considered impressment a violation of the rights of property and because he believed that the practice undercut public support for the cause. Evidently Washington felt that war profits were even more reprehensible than impressment, for in 1777 he used his authority to seize a supply of hides that he believed was being withheld from the market in order to gain a speculative profit.[58]

Congressional and army leaders also experimented with a number of minor changes designed to reduce war profits. Most of these were directed at the commissary and quartermaster department, which were constantly the object of suspicion. Officers were required to supply monthly and later weekly reports of supplies on hand. Proposals were offered to open the supply system to competitive bidding that would result from public notices. The most popular suggestion was to change the system of compensating procurement officers by a commission based on a percentage of purchases to one in which they received a fixed salary. This reform was accomplished in 1777.[59] (This anticipated the World War I decision to switch from contracts based on reimbursement of cost plus a percentage of cost to cost plus a fixed fee.)

The Revolution was peculiar in its deep distrust of military supply officers. Although administrative changes were often suggested, there was a fundamental assumption of the rapaciousness of supply officers that led to numerous demands for their replacement. No official, in fact, who was given a major responsibility in the supply system emerged from the war with his reputation entirely intact. Nathanael Greene, Joseph Trumbull, Robert Morris, Carpenter Wharton, Thomas Mifflin, and Stephen Moylan were all men who suffered a degree of ignominy. Ironically, near the close of the war Robert Morris redeemed his reputation during his term as superintendent of finance only by significantly shrinking the size of the supply system. The verdict of history, shared by the most careful students of the problem, is that the supply problems of the Continental army were not due to the greediness of its supply officers. Their chicanery was grossly exaggerated.[60]

Greene's case is revealing. Appointed Washington's quartermaster general in 1778, Greene supervised a corps of 3,000 men with authority to spend millions of dollars. Having ample opportunity to dip deeply into the public coffers, Greene was guilty of some minor indiscretions. He diverted some military business to a family enterprise managed by his brother and cousin in which he owned a share. In 1779 Greene also entered into a partnership with Jeremiah Wadsworth, the commissary general, and Barnabas Deane, as operating partner. This firm engaged in shipping and privateering, where it gained and lost a fortune. Neither of these enterprises was significantly profitable to Greene, but his commissions on war supplies were impressive. He earned about $170,000 in specie from his appointment, which he converted to extensive holdings in real estate and other ventures.[61]

Perhaps every war leaves behind a degree of bitterness, resentment, and re-crimination. The deep sense of indignation felt by the veterans of the American Revolution was approached in American history only by the dis-satisfaction common to veterans of the war in Vietnam. By 1780 Congressman John Paterson asked: "Where is the spirit of the year 1775? Where are those flaming *patriots* who were ready to sacrifice their fortunes, their all, for the public?"[62]

In the view of many officers and enlisted men, those who had espoused the republican concepts of virtue and disinterestedness in 1775 were now devoted to enlarging their own fortunes. "Those who had played a safe and calculating game," one veteran wrote, "were rewarded for it; pelf, it appeared, was a better goal than liberty; and at no period in my recollection, was the worship of Mammon more widely spread, more sordid and disgusting. Those who had fought the battles of the country, at least in the humbler grades, had as yet earned nothing but poverty and contempt; while their wiser fellow-citizens who had attended to their interests, were the men of mark and consideration." One of his comrades wrote half-apologetically (for a virtuous republican ought not to be bitter) to General Washington: "I candidly confess I feel a de-gree of resentment against the conduct of many."[63]

That resentment manifested itself in a number of ways. In 1780 a dispute arose in Connecticut over war taxes. As an incentive to enlistment, the state proposed that all new recruits would be excused from paying tax obligations. Soldiers already in the national service, however, would continue to be taxed, and much of the tax money would go to pay bounties to the new recruits. The experienced veterans, many of whom who had suffered through the darkest days of the war, resented paying taxes to men whom they considered to be lag-gards.[64]

The national controversy over postwar payments to servicemen was also related to the fundamental question of the equity of the war burden. Officers demanded payment of their back wages in full, plus half-pay for life. Congress agreed to grant unpaid back wages plus five full years of wages, to be paid in federal securities. This compromise brought protests from some taxpayers, who believed that officers of the Continental army had profited improperly during the war. These civilians objected to the permanent continuation of a situation they viewed as inequitable.[65]

The question of amnesty for Tories was also in part a war profiteering issue. Tories tended to be wealthy, at least in New England. During the war they often abandoned their estates to take up residence in politically more con-genial climates. While others were making sacrifices, the Tories had either sat out the war or had actively aided the enemy. When peace returned, there was a real possibility that the property of Loyalists would be returned to them; that was what the peace treaty implied. The thought of British sympathizers living comfortably on commodious American farms so outraged some Americans that mob action resulted.

To prevent another round of charges of extortionate pricing, many of the leaders of the Revolution believed that the government should own its own armaments factories. This was the recommendation of Alexander Hamilton in his famous Report on Manufacturing of 1791.[66] In this respect, Hamilton anticipated the position of Woodrow Wilson in 1919, who endorsed a League of Nations statement that private trade in arms was open to "grave objection." (This was perhaps the only issue upon which Hamilton and Wilson agreed.) In the balmy days of the Republic, the war profits issue seemed resolvable. At a farewell banquet given at the conclusion of the Revolution, comrades-in-arms drank to the belief that the characters who had defiled the cause "will be past us":

> The Merchant who Ven'r
> The Rich Brawling Lawyer
> Plush Coated Quack
> The Meagre Chopp'd Usurer.[67]

The toast, unfortunately, was premature. The nation was not entirely rid of the unvirtuous.

3

Left-Handed Trade

The speculators of the land
A great proportion as we see
Without one single blush demand
Four dollars for a pound of Tea!

UNITED STATES GAZETTE (5 JANUARY 1814)

This Mexican war—a perfect outrage upon every principle of both civilization and Christianity—was got up, and is kept up, mainly by these leaches, so that they may glut themselves on the spoils. Oh, my countrymen, be intreated to tolerate neither these evils nor this accursed war any longer!

AMERICAN PHRENOLOGICAL JOURNAL (NOVEMBER 1847)

When the Continental army disbanded, America's military establishment essentially disappeared. In late 1783 the regular army numbered but eighty soldiers, none ranking higher than captain. During the Articles of Confederation period, Congress increased the army to an infantry regiment and a small artillery unit, but military appropriations were trifling. As long as the army languished as a skeleton force, war profiteering remained a dormant issue.[1] In the 1790s the Federalists rebuilt the nation's military forces, but between the Revolution and the Civil War only conflicts with Britain in 1812 and with Mexico in 1846 resuscitated the war profits controversy.

In the 1780s contention about war profits dwindled to bickering about who gained and who lost during the Revolution. In the Philadelphia Convention in 1787, Pierce Butler of South Carolina sought to have the new U.S. Constitution discriminate between Revolutionary veterans who had been paid for their service in government bonds and those who had speculated in the issues. Butler "expressed his dissatisfaction lest [the Constitution] should compel payment as well to the bloodsuckers who had speculated on the distresses of others as to those who had fought and bled for their country."[2]

When George Washington was inaugurated as president in 1789, the army still counted only 694 soldiers in its ranks. The navy, scrapped after the Revolution, was nonexistent. Washington disapproved of a defense policy that relied heavily on a militia, and Federalist leaders gradually strengthened the nation's regular military forces. When the Federalists left office in 1801, the army's complement had reached 5,400 troops, and the resurrected navy had thirteen frigates in service and six ships-of-the-line under construction.[3]

This military augmentation was not cheap. Between 1790 and 1794, military costs were 40 percent of all federal expenditures, and during that time the defense budget increased by three and a half times to $2.5 million annually. Coastal fortifications costing about $1 million were erected during the Federalist years. To arm the expanded forces, the army had first relied on European suppliers. Washington had painful experiences with supply problems during the Revolution and with the frontier army, and he was concerned lest the French Revolution interrupt the European arms pipeline. He took an active role in establishing permanent federal armories first at Springfield, Massachusetts, and later at Harpers Ferry, Virginia. Besides spending on forts and armories, when war with France threatened in 1798-99 the Federalists budgeted support for twelve army regiments and an enlarged navy. In 1798 private foundries began casting cannon, and the War Department let contracts to private firms for the manufacture of small arms. The early rifle and pistol contracts were generally unprofitable, however.[4]

In 1800, with defense spending threatening to soak up half the federal budget, the Republican Party sensed a winning campaign issue. The $7 million earmarked for the defense budget could be obtained only by borrowing, and Republican leaders charged that interest rates would be exorbitant. After capturing the presidency and control of Congress, the Republicans blocked the bankers' potential gains by cutting the army by 39 percent and decommissioning most of the frigates. Albert Gallatin, Thomas Jefferson's tightfisted treasury secretary, was convinced that the military establishment wasted resources, led to debt, and was largely unneeded. During Jefferson's first years in the presidency, Gallatin held appropriations for the "Peace Establishment" consistently under $1 million annually. In 1802, to save money and wipe out fraud, Congress eliminated the quartermaster and commissary departments (a decision it would regret a decade later). There was little room for complaint about war profits, even though in 1808 a deepening feud with Britain impelled the Republican leadership to return the army's strength to Federalist levels.[5]

Mr. Madison's War

The War of 1812 was arguably America's most unpopular war. The Federalist Party was almost unanimously opposed to the cause, and there were substantial elements of dissent within other constituencies as well.[6] In an atmosphere of unpopularity, the cost of the war and the price of war goods might well have instigated discord, but in this instance they were of secondary importance. The Federalists, who owned a well-deserved reputation for friendliness toward defense spending, were on slippery ground when criticizing President James Madison's military expenditures. The army's muskets and cannon were being made in armories their party had built, and Federalists were the chief proponents of manufacturing at home.

The Federalists preferred a more secure footing on which to base their case against Madison's war policy. They disputed the worthiness of the cause, questioned Madison's war plan, and argued that the war played into the hands of the hateful tyrant Napoleon. Almost as an afterthought, they noted that the war damaged a profitable commerce, required burdensome expenditures, and imposed oppressive taxes and inflationary costs. The Federalists denounced injustices allegedly imposed by Republican war finance, but their reproach centered on regional inequities rather than on unfair individual treatment.[7]

Although questions about the economic conduct of the war arrived late, they were in earnest. In 1813 British seapower interdicted supplies of numerous commodities. Along with heavy military purchasing, blockade-induced shortages drove up prices abruptly. In Philadelphia and other cities along the eastern seaboard there was extensive speculation in such staples as coal, sugar, flour, salt, coffee, and tea. "The gambling is high," observed a Marylander. "Immense sums will be won or lost. It appears to extend from one end of the union to the other. The price of these goods is amazingly enhanced. Many bankruptcies may be expected from this wild business."[8] A Republican journal, doubtless assuming that many commodity speculators were Federalists, denounced price extortion as "criminal." In 1814 a Federalist who identified himself as "ANTI SPECULATOR" reported that the price of sheepskins had become so augmented that someone had butchered four animals for their skins, not even bothering to save the carcasses. Outraged, he demanded that hounds be set forth against the "Scalping Speculators."[9]

The sectoral shifts brought about by wartime conditions inflicted hardship on most Americans but offered opportunity to others. The shipping industry was heavily affected; an American embargo and the British blockade kept merchant and fishing vessels in port. Conversely, land transportation commanded a premium price. In 1814 salt could be purchased for fifty to seventy cents per bushel at Amelia Island, Spanish Florida, or Kanawha Springs, Virginia. Transportation to Baltimore, Maryland, or Charleston, South Carolina, added $2.50 to the cost, but since the price of salt in those cities was $5.00 per bushel, the business was surely lucrative. The blockade tripled the price of coal in Philadelphia in 1813, and the cost of transporting a barrel of flour from New York to Boston increased sevenfold by 1814. Early in 1814 a disgusted Federalist grumbled, "There is one description of persons who may cordially greet each other with 'A happy New Year!' We mean . . . the long list of army contractors . . . who hold profitable or sinecure places as a reward for slandering George Washington and other fine deeds."[10]

Those who were situated to take advantage of these conditions tended to be in the South and West. This deepened the animosity for the war in the North and East. In 1815 advancing prices for cotton, sugar, and tobacco brought anger to seaboard cities: "They have given money a south and southwestern direction," the coast complained. On the other hand, a group of North Carolina merchants became so attached to the lucrative business acquired when soldiers were stationed nearby that they were enraged when army commanders marched the regiments.[11]

Although the war closed off some markets to northeastern merchants, it opened others, at least to those unconcerned about compromising their republican virtue. During the War of 1812, trading with the enemy thrived as in no other American conflict. This was due in part to the unpopularity of the cause and in part to the ready market in British Canada. In the first three months of the war, Britain cut off trade with American customers, except for those granted a special license. The eagerness of Americans to trade with the enemy was made clear when these licenses sold openly for up to $5,000 in American cities. They were also often counterfeited.[12]

Traditional smuggling routes along the New England border reopened soon after the conflict broke out. Each summer of the war, herds of cattle from downstate New York and from Maine, New Hampshire, and Vermont trampled paths through the forest as they ambled northward to nourish the Royal Army in Canada. Not even severe weather could interrupt an American pastime that was as satisfying as it was shameful. In January 1813, smuggling was reportedly "as brisk as ever." In a single winter day federal agents in Vermont captured twenty-seven sleighs "laden with English goods," and charges of smuggling clogged the docket of the U.S. District Court at Rutland, Vermont. In 1814 every day twenty to fifty wagonloads of illicit goods crossed through Castine, Maine, allowing a royal governor to boast that two-thirds of the British army stationed in Canada lived on American beef. "The depravity of these creatures," reproved a Republican editor disgusted with his unvirtuous countrymen, "is beyond anything we ever expected to find in the United States."[13]

Much of the illicit trade was in goods that while militarily valuable would not be directly useful in battle—foodstuffs, cloth, and wire, for example. Nevertheless, trade with the British sometimes clearly crossed the indistinct boundary between subsistence and weaponry. Americans supplied spars for the construction of the British warship *Confiance* built at Isle-aux-Noix on Lake Champlain. Planks intended for the vessel were confiscated and found to be worth more than $5,000. Rhode Islanders living on Block Island maintained their well-earned reputation for disregarding trade restrictions by carrying intelligence and supplies to British blockaders on a daily basis. "Self, the great ruling principle," pronounced a British officer, "[is] more powerful with Yankees than any people I ever saw."[14]

Trade with the enemy was sufficiently extensive to damage the war effort. Numerous troops were sent north in a fruitless attempt to close the border. Armed skirmishes were occasionally fought between smugglers and American border guards—who were known in smugglers' parlance as "highway robbers." In 1814 near Burlington, Vermont, thirteen mounted infantrymen were dispatched to capture smugglers. They were attacked by a gang of thirty traffickers, who killed one soldier and took five prisoner. A mob of eighty armed smugglers including a merchant, a physician, and a grand juror sortied from Georgia, Vermont, and fell upon outgunned border guards, several of whom were badly beaten. "How superabundant is our country of scoundrels," sighed the *Niles' Weekly Register*.[15]

Smuggling was not only violent but expensive. A nationwide shortage of specie developed as money flowed abroad to pay for illegal imports. According to Donald Hickey, early in 1814 $2 million in gold went offshore, followed by another $1.8 million the following summer. Part of this money departed via Canada, but Boston and New York also served as outlet points. In 1814 illicit shipments worth $30,000 each were confiscated at New Haven and Stonington, Connecticut. "This nefarious business has greatly contributed to the scarcity of specie," complained a Marylander. "This left-handed trade is doing us serious injury, totally changing the relations of trade."[16]

Trading with the enemy was a lucrative but illegal way to profit on the war. Another convenient means of profiting was even more remunerative and also legal: privateering. This activity was comparatively expensive and quite dangerous, but these considerations discouraged few participants. In 1812 there were many shipowners still sailing who had hunted the seas for British booty during the Revolution. Others were sons of former privateers or otherwise related to them. Although ownership of a privateering vessel was a rich man's game, there was plenty of opportunity for a common seaman to earn prize money. Owners of privateers customarily split the prize money with their crews, and within each ship the money was normally split evenly between officers and tars. In addition to the value of the ship that could be confiscated, the U.S. government paid a bounty for each British sailor captured. This fee began at $20 in 1812, rose to $25 in 1813, and reached $100 in 1814.[17]

Prize money was equally available to sailors in the U.S. Navy, and whether obtained aboard a privateer or a frigate, the gain could be substantial. Early in 1813 a naval squadron under the command of Commodore John Rodgers captured the British packet *Swallow*, which had a cargo worth $158,000 aboard. When Rodgers returned triumphantly to port, he needed six wagons to carry his booty. Parading with colors flying, drums beating, and accompanied by a marine guard, Rodgers deposited his wagonloads of loot "amidst huzzas of spectators." Few Americans can have arrived at a bank in more spectacular fashion than did Rodgers, and when the news of his coup spread to other seaports the effect was electric. Rodgers' personal share was nearly $12,000, which was comparable to the bonus paid to Commodore Oliver Hazard Perry after his victory in the battle of Lake Erie.[18]

Privateering was a popular enterprise from the Republican South to Federalist New England. In Salem, Massachusetts, the *Essex Register* kept a running account of the profits obtained. In November 1813, as the second year of the war came to a close, Salem lost seventeen privateers worth $164,100. By contrast, Salem's adventuresome seamen had seized and brought to port enemy ships valued at $675,695.93, a return of better than four to one. The great disadvantage to the privateering business was the substantial risk of capture, but American shipwrights offered a solution. A sleek new ship design possessed an uncanny ability to point high into the wind, greatly improving the chances of eluding a pursuing English vessel. When war profits were high, technology responded.[19]

In December 1813, Congress belatedly imposed an embargo intended to block illicit trade with the enemy. Although comprehensive, the embargo of 1813 was not total, as it allowed privateers to clear American ports. This led to abuses, as "privateers" (particularly those from New England) sometimes packed their holds, declared they were seeking prizes, then sold their cargos to the enemy. When privateers or merchantmen were captured, their owners often paid ransom to recover the vessel and crew. When British ships off Cape Ann captured American coastal vessels, they customarily sent the captain ashore to obtain $200 in ransom to recover his ship. Occasionally they detained the captain himself pending payment of a $50 ransom. Some American captains therefore allowed themselves and their ships to be captured in return for a share of the ransom money. In February 1814, Congress attempted to prevent "collusive captures" by forbidding the ransoming of prizes.[20]

Besides ransoming, two other hoary old forms of war profiting that were waning were plunder and scalp bounties. When news of the declaration of war in 1812 reached the northern border, Americans living in Eastport, Maine, crossed into New Brunswick and plundered the village of St. Andrews, seizing all the goods they could find. But when the tides of war turned, the Canadians crossed back into Eastport and ravaged the American plunderers, revealing the disadvantage of the practice. Early in the war an American privateer, aptly named the *Midas,* plundered Harbour Island in the Bahamas, stealing 740 doubloons from one unfortunate British subject. President Madison revoked *Midas*'s letter of marque. The proscription against plundering had limits, however. In 1814 Captain William Patterson plundered a pirate settlement on Grand Terre Island in Barataria Bay in Louisiana. Patterson seized seven schooners, plus specie and dry goods, worth about $500,000. No objection was raised; plundering an English settlement might be ungentlemanly or imprudent, but plundering a pirate camp was sensible policy.[21]

The War of 1812 set the frontier aflame, and with it the issue of scalping arose once more. Disregarding their own customary practice of two centuries of forest warfare, Americans disingenuously professed outrage that the demonic English were reportedly offering a bounty of six dollars per American scalp. (As late as 1794, committees of private citizens in Cincinnati and Columbus, Ohio, were still offering bounties for Indian scalps. They paid according to a sliding scale of $136 for each of the first ten scalps, $117 each for the second ten, and $95 each for the third ten.) Nevertheless, the scalp bounty charge became a staple of American propaganda. Early in 1813 the semiofficial *Niles Weekly Register* reported that the British ship *Euretta* bound for London was carrying in its hold five hundred American scalps. Since the price the British were allegedly offering in 1812 was well below the previous market price, and since Americans regularly paid ransom to retrieve prisoners captured by Indians, this murderous form of war profit had probably ended or at the very least was mercifully close to its conclusion.[22]

Although the practice of paying scalp bounties had faded, Congress relied heavily upon other cash prizes to encourage military service. The sons of the minutemen still voiced support for the Revolutionary concept of republican self-sacrifice—the main formal qualification for an army commission was "virtuous conduct"—but in practice they depended on very attractive monetary rewards for enlistment. The cash bounty for enlistment was alone worth about a year's wages for a young man, and the total bounty was more than most people earned in two years. The Jeffersonian purist John Taylor of Caroline lamented, accordingly, that the virtuous soul of Revolutionary America was dimming. The need for enlistment bounties as well as the widespread practice of trading with the enemy offered clear evidence that the fervid individualism of the Jacksonian years was making its appearance.[23]

The problem facing the Madison administration was how to make use of monetary incentives without succumbing to avarice. To this end, various means for controlling war profits were employed, several of which originated during the Federalist years. There was good reason to steal the Federalist thunder, because as the months passed, criticism of the war became increasingly shrill. In December 1812, the "Friends of Peace" of Salem, New Jersey, denounced war costs: "ONE HUNDRED MILLIONS per annum will be required," the Friends exaggerated. "Who, fellow citizens, are [these expenditures] to benefit? None but office holders, army contractors and the favourites of a party administration." "Loans on loans, taxes on taxes, double and increasing prices," cried a Federalist. "What benefit is this war to any but those who live on the spoils of the people?"[24]

In 1812 Congress reestablished the army's supply agencies, casualties of Jefferson's cost-cutting campaign a decade earlier. Madison appointed a college friend and veteran of the Revolution, Morgan Lewis, to be quartermaster general. Callender Irvine became commissary general of purchases, and Colonel

Decius Wadsworth assumed the position of commissary general of ordnance. Although there was considerable waste and some serious shortages of supplies, these men were generally capable, and the supply system, while still inefficient, was improved from the woeful performance of the Revolution.[25]

The ordnance department had the most innovative management. The government armories at Springfield and Harpers Ferry, which the Jeffersonians much disliked, served them well during "Mr. Madison's War." Colonel Wadsworth implemented a system of controlling procurement costs by dividing production of both small arms and cannon among federal armories and private arsenals. When muskets were produced in various places with interchangeable parts, an accurate estimate of the cost of production was possible. The cost of producing weapons in a government armory served as a "yardstick" by which to measure the cost of production by private contractors. This was the origin of the distinctive feature of American arms production by which weapons development and production is carried forth by a mixture of federally owned armories and private firms. By contrast, militia units in such states as Massachusetts, Pennsylvania, and Kentucky were very poorly equipped.[26]

Supplies of provisions and clothing were less efficiently managed. The system of providing army food had suffered severely during the Jeffersonian frugality. It had become customary before the war for contractors to omit parts of a consignment, such as candles or soap, and if necessary to bribe commanders to take delivery of the shipment. To stop this practice, the commissary department attempted to force contractors to sign a binding agreement and post a performance bond, but this method proved ineffective. Numerous instances of the degraded quality form of profiteering were reported. Shipments labeled "prime beef" turned out to be horsemeat, army bread was too hard, and leaky barrels that spilled brine resulted in inedible pork. In perhaps the most repulsive example in the history of degraded supplies, a frontier doctor found excrement in a shipment of army bread.[27]

Improvements in the Quartermaster Corps were not sufficient to eliminate fraud. In 1814 Colonel James Thomas, quartermaster general of the northern army, and his assistant Michael T. Simpson overstated the cost of provisions they procured. Because quartermasters customarily received a fee in the amount of 2.5 percent of the sums expended, magnified prices resulted in substantial profits for the two men. Significantly, the discrepancy was discovered when their accounts were audited, suggesting an improvement in managerial techniques. On the day after Congress was notified of the discrepancy, a Republican member called for an investigation of supply contracts. Representative Lewis Condict, a physician from New Jersey, demanded to know if any funds had been "misapplied," if any supplies had been deficient or "unfit to use," what was lost, and what was being done to recover it. In requesting such an investigation, Condict anticipated numerous other legislative inquiries into arms procurement, including the Committee on the Conduct of the War of 1862-65, the Nye committee of 1934-36, and the Truman committee of World War II. Condict's demand was rejected, and his quest for legislative account-ability, like those of many of his successors, did not bear fruit.[28]

Congress attempted to even out the burden of the war through its system of war taxes. The first war taxes were imposed by the Republican-dominated Congress in 1812, and they sharply increased duties on imported goods. This had a heavy impact on the Federalist cities of the North and East, who protested loudly. The Republicans had traditionally been opposed to direct taxation, but the war emergency, like so many others in American history, forced them to revise their doctrine. At least two of the taxes were regional leg-islation: a twenty cent per gallon tax on distilled spirits (heavier than the "Whiskey Tax" that prompted rebellion in 1794) and a twenty cent per bushel tax on imported salt. Both of these fell most heavily on southerners and west-erners. The liquor tax was discriminatory because whiskey was the drink of choice in their region, and the salt tax was discriminatory because interior farmers had great need for salt but little access to it. The Republicans also made a clear if rudimentary attempt at progressive taxation. Several luxury taxes were imposed. These taxes would apply only to those who owned gold, silver, and plated ware, gold and silver watches, fancy carriages, and expensive furniture. In 1815 Treasury Secretary Alexander J. Dallas proposed adoption of an income tax that would heavily impact the North and East, but he met with no success.[29]

The tax structure made a halfhearted attempt to distribute the burden of the war evenly, but the system of war finance offered some considerable gains to its financiers. In March 1813, the Treasury Department sold notes in the amount of $16 million. Unfortunately, these could not be sold at face value, as they brought only $88 for every $100 bond. The 12 percent discount resulted in a postwar gain of nearly $2 million to the purchasers. The primary beneficia-ries were wealthy merchants who subscribed to two-thirds of the loan: David Parish, Stephen Girard, and John Jacob Astor.[30]

A year later, Congress authorized loans in the amount of $25 million. In May 1814, the Treasury sold $7.9 million in 6 percent bonds having a face value of $100 but discounted to $88. The bankers extracted a promise that if bonds were sold at a more favorable price later, the lower price must be made retroactive. In August, the Treasury sold $2.5 million in 6 percent bonds dis-counted to $80. According to the agreement, the Treasury was compelled to give a rebate to the first purchasers. Therefore, the bankers loaned $10.4 mil-lion for government obligations carrying a 20 percent discount. When the notes were repaid after the war, the bankers realized a capital gain of $2.6 mil-lion.[31]

The War of 1812 marked the climb to banking prominence of Stephen Girard of Philadelphia, who used the occasion to achieve a place in the thriving ranks of great financiers. In early 1812 Girard formed a private, un-chartered bank that met severe difficulty in gaining recognition in the nation's banking community. Shrewdly perceiving an opportunity presented by the Treasury Department's need to place war loans, Girard agreed to act as the government's broker. This agency brought recognition and good profit to Girard's bank, which continued to record comfortable profit margins for many years after hostilities ceased.[32] Girard thus pioneered a path that would be followed eagerly and gainfully by later opportunists—W. W. Corcoran on the next occasion and Jay Cooke during the Civil War.

Mr. Polk's War

America's war with Mexico was hotly debated, but its financial burden formed only a minor part of the quarrel. In the euphoric days of Manifest Destiny, war supporters believed that America was upholding revolutionary republicanism against the menace of benighted monarchism. Albert Gallatin, the eighty-five-year-old former secretary of the treasury, was a lonely voice when he decried the Mexican incursion on the ground of its economic cost. Struggling against an impassioned tide of expansionism, Gallatin's pleas for frugality fell on deaf ears.[33] The war was cheap in relation to its predecessors, and when the nation's gains were measured against the economic costs the war imposed, it was among the most highly rewarding of America's many successful military ventures. The 1840s were an inopportune time for the war profits argument to thrive.

There were several reasons why the war profits question lay dormant during the Mexican War. Although the war aroused vigorous disapproval, the opposition was divided into two camps. Both groups condemned the aims of the war, but they differed on its intrinsic evilness. The opponents generally concurred that the United States was engaged in what the Rhode Island legislature called a "war of conquest." The more moderate Whig critics simply asserted that the war was unwarranted aggression, unlawfully precipitated by the Democratic president, James K. Polk. Speaking at Faneuil Hall in Boston in November 1846, the Whig spokesman Daniel Webster contended (inaccurately) that "the Mexican war is universally odious throughout the United States." By censuring it as a "presidential war," Webster disclosed the basic Whig strategy of emphasizing Polk's culpability as a bringer of war, as opposed to accentuating the cost of the war.[34]

The more extreme critics asserted that Polk was carrying out a secret but nonetheless recognizable plot to seize territory in order to facilitate the spread of slavery. The "unhallowed object" of the war, declared a Presbyterian divine from Ohio, was deeply sinister—the acquisition of land for the expansion of slavery.[35] Thus while both groups alleged that the war's ultimate objective was economic gain—the conquest of territory and/or the preservation of slavery—

neither alleged that the war's immediate purpose was direct personal profit. The more moderate, politically astute Whig critics recognized the broad support the war attracted. They elected to focus their criticism on Polk's actions at the inception of the war and generally to avoid attacks on the conduct of the war itself. This shrewd choice was undoubtedly influenced by the presence at the front of two commanders who were being considered for the Whig presidential nomination: Generals Zachary Taylor and Winfield Scott.[36] The cost of the Mexican War thus became of secondary concern to the war's critics; the political opponents preferred not to challenge the execution of Polk's war plan, and the abolitionists had bigger fish to fry. Slavery was far worse than war profiteering, and the abolitionists would not fail to exploit any opportunity to vilify the greatest evil of their day.

Although a matter of secondary concern during the Mexican War, war profits nevertheless attracted some attention, largely but not exclusively from the more radical war opponents. "War, wasting a nation's wealth, depresses the great mass of the people," observed the transcendentalist reformer Theodore Parker. "But [war] serves to elevate a few to opulence and power. . . . There is one class of men who find their pecuniary advantage in it. I mean army contractors, when they chance to be the favourites of the party in power." Other abolitionists used the existence of war profits as an opening to undercut the moderation of the Whig position. One line of argument held that so many Americans were making money out of the war that the profits undercut any inclination to consider and spurn the war's sordid purposes. Another line of attack was that war profits showed that Polk alone could not be responsible for the war's existence and continuation. "Did nobody vote for the supplies, and are not supplies the sinews of war?" demanded another New England clergyman. A final abolitionist charge was that the conquest of additional slave soil was motivated by an avaricious desire to drive up the price of slaves.[37]

If they had been so inclined, war critics would have had little difficulty finding ample grist for controversy in the mobilization effort. The Mexican War was America's first foreign campaign, and maneuvers at great distance presented unusually disruptive supply problems. Until Congress approved a declaration of war in May 1846, no special funds had been appropriated. The War Department was then expected to supply no fewer than three armies of invasion simultaneously. Each army was stationed about a thousand miles from Washington, and each expected to stretch its supply line much farther. Transportation therefore became a major part of the war's cost. The War Department purchased massive quantities of ships, draft animals, and wagons, and these caused sectoral shifts in the American economy that were potentially controversial.[38]

Estimating the price of campaigning at great distances was extremely difficult. In the first eighteen months of the war, the Quartermaster Corps bought 11,529 horses, 22,907 mules, 16,288 oxen, 6,886 wagons, and 72 sailing and steam vessels. It also hired 300 wagons, 5,000 mules, and several hundred other watercraft. Purchases of this magnitude inevitably drove up prices. Surf boats needed for General Winfield Scott's attack on Vera Cruz were budgeted to cost $200 but eventually cost $950. Trading in used wagons became very active. In Baltimore, used wagons that had cost $90 new could be sold to the army for $150. There were reports of persons who speculated in used wagons making as much as $600—a year's wages—in a single day.[39]

Along the Gulf Coast, military needs affected economic conditions dramatically. The quartermaster corps purchased or leased almost every vessel available in New Orleans and Mobile, and some were "seaworthy" only by a loose wartime definition. Although government agents tried to obtain the lowest possible prices, the seller's market produced rates that were favorable to the shipowners. By February 1847, more than 100 merchant vessels were plying the Gulf under lease to the government, and while this development was very advantageous to the shipowners it was unwelcome news for shippers. New Orleans traders engaged in civilian commerce were throttled by the shortage of shipping capacity, which caused shipping rates to double.[40]

Food prices fluctuated dramatically. In addition to the heavy military demand, there were crop failures in Europe. Foreign sales of American agricultural commodities, particularly in Britain and Germany, were exceptionally strong. In Corpus Christi, a supply depot for the invasion of Mexico, corn reached the unheard of price of $1.50 per bushel. Prices of most other provisions also shot upward, but beef did not. With ample herds of beef cattle available in Texas, beef prices held stable at one cent to one and one-half cents per pound. Salt pork, a staple of the military diet, was considerably more expensive. A barrel of pork weighing 320 pounds sold for only $10 in St. Louis, but shipping it to Santa Fe added $40 to the price, resulting in a final price of 15.63¢ per pound. When army supply officers entered these unstable markets, they became easy targets for critics. A Whig commentator maintained that "utter ignorance, waste and extravagance have marked all the arrangements." Most war foes, nevertheless, were preoccupied with attacking Polk and the expansion of slavery and did not pick up the scent of scandal.[41]

The Polk administration deftly skirted criticism of the procurement process. Because he distrusted the bargaining abilities of naval officers, Polk appointed Gideon Welles (who would later become famous as secretary of the navy during the Civil War) to head the Naval Bureau of Provisions and Clothing. Welles served efficiently as the navy's chief purchasing agent, and the army was well served by capable officers in the quartermaster, commissary, and ordnance departments.[42] Quartermaster General Thomas S. Jesup demonstrated his resourcefulness by relying heavily on army arsenals for the production of war matériel. The federal arsenal at Watervliet, New York, hired 400 boys to manufacture musket cartridges, of which a million were under order. An even more dramatic expansion of production occurred at the Philadelphia supply depot. The number of workers employed at Philadelphia increased from 400 to 4,000 (many of them seamstresses), making it very likely the largest manufacturing complex in the nation. Producing 500 different articles, at peak production the depot sent the troops 12,000 pairs of shoes, 700 tents, and 35,000 garments per month.[43]

With production under the control of the army itself, there was little room for criticism in respect to extortionate pricing or quality degradation. The army admitted that it paid its temporary workers a premium wage—which the quartermaster general estimated at about one-third "too high"—but this stirred no uproar. Still, the system was by no means foolproof. A shoe manufacturer holding a contract to supply the army with shoes at $1.05 per pair had a shipment rejected for poor workmanship by a government agent at Philadelphia. The supplier then sold the unsatisfactory shoes to an intermediary, who cleverly ferreted out another government buyer who agreed to accept the entire lot at a price of $1.50 per pair.[44]

The peculiarities of a war with Mexico affected the war profits problem in other ways. Since the Mexican ocean fleet was of negligible size, there was little chance for profit in privateering. U.S. naval officers were able to relieve the boredom of blockade duty and to find financial opportunity by enforcing the ban on slave trading. In 1846 the USS *Yorktown* captured the slaver *Pons,* resulting in a lucrative distribution. Half the prize money went directly to the officers of the *Yorktown,* and the other half was set aside for their pension fund.[45]

President Polk elevated and semilegitimized the ancient art of plunder by announcing that "it is the right of conquerors to levy contributions." He then directed the army to commandeer Mexican property to offset war costs. An early report by Secretary of War William L. Marcy estimated that "captured property" confiscated from the Mexicans carried a value of $3,840,000. (The money went into the army's general purpose fund.)[46] The flow of goods was not entirely one way, however. Since most of the supplies for General Scott's invasion force were purchased in Mexico, the Mexican economy received a welcome influx of specie. When the American army withdrew from Vera Cruz in 1848, it disposed of many of its animals and much of its supplies at public auction. A number of criticisms were expressed about the prices obtained for the American equipment—one ship, the *Saint Louis,* was sold for scrap for $500—but the Mexican markets were too far off to stir much contention in the United States.[47] The war had been far too successful to permit much quibbling about a few disadvantageous sales to the conquered.

Although little was made of it at the time, the Mexican War was lavishly valuable to its financiers. When the war commenced, the U.S. Treasury was running a hefty surplus resulting from income from a protective tariff. The Democratic Party was committed to reducing the tariff, which it accomplished in 1847. (The Walker tariff, the only major wartime tax cut in American history, probably helped neutralize complaints about wartime gains.) Having cut taxes, in order to finance the war the Polk administration was forced to borrow. In 1847-48 the Treasury sold bonds totaling $34 million in two issues. The major player in the bond purchase was the Washington banking firm of Corcoran and Riggs, which had close ties to Secretary of the Treasury Robert J. Walker and other influential Democrats. Smaller shares went to the banking house of E. W. Clark and Co. of Philadelphia and to other firms.[48]

After the bankers contracted to purchase the Treasury bonds, their role was to resell them on the open market. As they anticipated, they were able to resell most of the bonds at a higher price, turning a tidy profit. In 1847, W. W. Corcoran, the senior partner in Corcoran and Riggs, received $281,000 as his share of the firm's profits on the war loans. In 1849 his earnings were another $130,000. Jay Cooke, then a junior partner in the Clark firm, did less well, but he would go on to augment his fortune and gain notoriety as the financier of the Civil War. By contrast, an army private received $8 per month, or $96 per year (plus subsistence). Even a colonel's pay was only $75 per month, or $900 per year.[49] Corcoran, the war's financier, thus achieved an income in 1847 nearly three thousand times as large as that of a combat soldier. That he could do so without attracting critical comment exemplified the dedication of the Jacksonians to the concept of unfettered economic individualism.

4.

The "Shoddyocracy"

Shoddy coats, shoddy shoes, shoddy blankets, shoddy horses, shoddy arms, shoddy ammunition, shoddy boats, shoddy beef and bread, shoddy bravery.... We can vie with any people who ever practised the great art of knavery.... Men ... should carry a scarlet 'S' upon the forehead, much more prominent than the 'A' on the breast of poor Hester Prynne.

HENRY MORFORD, *THE DAYS OF SHODDY: A NOVEL OF THE GREAT REBELLION OF 1861* (1863)

Within weeks after Confederate siege guns forced the capitulation of Fort Sumter, rumors of scandal in the War Department began to circulate. By the first Christmas of the war, these accusations reached a crescendo. "The record is a sad and gloomy one," pronounced the *New York Herald,* the nation's largest newspaper. Of the first $200 million spent on the war, the *Herald* maintained, $50 million had been "dishonestly pocketed."[1]

For a century thereafter, the history of Northern mobilization has been presented as a kind of morality play. Saving the Union and ending slavery were noble accomplishments, the story went, but providing the tools of victory gave rise to rampant corruption. Greed on the home front became an inglorious but oft-repeated counterpoise to the sacrifice of the troops at the front. Partisan charges taken from a deeply divided Romantic era substituted for a measured historical account, with Progressive historians being notably meticulous in repeating the wartime allegations.[2] In his muckraking 1907 account, *History of the Great American Fortunes,* Gustavus Myers told of "indiscriminate plundering," "stupendous corruption," and "shameless frauds." "The Federal armies," Myers charged, "were ... the helpless targets of the profit mongers of their own section who insidiously slew great numbers of them—not, it is true, out of deliberate lust for murder, but because the craze for profits crushed every instinct of honor and humanity, and rendered them callous to the appalling consequences."[3] In the aftermath of the Great War, Myers's charges were echoed and expanded by Fred A. Shannon, A. Howard Meneely, and others. The sparse rank of dissenters included R. Gordon Wasson, Russell F. Weigley, and J. Matthew Gallman.[4]

Steeped in controversy, like the conflict itself, the issue of war profits during the Civil War remains unresolved. A careful investigation of the Union mobilization effort (the Confederate side is not a part of the present discussion) discloses a degree of complexity that has not been sufficiently examined. Most writers who have described the scale of corruption have based their assessment on the improvised, haphazard procedures of the early days of the war rather than on an examination of the full record of the conflict. Although chicanery was never absent, profits accruing to military contractors and others derived largely from the massive, unprecedented disruption of the civilian economy at the commencement of hostilities; from the unplanned, hell-for-leather mobilization; from the rudimentary nature of a state-based arms procurement system; and from the war's unexpected and unpredicted duration. "Few . . . know or can appreciate the actual condition of things . . . in those days," remembered Secretary of the Navy Gideon Welles of the desperate situation in 1861. "Nearly sixty years of peace had unfitted us for any war, but the most terrible of all wars, a civil one, was upon us, and it had to be met."[5]

The War for the Union was a transitional experience in respect to technology and tactics, but it also modified war mobilization. The groping methods of the early months gradually gave way to more systematic procedures as a corps of able managers rose to power. This expansion of administrative capacity—what Frank E. Vandiver called the ascendancy of "democratic militarism"—was part of an organizational augmentation that was under way throughout the Union government.[6] The resourceful administrators included civilians and soldiers holding key executive posts in various departments (such as Secretary of War Edwin M. Stanton, Quartermaster General Montgomery C. Meigs, and de facto Chief of Ordnance Captain George T. Balch).[7] Aided by increased congressional scrutiny, they boldly streamlined the procurement process, most by shifting contracting from the states to the federal government. These measures reduced, but could not eliminate, the incidence of profiteering.

The Civil War was also a transitional episode in respect to opportunities for financial gain. While the principal forms of profit seeking that stirred controversy continued to be extortionate pricing and degraded quality, the war also produced contention over several older types of profiteering that would decline or disappear in the twentieth century: plunder, privateering, trading with the enemy, and bounty jumping.

The transitional nature of the war is revealingly illustrated by inconsistent policies governing plundering. Naval officers were allowed to receive a monetary prize for capturing enemy ships or property, as was traditional, but army officers were banned from these gains. In another striking example, in 1863 Lincoln's cabinet debated whether to recommence the hoary practice of privateering. Secretaries William H. Seward and Salmon P. Chase of the State and Treasury Departments wanted to issue letters of marque to permit private vessels to interdict Confederate ocean commerce, but Gideon Welles and Attorney General Edward Bates objected, arguing that privateering would antagonize Britain and would not significantly dent Confederate trade. They eventually compromised by deciding to commission private vessels into the navy.[8]

As the war lengthened, these traditional varieties of profiteering were modified and complicated by the implementation of advanced technology. The industrial revolution introduced to the military environment such technological advances as railroads, steamships, and machine tools. Along with several new and more complex weapons, these machines opened fresh opportunities for profit making that would remain at the center of the profiteering controversy in the twentieth century. These modern modes of war profiteering included patent royalties, stock market speculation, and gains in executive income. To reduce the new and numerous manifestations of war wealth, the War Department and the Civil War Congress devised control mechanisms that were experimental then but that would flourish in later wars: a federal income tax, renegotiated pricing, and prosecution for fraud in war contracts.

Besides its military-political qualities, the War for the Union was also a transforming cultural event, and it denoted the emergence of profiteering as a minor but enduring theme in American fiction. Henry Morford, an author whose goal was to "stamp the human vipers with infamy," became the first American novelist to employ profiteering as a major theme. In an otherwise unmemorable 1863 yarn, *The Days of Shoddy*, Morford invented Charles Holt, a conniving New York merchant, as the prototypical fictional profiteer. While daydreaming during a church service immediately following the assault upon Fort Sumter, Holt plots to sell shoddy cloth to the army. His scheme succeeds spectacularly, and he becomes vulgarly opulent. Olympia, his alcoholic and adulterous wife, personifies by her debauched morality and coarse manners the commercial class the war allegedly spawned. This group of loathsome profiteers would become known as the "shoddy aristocracy" or, more simply, the "shoddyocracy." It was a metaphor for Civil War business itself.[9]

During the first three months of the Civil War, effective planning of the Union war effort was crippled by a fundamental misunderstanding about the enormity of the conflict.[10] When the chief of the Bureau of Ordnance, only a month after the beginning of what would turn out to be a major war, disposed of 5,000 carbines at a bargain price, he made a costly blunder. For this mistake and others General James W. Ripley was dubbed "Ripley Van Winkle." In Ripley's defense, he was acting under a widely shared assumption that the rebellion would be short-lived and that the army would not grow beyond 250,000 men.[11] In April 1861, optimists estimated that the rebellion would be crushed within ninety days, whereas pessimists thought that victory might take six months or a year. The *New York Herald* placed the expected cost of the war at $150-$200 million.[12]

Of course, all these estimates were grossly inaccurate—low by a factor of eight. In the early weeks of the war, this misapprehension seriously hampered defense procurement, as there is a great difference between preparing to fight a three-month rebellion and organizing for a major war that might last for several years. The U.S. Army eventually grew to 1,700,000 men, not a quarter-million, and the Civil War ended up as the most rushed mobilization in American history.[13]

Because the rebellion was expected to be short-lived, the War Department delayed sponsoring new arms factories, and the Navy Department elected to lease ships rather than to build them. Forecasting the length of wars is a notoriously inexact art, and mistakes continued throughout the Civil War. In December 1861, when the war had already far outlasted early predictions, Salmon P. Chase estimated its cost at $475 million based on an expected further duration of one more year. In September 1863, a year and a half early, the *New York Herald* confidently announced that "the rebellion is now drawing to a close."[14]

A second problem originated in the peculiar location of the rebellion. The city of Washington was nestled in an area harboring numerous disloyal citizens, and for good reason the Lincoln administration presumed that some civil servants were disloyal and might give military secrets to the rebels. (In June 1861, when an oath of loyalty was administered to one hundred clerks in the War Department, twenty refused to take it.)[15] To prevent the transmission of military information to the enemy, the Union government preferred to locate defense procurement in the more loyal states until the capital could be cleansed of treason.

During the colonial period, and particularly during the Revolution, military procurement had been largely a state function. State militias remained in existence, and each had a quartermaster whose responsibility was purchasing war matériel. Most were located closer to the centers of defense production than was the War Department in Washington. Communication lines from Washington to the loyal regions necessarily passed through rebel territory, where they were subject to disruption or interception. For all these reasons, the Union war effort initially concentrated on state-level purchasing. There was a distinct resemblance to the mobilization for the Revolution, in which states played the major role.

Shoddy Uniforms

State-level purchasing led to the first major procurement scandal of the Civil War, an episode that caused the conflict to gain a reputation for having an unusually high level of profiteering. This was a case of the degraded quality type of profiteering, and it involved the purchase by the State of New York of uniforms for its volunteer troops. Some of these uniforms were hastily sewn of poor-quality cloth that deteriorated quickly, causing much embarrassment to New Yorkers but no significant damage to the war effort.

The story begins on Friday, 26 April 1861, just two weeks after the fall of Fort Sumter and President Lincoln's subsequent call for troops to suppress the rebellion. The New York Military Board, which included such prominent Republican politicians as Governor Edward D. Morgan and Treasurer Philip Dorsheimer, met in Albany and awarded a contract for the manufacture of 12,000 army uniforms to the clothing firm of Brooks Brothers of New York City. On the next day, Brooks Brothers hired 125 cutters and 5,000 hands. By the following Friday, 5,000 uniforms were completed, and the rest were finished soon afterward.[16]

There were numerous problems with the Brooks Brothers contract. Although the contract was opened to competitive bidding, potential suppliers were given only twenty-four hours to prepare and submit their bids. Robert Freeman, the agent for Brooks Brothers, improperly obtained prior knowledge that gave his firm an unfair advantage, and the wife of the state treasurer who signed the contract received a new dress as a gift from the contractor. Although the contract was for a large number of uniforms, Brooks Brothers did not divide it among other clothing firms, which would have hastened completion and avoided jealousy. Brooks Brothers submitted its winning bid at $20.00 per garment, but Governor Morgan forced a reduction only to $19.50, even though subsequent contracts would be made at $18.00.[17]

The most telling charges concerned the quality of material and workmanship. When Brooks Brothers discovered that the supply of regulation army woolen cloth in New York City was insufficient to complete the order, the firm requested and obtained permission from the Military Board to substitute other available materials—petersham satinet, felt, mixed cassimere, and mixed coating. Some of this cloth was of inferior quality, even partially decayed, and it deteriorated rapidly. The lower-grade material was known in the garment industry as "shoddy," but only a trained eye could tell it from superior grades. The construction of the uniforms also bore the mark of great haste. Some garments were badly cut and ill-fitting; others were missing buttons, and some lacked button holes. Many wore poorly: some ripped open when first put on, and others lasted only a few days or a few weeks. This forced combat-eager recruits to spend a great deal of time mending their clothing. The New York Volunteers presented a ragged, unsoldierly appearance, and they were mocked by men from other states. This was a great embarrassment both to the recruits and to their patrons, many of whom imagined that the Civil War would amount to little more than a dress parade, a show of force, and a great victory celebration. Because each soldier was required to pay for his own uniform from a clothing allowance, the gallant volunteers of '61 would also be forced to bear the shameful cost themselves.[18]

The actual extent of the scandal was a matter of dispute. Brooks Brothers claimed that of 36,000 uniforms the firm manufactured in 1861, only 500 were damaged goods. The lieutenant governor of New York claimed that the "greater part" were substandard. After an extensive investigation by the New York state legislature, Brooks Brothers agreed to replace 2,350 uniforms, worth about $45,000. This action partially compensated the state for the firm's two discreditable actions. Brooks Brothers should never have shipped thousands of damaged garments, and when it substituted a cheaper grade of cloth, it should have granted a reduction in price. Asked by a state legislator how much money he had made by the substitution, Elisha Brooks replied evasively and revealingly: "I think I cannot ascertain the difference without spending more time than I can now devote to that purpose."[19]

Although Brooks Brothers had behaved improperly in its discharge of the uniform contract, the matter hardly deserved the notoriety it received, as some contemporaries understood.[20] The terms *shoddy aristocracy* and often simply *shoddy* became generic terms representing all who made profits on war contracts. The first use, and certainly the most extensive use, of the word *shoddy* as meaning defense contractors rather than a kind of cloth was by the large and influential *New York Herald*,[21] which had a partisan reason for doing so. Its editor, James Gordon Bennett, had a strong dislike for the merchant classes of New York City, for the Republican Party, and for the early war policy of the Lincoln administration. The *Herald* implied that *shoddy* was synonymous with the Lincoln wing of the GOP and that Lincoln was a "shoddy candidate." The newspaper also disliked the use of shoddy in cloth, because it allegedly drove up the price of newsprint, which was then made of rags. The price of paper rose 100 percent in the first year of the war, and Bennett, who was probably the largest purchaser of newsprint in the country, said that papermakers were among the worst of the war profiteers.[22]

The Brooks contract offered an ideal opportunity to lambaste the alleged rascals. The contract was let improperly by the highest officers of the state, Republicans all. Besides Governor Morgan and Treasurer Dorsheimer, Thurlow Weed, a high Republican leader and rival editor whom Bennett regarded as personifying the antichrist, played an important but hidden role in securing the contract for Brooks Brothers. The inspectors who approved the uniforms were also influential Republican merchants. The chief of these was George Opdyke, who was to run successfully (against Bennett's strong opposition) for the office of mayor of New York in 1862. Before the war the *Herald* had routinely denounced New Yorkers who had been born to wealth (the "Knickerbocker Aristocracy") and who had made their money in fishing (the "Codfish Aristocracy") or in fiber (the "Cotton Aristocracy"). Now it poured vitriol on the "Shoddy Aristocracy."[23]

There were few limits to the *Herald*'s scorn. It reported that because of shoddy uniforms, New York troops were "half-naked" and that "it is by such contracts as these that the life-blood is being sucked out of the nation by the vampires." It claimed that "politicians . . . have been delving in the shoddy pool, bathing in its filthy waters, . . . but instead of cleansing them of their leprous spots it has covered them with slime that will hang to them during their natural lives." The *Herald* charged that Fifth Avenue brownstones that had become vacant were now filling with families fattened by war profits. Their homes contained shoddy carpets ("brilliant colors and little wool"), shoddy pianos ("all case and no music"), shoddy portraits ("all paint and no likeness"), and even shoddy toys ("dead, pink-eyed rabbits"). The shoddy aristocracy allegedly employed shoddy cooks ("more French than skillful"), served shoddy wines ("with all their excellence on the label"), and drove shoddy horses ("prance more than they go"). When the shoddy aristocracy attended the opera, they supposedly applauded at the wrong time, wore excessive makeup, and even looked through the wrong end of their opera glasses. In justice, the *Herald* demanded, the shoddy aristocracy "ought to be repenting in State prisons instead of living in brownstone fronts."[24]

Despite the extremism of the charges, the label stuck. "Far and wide," reported a New York investigating committee, "[the Brooks Brothers contract] has become the subject of grave discussion, and the clothing which was furnished under it is now only known to the people by the term of 'shoddy.'" A private wrote home from the front that "shoddycracy is pretty large in New York, they say, the hideous offspring of the monster war." And *Harper's Weekly* reported a visit to a Fifth Avenue brownstone inhabited by "Mr. Shoddy" and "Mrs. Shoddy," who supposedly had made $195,000 in two weeks. The former was "a huge Titan in dancing attire—a flabby villainous countenance—diamonds flashing from the center of a wall of ruffled linen—an atmosphere heavy with pomade." Mrs. Shoddy wore "intensely blue silk, and a huge coronet of pink and purple artificial flowers." She used *ain't,* and she substituted *figger* for *finger, parler* for *parlor,* and *reg'lar* for *regular.* Their home was decorated by "abominable, showily framed paintings" by a cheap "Western artist" whose name the Shoddies could not remember. Although the "Shoddies" were fictional, the image of Civil War defense contractors as vulgar *nouveaux riches* was real and lasting. The Brooks Brothers contract incident was paralleled by a similar episode in Philadelphia that led to charges and investigations of cronyism, inferior quality, and fraud.[25]

Railroads

A second early war scandal that received great notoriety concerned the transportation of troops by rail. Again, the unfortunate location of the nation's capital played a role. In early 1861 it was vital that the government rush reinforcements into Washington to prevent the Confederates from capturing the capital. There were only two rail connections to the North: the Baltimore and Ohio ran east of the city to Baltimore and then north, and the Northern Central ran more directly north to Harrisburg, Pennsylvania, where it connected to trunk lines. There was an emergency need for passenger transportation on these lines, and Secretary of War Simon Cameron, whose family owned the Northern Central, had the responsibility of deciding how much of it would flow over his family's track and at what price. Under the emergency conditions pertaining, a conflict of interest was unavoidable.

The transportation of soldiers during the Civil War proved to be a lucrative and controversial business. Three years before the outbreak of the war, officers of western roads had met secretly and agreed not to transport troops for anything less than first-class fares and not to allow the troops to carry with them any more than eighty pounds of baggage. In June 1861, there was further evidence of overcharging on troop transportation. Governor Andrew G. Curtin of Pennsylvania discovered wide variation in the price of transporting his state's volunteers: some roads were charging three cents per passenger per mile (the "local" rate), and others were charging only two and one-fourth or two and one-half cents (the "through" rate).[26] On June 4, Governor Curtin assembled delegates from twenty-one lines in Harrisburg in an attempt to settle on a standard rate. Against the opposition of Cameron's Northern Central, the convention set the rate at two cents per passenger-mile. The roads also agreed that military freight would be charged "local" rates, which were also more expensive than "through" rates. This was to the railroads' advantage because freight transportation was generally more profitable than passenger transportation.[27]

Governor Curtin's convention, which reached a voluntary, government-encouraged agreement establishing standards for an entire industry, set a precedent for federal policy. It anticipated the practices of the War Industries Board during World War I and the associationalism advocated by Herbert Hoover during the 1920s. In another harbinger of the controversies of the World War I era, the Lincoln administration initiated the employment of businessmen on a temporary, unpaid basis for the period of the war emergency— a kind of prototype of the "dollar-a-year" man.

Secretary Cameron appointed as his assistant for railroad matters a vice president of the Pennsylvania Railroad, Thomas A. Scott, who only took a leave of absence rather than resign his position because the rebellion was not expected to last long. Scott thus incurred a direct conflict of interest: while still a railroad executive, he was also serving as an agent of the government. Although the term *conflict of interest* had not yet gained currency, Scott became a target of criticism for bearing competing responsibilities. The *New York Times* protested that "a public official should be . . . unsuspected and above suspicion." Noting that Scott received a salary of $9,000 as a railroad executive but only $3,000 as a government official, the *Times* loftily declared that "the virtue of the public officer should never be subjected to unnecessary temptation."[28] Although this rhetoric was reminiscent of the American Revolution, such an arrangement would not have been protested then. During World War I such arrangements were common, but by World War II they would not be acceptable.

Disregarding his critics, Scott assumed responsibility to work out a standard rate schedule for military rail transportation. The first consideration was the price of shipping into the capital. The Baltimore and Ohio, whose officers were widely suspected of having Confederate sympathies, held a monopoly of traffic from Baltimore to Washington. The B&O charged the government three and three-fourths cents per mile for transporting troops into the city, which was well above the two-cent rate charged by the Pennsylvania roads. For the entire New York to Washington trip, the B&O price was six dollars per soldier. Cameron and Scott forced the B&O to cut this to four dollars, but only by rerouting traffic along the Pennsylvania Railroad to Harrisburg, where it picked up the Cameron-owned Northern Central. Of course, this opened both Cameron and Scott to charges of nest-feathering at their competitors' expense. The B&O did not suffer; before the war commenced, it ordinarily ran eight cars per day into Washington. Within a year, traffic increased to 400 cars per day.[29]

On July 7, 1861, Assistant Secretary Scott issued a directive that would establish the standard nationwide rate schedule for rail transportation for the entire war. Scott's letter set the price that the government would pay at two cents per passenger per mile (the same as the state of Pennsylvania rate). In the manner of countless other bureaucrats, Scott attempted to soften his edict, testifying that he intended the two-cent figure to be "a guide and maximum amount." His letter belied his claim, as it simply ordered War Department agents to "please observe the following as a general basis" when it settled upon two cents.[30]

At this price, troop transportation was very attractive to northern railroads. The cost to the roads of troop transportation was not commonly revealed, but one executive of Cameron's Northern Central testified that the figure generally used in estimating the cost of passenger traffic for long distances was one and one-third cents per mile. However, if freight cars were used, as was often the case with troops, the cost was customarily estimated at nine mills per mile.[31] Therefore, the markup for army traffic was generous—from 50 to 122 percent.

The railroads maintained that there were certain disadvantages to military business that justified these prices. Because troop trains left on short notice, extra conductors and telegraphers had to be employed. The young warriors were allegedly hard on equipment: their bayonets tore headliners, their knives cut cushions and upholstery, and if they were carried in freight cars, they gouged holes in the wall for ventilation. Soldiers often pilfered from other freight cars, especially if they carried tobacco or cheese. Troop trains had priority, which delayed freight trains, and troop transfers were usually one-way journeys, which required deadheading on the return trip. Although these arguments undoubtedly had some merit, their validity was doubtful.[32]

Proof that these nuisances were not overly burdensome materialized when railroads moved aggressively to obtain as much military business as possible. Because the War Department paid a flat rate per soldier per mile, the most profitable contracts were those that carried large numbers of soldiers long distances. In practice, this meant the transportation of western regiments to Washington. The competition for western military business was very active.

To carry a 1,000-man regiment in the Union army required approximately eighteen passenger cars and seven luggage cars, or about two trains. A railroad would charge the government about $14,200 for the 714-mile trip from Chicago to Harrisburg, and about $4,700 of this would be profit. In order to obtain this payment, railroad executives were willing to pay generous bribes to regimental officers—approximately $1,000-$1,500 per regiment. The Sixth Wisconsin, which would distinguish itself in combat at Antietam, Fredericksburg, and Gettysburg, had a rather inglorious arrival. Its commander, Colonel Lysander Cutler, admitted that he accepted a $400 payment from the Pittsburgh, Fort Wayne and Chicago to route his regiment along that line. Other colonels accepted substantially more. The Michigan Southern paid between $5,000 and $8,000 to obtain military business.[33]

Swollen railroad profits in the early months of the war prompted the government to take action in February 1862. It called a convention of the major lines to work out a freight rate schedule. The contest again centered on "local" vs. "through" rates. Scott's schedule of July 1861 required only a slight reduction in rates for the longer distances—about 25 percent. A number of critics charged that the difference should have been much greater. Scott disagreed, but at least one railroad general superintendent testified that "through" prices were about 50 percent too high.[34]

The result was a compromise: passenger rates would continue at two cents per person per mile, and government freight would be carried at prices that were 10 percent below published schedules. In return, the government agreed not to encourage "harsh competition." In this matter, the Lincoln administration anticipated the decision of President Franklin D. Roosevelt to set aside antitrust prosecutions for the duration of World War II. In 1862 Congress approved the Railroad Act, which empowered the government to take over railroad lines if the public safety required. This gave the government a powerful weapon that anticipated the takeover of railroads during World War I and the control of business by the Office of Price Administration during World War II.[35]

Owing to these congenial agreements, nearly all Northern railroads were quite profitable during the early period. The war-stimulated economy produced huge volumes of business for the east-west lines, which benefited most firms. There were few north-south lines in 1861, and while they were more likely to be ravaged than east-west lines, they were still able to remain healthy. In the first six months of the war, profits of most businesses fell, but railroad profits improved, in some cases dramatically. The total earnings of the eighteen northern railroads that issued monthly reports improved from $26.8 million in 1860 to $29.5 million in the early months of the war. This was an overall increase of 10 percent, although four roads lost money. Profits of Secretary Cameron's Northern Central increased 44.5 percent in the first year of the war, a fact which did not escape public notice and which certainly contributed to his replacement. In Cameron's defense, however, he was placed in an impossible position—the safety of the nation and his personal financial interests were inextricably connected. Although his performance was less than sterling, he did not quite deserve the opprobrium he received.[36]

Throughout the war, railroad profits were similarly impressive. The Pennsylvania Railroad increased its earnings from $5 million to $17 million (unadjusted for inflation) over the period 1860-1865, and the New York Central doubled its profits. Indiana lines that had been near bankruptcy recovered so well that they could make repairs and also pay generous, even handsome, dividends. In 1864 the Indianapolis to Terre Haute paid a stock dividend of 25 percent, a regular cash dividend of 5 percent, and a special dividend of 5 percent. Even the Baltimore and Ohio, whose location placed it in great jeopardy of destruction, recorded a total increase in investment from $31.6 million to $43.1 million. The B&O dividend, which its directors set at 6 percent from 1861 to 1863, improved to 7 percent in 1864 and 8 percent in 1865. Perhaps no company was more fortunate, however, than the strategically located Philadelphia, Wilmington and Baltimore, whose profits increased from $236,000 in the last full year of peace to $1,645,000 in the first full year of war—an increase of 452 percent. Railroad express companies, which had the important morale-building role of carrying packages from home to the soldiers and the grisly role of returning corpses from the front, were strategically situated to profit from the conflict. American Express paid a 35 percent cash dividend and a 50 percent stock dividend in 1864.[37]

Shipping

The early war period produced confusion and controversy in the supply of another kind of transportation services: shipping. The Union strategy was first to impose a blockade of the Confederacy and later to invade it from the sea and along the Mississippi River. This required a large and rapid enlargement of the government's shipping capacity. This expansion was at first carried out in the fashion that had been used during the Revolution, but this means proved too costly and too controversial for the nineteenth century.

In the summer of 1861, the United States was a great maritime nation, owning some four million tons of shipping. Nevertheless, the Navy Department lacked the ships needed to enforce the blockade and to transport troops. Navy Secretary Gideon Welles moved quickly to expand the fleet, and like Secretary Cameron in the War Department, Welles relied upon New Yorkers to carry out the expansion. Welles's chief agent was George D. Morgan, an exceptionally well connected individual who had ties both to Governor Edward D. Morgan and to the secretary himself. George D. Morgan was the governor's cousin and Welles's brother-in-law.[38]

The system that Welles employed for the purchase and lease of naval vessels was borrowed from methods used in the ship brokerage business. It was also quite similar to practices used in the American Revolution. The government designated a person to be its purchasing agent, and he in turn received a commission on each purchase or charter, in most cases 2.5 percent. Commissions were irregular in the peacetime ship brokerage business, and 2.5 percent was the customary rate. In 1861 the Navy Department's purchases were huge by prior standards, and a commission of 2.5 percent was a very attractive plum. In four and a half months George D. Morgan purchased eighty-nine ships for the navy, receiving as his commission a total of $95,008—roughly as much as President Lincoln earned in all four years in office. The government did realize a substantial benefit from Morgan's service, since the total cost of the ships he purchased was about $900,000 less than the owners' asking prices. While Gideon Welles exercised good judgment when he employed an astute ship broker as the navy's agent, it seems probable that knowledgeable advice could have been obtained at lower cost. Following heavy congressional criticism, Morgan left for an extended vacation in Europe.[39]

Several unfair charges were hurled at shipping agents. A congressional committee investigating the Burnside expedition of 1862 charged Captain R. F. Loper with extracting exorbitant fees of 5 percent for chartering ships. In fact, Loper charged no commission; he only charged 5 percent interest on his own money, which he advanced to the lessors. Critics also often overlooked the extremity of the emergency conditions. In 1862 John Tucker, the transportation agent of the War Department, chartered shipping for General George McClellan's peninsular campaign. Tucker was later criticized for failing to advertise contracts. In rejoinder, Tucker described how he had been called into Lincoln's office and told that each day's delay cost the government $1 million. The president told Tucker directly that every hour of delay was more disastrous to the nation than was the loss of the money. Tucker pointed out that he had notified shipowners verbally rather than publicly because verbal notice protected military secrecy and was nearly as effective at spreading the news as a written announcement. If the size of the fleet needed to transport McClellan's invasion force—301 ships and 88 barges—had been made public, the enemy would have been alerted to the plan. When McClellan's attack failed, Tucker was directed to bring the army back from the peninsula so that it could be placed between Lee and Washington. "Go," ordered Secretary

Stanton, "and make the whole power of the War Department bend to bring-ing that army away in the shortest possible time." In six days, 80,000 men, 27,500 animals, and 2,600 wagons and batteries were moved back to Wash-ington. To accomplish this feat, Tucker chartered every ship available, all at fixed prices.[40] If Tucker had paused to negotiate prices, he would have been highly irresponsible.

Another aspect of the shipping conundrum involved the price of charter-ing merchant ships for military use. In 1861 a customary rate prevailed in New York and Boston for chartering ships for commercial use. This was three to four dollars per ton per month for a four- to five-month charter, with the owner accepting the ordinary risk of loss at sea. A ship displacing 1,000 tons would thus normally rent for $3,000 to $4,000 per month. In the summer of 1861, there was no shortage of ships in northern ports because business condi-tions were depressed and because owners were unwilling to risk loss of the ships to Confederate raiders. Nevertheless, agents for the Union government paid five to six dollars per ton per month for three-month charters, with the government accepting the risk of loss due to military action. This was about 40 percent above the market rate.[41]

Some of this excess cost resulted from bribes paid to government officials. Lessors of ships sometimes surrendered part of their customary 2.5 percent commissions to naval agents. Washington Libbey, a Boston ship contractor, testified to a congressional investigating committee that he chartered his ship *Eliza and Ella* to the navy for $3,600 per month. At 800 tons, the *Eliza and Ella* normally brought $2,700. Libbey admitted that the extra $900 went to bribe the naval agent. When he was asked, "Doesn't it look very much like cheating the government?" Libbey replied, "Yes, sir."[42]

Another fraudulent stratagem involved the surreptitious sale to the navy of ships personally owned by naval agents themselves. W. H. Starbuck, a ship broker of New Bedford, Massachusetts, was employed by the navy to purchase two whaling ships for a total of no more than $15,000. Starbuck located the *William Badger* selling at $2,500 and the *Roman* selling at $4,000. He had an accomplice purchase the two for the total of $6,500. Acting as the government's agent, Starbuck then bought the two ships from his friend for $14,550, with the two partners sharing the $8,050 gain. In addition, Starbuck pocketed a com-mission of 7.5 percent on the purchase. The navy forced him to refund $6,166.[43]

The extravagance in shipping costs in the early months of the war was ac-tually more waste and inefficiency than it was theft. When officials proceeded under the assumption that the rebellion would be short-lived, it made sense for the government to lease ships rather than to purchase them. Frequently ships were rented at expensive daily rates, and frequently there were delays in sailing orders, as military strategy lacked firm direction. As a result, ships sometimes accumulated embarrassing rental charges even before they left port. In April 1861, the navy chartered the steamer *Cataline* for a fee of $10,000 per month. Since the owners had only recently purchased the vessel for $18,000 plus $5,000 for refitting, the fee was generously remunerative. In addition, the *Cataline*'s owners extracted a promise from the government to pay them $50,000 if the ship were lost. When the ship burned, rumors of arson circulated. Whatever the cause of the fire, the owners turned a tidy $20,000 profit in less than three months. The following year, the government charted the steamship *Marion* for $1,000 per day. Since the ship had a value of $100,000, this was reasonable for a short cruise. But the navy did not return the vessel for 168 days, thereby running up a rental bill that far exceeded the value of the property.[44]

Much of this might have been corrected if the nature of the conflict had been clear at the beginning. Once it became evident that the war would be long and large, the government was in a better position to limit abuses. The disastrous battle of Bull Run revealed the full extent of the struggle, and thereafter officials planned for a more extended period of strife. By November 1861, the navy was driving a harder bargain. When it purchased the *Mercedita*, the Navy Board forced a reduction in price from $135,000 to $100,000 and successfully insisted upon a $15,000 refitting. Assistant Secretary of the Navy Gustavus Vasa Fox gloated privately over the new circumstances: "They came at us with the belief that we were prepared and willing to buy every steamboat at anybody's price, and they have found their error, and gnash their teeth in bitterness and disappointment."[45] If the navy could have proceeded more deliberately early in the war, this reversal might have been accomplished sooner.

More serious than the charges of exorbitant prices for shipping was the charge that ships sold or leased to the government were not seaworthy. This was a most dangerous example of degraded quality in military supplies as it could lead to defeat at sea and mass death. In 1862 the navy chartered fifty-one ships to transport an expeditionary force commanded by General Nathaniel P. Banks to New Orleans. One of these, the *Niagara*, nearly foundered, forcing the crew to make an emergency port call. A full-scale investigation by the Senate followed. Since Cornelius Vanderbilt, the largest shipowner in the country, had chartered the ships, the controversy attracted much attention. Nautical opinion on the seaworthiness of the *Niagara* differed dramatically, with some witnesses pronouncing it safe and others finding it unsound. The investigation did establish certain incontrovertible facts. The ship was twenty years old and had been designed for passenger service on inland lakes. Government inspectors discovered serious rotting of its main beams. Although there were 500 passengers aboard, the *Niagara* carried only 100 life preservers, a clear violation of safety rules. Despite these defects, by the standards of the 1860s the ship was probably safe for lake voyages, except for the shortage of life preservers. The vessel was not fit, however, to round Cape Hatteras in winter, particularly when overloaded with soldiers and their gear.[46]

The *Niagara* affair illustrates the difficulty of assigning blame for degraded quality. The ship was certainly no bargain: its purchase price in 1862 was only $10,000, although its owner claimed he spent another $8,000 to $10,000 to fit her out. This included a coat of paint, leading to an allegation that he simply painted over rotten beams. He leased her to the government for $400 per day, a price that exceeded operating costs by about $300 per day. At this rate, the owner would recover his investment in two months, assuming the vessel continued to float. Nevertheless, the greatest abuse arose from the danger to life deriving from the ship's usage. The navy surely knew that it was dangerous to send an aged and overloaded lake steamer to sea. But war entails risks, and in this case, luckily, no lives were lost. Of fifty-one ships chartered for the peninsular expedition, none sank, and only one was investigated, despite numerous charges of sinister behavior. Commodore Vanderbilt was exonerated of blame, although some historians have doubted his innocence.[47] Vanderbilt went on to serve the Union faithfully, even presenting an expensive steamship (christened, predictably, the USS *Vanderbilt*) to the navy.[48]

There are no verifiable cases of the loss of life due to unsafe ships in the Civil War. There are, however, other verifiable cases of the misuse by the navy of otherwise safe vessels.[49] Insurance carriers were understandably averse to underwriting voyages through narrow, crooked rivers guided by inexperienced pilots. Rates were sometimes nearly astronomical: 33 percent of the ship's value for the peninsular campaign of 1862, and as much as 10 percent per month for other expeditions.[50]

Fortifications

Many of the charges of fraudulent spending in the early war period were directed toward the Department of Missouri, commanded by General John C. Frémont. Missouri was vital to the Union cause: a slave state which controlled important waterways, its loss to the Confederacy would have had a devastating, perhaps decisive, effect. Frémont was the most famed of the "political generals." The Republican nominee for president in 1856, he had a vast following in the abolitionist wing of the party. In late July 1861, following the Union defeat at Bull Run, Frémont believed that he was in great danger of losing the state. "The rebels are advancing in force from the south," he wrote Secretary of State William H. Seward. "We have plenty of men but absolutely no arms, and the condition of the state critical."

Frémont's state of mind goes far to explain why charges were levied against him. He believed that in order to save the state he had to hold St. Louis. To protect the city he ordered that fortifications be built at any cost. A contractor, E. L. Beard, estimated that the city could be fortified for $315,000. Frémont accepted this offer on 4 September 1861, and Beard received an advance payment of $171,000. He hired 4,000 men immediately by doubling prevailing wages.

After spending $3,000 on lights, work went on night and day, and the fortifications were completed in thirty days. In a controversial decision, Frémont released an additional $151,000 to Beard only two days after construction commenced, rather than waiting to pay upon completion. Beard later claimed that his prices were "very fair," but he admitted that the contract "gave a good profit" and cost two to three times as much to complete than if the work had not been done in haste.[51]

There were other examples of needless waste in Frémont's command. Colonel John Reeside, Frémont's inspector of horses, was paid a commission of 2.5 percent of their price for inspecting them, or about three dollars a head. In New York City, by contrast, the inspector of army horses received only fifty cents. Since Reeside inspected about five thousand horses in the last five months of 1861, he received about $15,000 instead of the $2,500 which the New Yorker would have charged (a generous sum in itself). There were also many charges of fraud, some of which were probably true, although this is difficult to determine. For example, Frémont's chief quartermaster, Major (later Brigadier General) Justus B. McKinstry, was court-martialed and convicted of twenty-six violations of military regulations in connection with army contracts in Missouri. Allegedly, McKinstry let contracts without advertising, paid excessive prices, and dealt with middlemen. Most historians have accepted his guilt,[52] but there is reason to doubt it. McKinstry incurred the animosity of the powerful Blair clan, which early sought his removal for failing to purchase from their cronies. He came under heavy pressure to purchase from Republicans alone, receiving a letter to this effect from Abraham Lincoln himself, another from Secretary of War Simon Cameron, and six from Congressman Frank P. Blair's son. When General Frémont ordered McKinstry to purchase $750,000 worth of supplies without advertising, the quartermaster refused. McKinstry was given little opportunity to prepare his defense for his trial, and he had little support from Chief Quartermaster Montgomery Meigs, a Blair ally.[53]

In any case, the expenditures were not sufficiently wasteful or corrupt to merit the tone of criticism from Frémont's political opponents, who were far from the scene and could hardly understand the depth of the emergency. Unlike the situation elsewhere, the Missouri state government was unwilling to assist the federal troops. On one occasion, Quartermaster McKinstry was ordered to buy 500 sets of cavalry equipage "tomorrow." On another, he was directed to purchase clothing and equipment for twenty-seven regiments in two weeks. Unadvertised purchases were also perfectly legal under some circumstances. An act of 1809 allowed unadvertised purchases "when immediate delivery is required by the public exigency."[54] If ever there was such an exigency in American history, it was certainly in Missouri in 1861. In June 1862, congressmen approved legislation requiring written contracts, but one month later they suspended their decision for the remainder of the year.[55]

Nevertheless, one congressman declared that "a horde of pirates" was ruining the credit of the government. The *New York Herald* called Frémont the "Great Mogul" and denounced him for "personal extravagances, rivaling that of princes." (Frémont's headquarters were in an opulent mansion renting for $6,000 per year.) The *Herald* claimed he wanted to establish the "kingdom of Frémont the First." A more measured response came from the congressional Committee on the Conduct of the War. "The exigencies of the department," it reported, "were such that much should be pardoned in one compelled to act so promptly, and with so little at his command." Frémont certainly spent extravagantly, but had he spent excessively? His total expenditure was $12 million. Was this too much to hold Missouri for the Union in 1861? Although some contemporaries and some Progressive historians deplored the waste, President Lincoln was not greatly concerned. Secretary of War Cameron visited Frémont and cut his spending, but when Lincoln replaced his commander in Missouri, it was for military reasons. In a message to Congress, Lincoln declared, "I am not aware that a dollar of the public funds . . . was either lost or wasted." In 1865, when the emergency had passed and the political controversy had cooled, a Senate committee quietly recommended payment for wagons ordered three years earlier by Frémont without competitive bidding.[56]

Firearms

When the Civil War began, there were not enough modern weapons available, either in the United States or anywhere else in the world, to fight a war of its magnitude. The mobilization of Northern manpower was so rapid that American and European manufacturers could not possibly meet the demand for arms. Within four months, the U.S. Army grew from 17,000 men to 500,000—an increase of twenty-nine times.[57] In these months, the acquisition of rifles became a limiting factor on the extent of the slaughter. There was little point in inducting more men until world armaments factories could tool up to equip them. In 1861 and 1862, the Civil War occupied the entire capacity of the American armament industry and much of the world's as well.

A year before hostilities opened, the War Department had shipped 105,000 rifles from Northern arsenals to Southern depots. When the fighting began, Northern inventories of modern weapons were insufficient—there were only about 60,000 up-to-date weapons among the 500,000 on hand.[58] Before the war, the government-owned Springfield armory produced about 10,000 rifles per year. By doubling employment and operating almost round-the-clock, ordnance chief James Ripley correctly estimated that production could be increased to 3,000 weapons per month.[59] Yet even at this rate, it would take the Springfield armory an impossible twenty-eight years to arm a million-man army. Ripley had no choice except to enlarge the army's procurement system, which had traditionally relied on a mixture of public and private enterprise.

Civilian sources, both at home and abroad, would now become the corner-stone of arms production.

Thousands of rifles were imported from Europe, with both state and fed-eral governments pursuing the overseas arms supply. In August 1861, the State of New York contracted with the firm of Schuyler, Hartley, and Graham for the purchase of 10,000 English-made Enfield rifles "at the lowest practicable price, and in the shortest time, and of the best quality."[60] The difficulty was that sev-eral American agents (and sometimes the enemy) might be bidding upon the same lot of weapons simultaneously. This led to charges of price extortion. Because arms agents customarily received a commission based on a percentage of the funds they disbursed, it was also alleged that they sustained the bidding war for selfish reasons.[61]

In November 1861, after the immediate emergency had diminished, Secretary Cameron attempted to take control of the situation by asking war governors to stop sending purchasing agents to compete with federal officials. The governors agreed, but coordination remained imperfect. Nearly a year later, an auction of firearms included as bidders agents of the War Department and the states of Missouri, Indiana, Massachusetts, and New York. Yet even at the peak of the tight market, the ability of foreign arms dealers to play off a state government against federal authorities was never unlimited. An importer of Enfield rifles attempted to extract a better price from the federal govern-ment by asserting that Pennsylvania had bid 25 percent more. Governor Curtin disavowed this offer, and the federal agent held firm.[62]

The U.S. government gradually became shrewd in dealing with foreign suppliers. Even during the tumultuous conditions of 1861, the War Depart-ment did not lose all ability to negotiate prices; Assistant Secretary Tom Scott drove down the price of 10,000 French muskets by $6 per weapon by refusing to meet a dealer's asking price.[63] In 1862 Secretary of War Stanton sent to Europe Marcellus Hartley, a thirty-five-year-old New York dealer in sporting arms who knew the overseas markets intimately. Upon arriving in England, Hartley found that a cartel of gunmakers formed as the Small Arms Company had nudged the price of rifles to monopoly levels. Under normal market con-ditions Enfields would sell at prices of 42 to 45 shillings ($12.13 to $13.00),[64] but the Small Arms agreement had nearly doubled the price to 80 shillings. By shrewd maneuvering and by declaring that he would pay no more than 50 shillings, Hartley broke the ring and eventually cut down the price to an aver-age of 45 shillings. By December 1862, Hartley had purchased 204,848 rifles worth about $2.7 million. "In London and Liege I cleaned the market out," he reported to Stanton. "It has required care and caution to push the manufac-turers to this unusual quantity, without materially advancing the price."[65] According to accepted custom, Hartley could have expected a 2.5 percent com-mission on his purchases, which would have amounted to about $66,000.[66] Hartley asked compensation only for his expenses, however.

A factor that affected the price of foreign arms to the great advantage of European arms dealers was the desire of the Union leaders to purchase weapons at whatever price was necessary to keep them out of the hands of Confederate agents. Since the Confederates had limited ability to manufacture weapons themselves, their only hope of obtaining them was by foreign purchase. The North had much more ready cash than the South, and it could use this effectively to buy up the world supply of military arms and thus deny weapons to the enemy. The War Department attempted to keep this purpose secret in order not to enhance the world price of guns, but its ability to conceal its intentions was undoubtedly meager.[67]

To obtain arms from private establishments, the army devised an innovative form of contracting: incentive pricing. Under this system, and in accord with established military practice, a contractor's ability to deliver quickly took precedence over price. Ripley's first step was to make a baseline estimation of the cost of manufacturing standard military weapons—Springfield rifles or cavalry sabres, for example. Since the army manufactured rifles itself, this was not difficult, and Ripley calculated the expense of producing a rifle at $13.93 and of forging a sabre at $4.53 (exclusive of charges for the cost of land, buildings, and machinery). After establishing baseline prices, Ripley recommended that "a liberal profit on the cost should be allowed."[68]

Translated, this meant that the government would pay twenty dollars per rifle or seven dollars per sabre. This price was intended to be sufficiently attractive to encourage new firms to enter the market and old firms to tool up for production. The Bureau of Ordnance let dozens of contracts to produce rifles at the twenty dollar figure. These contracts typically were for 50,000 weapons to be delivered by 1 January 1862. In September 1861, Ripley also awarded a contract to produce 1,000 cavalry sabres per week at seven dollars each. If the contractor could produce 20,000 in twelve weeks, he would receive a fifty cent bonus on each blade.[69] Ripley certainly knew that not all its contractors could possibly deliver on their promises, but he evidently hoped that some of them could do so.

In practice, none of the rifle contractors succeeded in fulfilling their contracts. The reason was that these contracts created a massive run on machine tools used to manufacture firearms. One contractor alone, the Trenton Arms Company, ordered forty-four milling machines, eleven boring machines, four screw machines, ten lathes, and miscellaneous other equipment. On short notice American machine tool producers could not possibly fill all the orders they received. The slaughter on the battlefield had to be postponed until new machines could be built that could manufacture mechanical killing devices. The army therefore extended the contractual deadlines for rifle delivery to May 1862.[70]

The 1861 allowable price of twenty dollars per rifle was clearly favorable to established firms. The Remington Arms Company, a very efficient producer, notified the war department that it could turn a suitable profit at sixteen dollars, and over the course of the war it received $3 million in military business.[71] For smaller, newer producers, rifle manufacture was not a bonanza. These companies had to tool up after the war began, when machine tools commanded a premium price. Almost continual usage wreaked havoc on the machines, and depreciation was rapid. In two wartime years, the value of gun making machinery typically depreciated about 50 percent. The cost of one small, new factory (owned by the mayor of New York, George Opdyke) was $97,000 for machinery, tools, and fixtures. Other new gun factories cost up to $500,000.[72]

In 1863 Mayor Opdyke's firm produced about 15,000 rifles per year at a cost of $16.48 each for materials, labor, and an allowance for wastage of 12.5 percent. When the cost of machinery depreciation is added ($1.62 per gun), the total cost becomes $18.10. A $20 price would yield a profit of only slightly more than 10 percent above the cost of production, which would probably not cover the cost of rent and insurance (Opdyke's factory was burned in the New York riots of July 1863). The price actually charged to the army, $24.70, was thus attractive to the manufacturer but not exorbitant. In 1861 a producer of intermediate efficiency could sell rifles profitably at $18.[73]

In the spring of 1862 Secretary of War Edwin M. Stanton, who replaced Simon Cameron, determined that there were outstanding contracts for 1,164,000 Springfields, whereas army needs were estimated at only 500,000. An outburst of public opposition demanded that these contracts be cut back, on the grounds that the rebellion would collapse before they could be delivered. A defense contractor, Robert H. Gallagher of the Union Arms Company, pleaded against cancellation on the grounds that the rebellion might not subside. He pointed out that Britain might intervene or that France might invade Mexico. He even suggested a possible British invasion of New England. Although Gallagher was plainly arguing in his own self-interest, he was certainly much more farsighted than soothsayers who were predicting an early end to the war. Nevertheless, the Senate directed Stanton to appoint a commission to reduce the number of contracts. The Commission on Ordnance Contracts, which consisted of former Secretary of War Joseph Holt and Robert Dale Owen, the son of the famous reformer, condensed the orders to 600,000. There were thirty-six rifle contracts outstanding, and the Commission cut thirty-two of them. The four that were not reduced were the very first ones let, all of which included a provision that no member of Congress was an interested party.[74]

In 1862 the Commission on Ordnance Contracts reviewed 107 outstanding defense contracts to determine whether prices were excessive. These contracts covered rifles, pistols, carbines, cannon, ammunition, saddles, sabres, machine tools, and such raw materials as lead and salt peter. Seven contracts were explicitly modified because of "unreasonably high" prices; these were principally for technically advanced weapons produced in relatively small quantities. The commission objected to the purchase of 10,000 breechloading rifles from the Spencer Repeating Rifle Company at $40.00 each and to 2,500 breechloaders from the Burnside Rifle Company at $38.50. It also rejected all but one contract for revolvers at $25.00. Twelve contracts were modified without comment; in most cases, this was so that the government could benefit from economies of scale. Instead of paying twenty dollars each for an entire order of 50,000 rifles, the commission cut the price for the second 25,000 to sixteen dollars. The commission also routinely reduced the price paid for cavalry sabres. In eight other cases, however, the commission explicitly declared that "prices do not seem unreasonable," or words to that effect. Thus after reviewing 107 contracts suspected of being exorbitant, the commission cut prices in nineteen cases and left prices unchanged in eighty-nine. In seven cases it directly stated that prices were unreasonable, and in eight it directly stated that they were fair. The new technique of controlling the cost of military supplies would be used more extensively during and after World Wars I and II, when it would be termed "renegotiation."[75]

The War Department continued to purchase huge quantities of muzzle-loading rifles throughout the war. In 1864 the rate was about 2,000 guns per day or nearly three-quarters of a million per year.[76] For experienced companies with comparatively low production costs, the incentive prices paid to attract new producers were a dream come true. The Ames Manufacturing Company of Chicopee, Massachusetts, exemplified the kind of profits that could accrue. In the last three years of peace, the Ames firm employed about 200 workers and paid profits of 8, 8, and 10 percent on its outstanding stock. As the war widened, employment reached 1,000 and dividends increased dramatically: first to 12 percent in 1861, and then to 32, 44, 25, and 25 percent for the years 1862 through 1865. When peace returned, dividends fell back to the 1860 level of 10 percent.[77] Although established, efficient producers like Remington or Ames did very well, the overall record does not support a charge of widespread price extortion. The problem of large, efficient firms earning handsome profits due to a pricing schedule sufficiently generous to keep small, inefficient firms in production would return during World War I.

Advanced Weapons

Most Civil War rifles were uncomplicated, single shot pieces that could be produced by several different manufacturers according to patterns developed at the Springfield armory. Repeating firearms were recent developments whose design was much more complicated. These technically advanced weapons—mainly the Colt revolver and a variety of repeating shoulder weapons—commanded a premium in the marketplace. The justice of these higher prices became a matter of considerable disagreement. Did these weapons cost more simply because they were more expensive to manufacture or because their makers were improperly extracting a monopoly profit based on their patent claims?

The Colt's Patent Fire Arms Company was the first firm to receive close public scrutiny. Despite his residence in Hartford, Connecticut, Colonel Samuel Colt was widely suspected of having Confederate sympathies. The basis for this suspicion was that shortly after secession he had contacted Southern officials in hopes of selling sidearms to the Rebel army. After hostilities erupted, however, Colt moved quickly to solidify his ties to the Union. In May 1861, Colt offered the use of his factory to the War Department, and he presented $50,000 worth of breechloading rifles to Connecticut for use by its volunteer forces.[78] Nevertheless, his company was soon accused of overcharging on the sale of pistols to the government.

For several years before the war, Colt had sold revolvers to U.S. military units at a price of twenty-five dollars. Estimates of the cost of production vary; Colt revolvers may have cost as much as nine dollars to make, or they may have cost only $4.00 or $5.00. In either case, when sold for twenty-five dollars, they returned a handsome profit. This was particularly annoying when it was discovered that Colt had sold revolvers simultaneously to the British government (a potential enemy) for $12.50 and to American civilians for $14.50. In 1860 Colt cut the American military price to twenty dollars, but this hardly satisfied his critics, who pointed out that a disparity between military and civilian rifles was more justifiable than between military and civilian handguns. Civilian customers willingly accepted the military style pistols, which were only slightly different from military versions, but civilians had little desire to use military rifles, which were generally too heavy for hunting. Revolvers of comparable quality to the Colt models were available from his chief competitor (the Remington Arms Company) for fifteen dollars in small quantities and thirteen dollars in large numbers.[79]

In 1861, to meet wartime demand, Colt added three new buildings to his factory. He agreed to produce Springfield rifles at the army incentive price of twenty dollars. His firm eventually employed 1,100 men and met a payroll of $50,000 per month. Upon his death in 1862, Colt was able to bequeath $2 million to his nephew. Although his wife successfully overturned the will, Colt's considerable fortune clearly accumulated substantially from military sales. The machinery at the Colt works was worth more than $500,000 in 1862, and when the plant was consumed by a great fire in 1864 the direct loss was between $1.5 and $2 million.[80] Samuel Colt was probably the first American to become a millionaire by making weapons. Although his fortune was built upon indisputable technical excellence, it exemplified the doubtful ethical rectitude of riches gained from war.[81]

High prices demanded by inventors were evidently an important reason that the army was slow to adopt breechloading rifles. In August 1861, the Ordnance Bureau advised against the purchase of 10,000 breechloading carbines offered at $35. "The price," General James Ripley said, "I consider too high." Even at $28 to $30 Ripley resisted the purchase of breechloaders. Yet despite Ripley's indifference, contractors were able to sell enough breechloaders to earn very healthy profits. During the third quarter of 1862, the Sharps company declared dividends that equaled 24 percent of the market value of its stock.[82] William F. Brooks, who had patented a carbine design, received a royalty of $6.50 per weapon. In 1864 he collected the ample if not lavish sum of $16,068.[83] Although major blame for the delay in adopting breechloaders must rest with General Ripley, excessive demands by their manufacturers also played a part. This was probably the greatest damage that defense contractors did to the Union war effort.

A third technically sophisticated weapon that stirred dispute was the more powerful cannon. The controversy centered on the actions of Captain Thomas J. Rodman, a young army officer who discovered an improved process for casting large guns. Troubled by the explosion of the great gun "Peacemaker" aboard the USS *Princeton* in 1845 (which killed Secretary of State Abel P. Upshur and several other officials), Rodman began to rethink cannon casting. He invented a process in which cannon were hollow-cast, allowing the barrel to be cooled from the inside out by pumping water down the hot muzzle. Rodman guns were much stronger than solid-cast cannons, which cooled first along the outside perimeter and slowly to the center before being bored out. Rodman guns would fire three to ten times as many shots as conventional types before they failed.[84]

Despite the promise of Rodman's theories, his senior officers refused to test his methods, doubting their value and claiming a lack of funds. Because testing was very expensive, Rodman assigned a half-interest in his patent to the firm of Knap and Totten, a large gun foundry. In return, this company agreed to cast a number of guns and test them. The tests, which cost $60,000, proved the superiority of the Rodman process, and the patent became extremely valuable. Rodman and Knap asked a royalty of 20 percent of the production cost, which would raise the price of cannon from 6.5¢ per pound to 7.8¢ per pound. In 1860 this was of no great importance, since only twenty-six guns were cast and the royalty amounted to $3,037.68. During the Civil War, however, 7,892 cannon were furnished to the Union army. Rodman and Knap agreed to license their process to other makers for one cent per pound. If all the guns were manufactured according to the Rodman process, the patentees would become very wealthy. To prevent this from happening, the War Department's Commission on Ordnance Contracts ruled that Rodman's assignment of his patent to a civilian was a violation of military regulations and nullified the action. The commission noted, however, that nothing in the regulations prevented the assignment if Rodman first resigned.[85]

In its ruling, the War Department conveniently ignored its own malfeasance. Before patenting his invention, Rodman had consulted his superior officer and inquired whether there would be any impropriety in doing so. The commander's reply was unequivocal: "Certainly not." If Captain Rodman had not sold a share of his patent to raise money for testing, the process would never have been developed, and the defense of the United States would have been seriously impaired. Although Rodman was severely criticized, his actions were thoroughly honorable, and he deserved every royalty he received. These were considerable but not outrageous. In 1864 the total weight of Rodman guns purchased by the army was six million pounds. At a penny a pound, this amounted to $60,000, half going to Knap and half going to Rodman. Cannon forged by Knap sold for 10 percent less than those of his competitor, the Dahlgren gun. Nevertheless, the War Department reduced his orders on grounds of excessive prices.[86]

The incomes of other cannon makers were not so defensible. One of the largest foundries in the country was the West Point Foundry at Cold Spring, New York. This firm produced at its peak twenty-five cannon and 7,000 projectiles per week, fulfilling about a fifth of the needs of the Union army for cannon. In 1864 its superintendent, Colonel Robert Parrott, reported a taxable income of $278,861.07, and his partner, Governeur Paulding, received $95,983.55. They paid federal income taxes in the amounts of $13,943.35 and $4,609.18, respectively.[87] Parrott's income, eleven times that of President Lincoln's salary, was nothing if not excessive, and the tax system was ineffective at relieving the abuse. In his defense, Parrott pointed out that despite inflationary pressures, he did not increase the price of cannon shells throughout the entire war. Still, his claim that he accepted only "a fair manufacturing profit" rings hollow.[88]

Provisions

Every war causes economic disruptions as the economy shifts from civilian to military production. The Civil War was very large, like World War II, and like the American Revolution, it was fought on American soil. In economic impact, the Civil War exceeded both, and in its circumstances of utmost emergency there were inevitably economic gainers and losers. Railroads that ran north and south lost business, while east-west lines benefited. Few, if any, Northern businesses were not affected. As early as April 1861 the randomness of the war's effect was apparent. When rebel forces cut off supply lines to Washington, the War Department entered into a verbal contract to purchase 2,000 head of cattle at eight cents a pound. This was well above the customary rate and reflected the unusual risk the contractor accepted in bringing the animals to the city. By a stroke of good luck, rail connections to the capital were restored just in time to transport the cattle safely, and the contractor collected a very generous sum.[89]

Other contractors enjoyed windfall profits. Among them, surprisingly, were cotton textile manufacturers. Although the Southern embargo on cotton might have been expected to cause chaos, in fact it produced bulging bank balances. The value of inventories of raw cotton shot upward. At the outbreak of the war, textile firms held about a three-month supply of cotton, which rapidly appreciated by at least 50 percent. Of twenty-four textile firms reporting in 1861, eighteen increased dividends, four paid the same dividend as the previous year, and only two reduced them. Total dividends rose from $618,000 to $1,013,000.[90] As the embargo tightened, some firms turned to arms making, but others continued to find the textile business quite profitable. In 1865 alone a Rhode Island textile firm recovered profits of $97,000 on an investment of $200,000.[91]

For suppliers that were placed in a fortunate position by the changed circumstances, benefits could be handsome. Kentuckians were lucky in that the government hoped to induce their loyalty by spending generously in their state. Newspapers prospered as they contentedly tapped a market eager for war news. The *New York Herald* saw its circulation improve from a daily average of 65,694 in 1860 to 92,158 in 1861, and the *New York Times* enjoyed a similar circulation gain. Some of the profits were absorbed by price increases for newsprint, and the *Times* uncharitably denounced "combinations of speculators" who had allegedly driven up the price of paper. The newsmen congratulated themselves for breaking the ring by importing paper from Europe, but their motives may be presumed to have been other than entirely patriotic.[92]

Ship chandlers found a ready market for almost every marine accessory. Rosin sold at $35 per barrel and turpentine at $200. Almost every kind of naval or military hardware required an abundance of screws, and this industry, centered in Rhode Island, prospered even more than textiles or iron manufacture. The large Boston firm of Smith Brothers and Co., a supplier of naval hardware, had annual sales of $150,000 in 1860 and an average of $567,000 in the following three years. This was a sufficiently large increase to attract a Senate investigation, and Franklin W. Smith, a partner in the firm, admitted under oath that he charged the government more than the "lowest market cash prices." This surcharge was justified, he protested in a complaint that echoed gripes nourished by countless legions of government suppliers, because payment was always delayed one to six months after delivery. Smith also carped that he was obliged to take discounted certificates that were worth only ninety-six to ninety-eight cents on the dollar and that banks would not take navy vouchers as collateral.[93]

Speculation

Uncertain wartime conditions prompted wild speculation in commodities and stocks. Gold prices rose dramatically after the government suspended specie payment on 1 January 1862. This had a great effect on international trade. Americans who held bills of exchange saw values appreciate by as much as 25 percent in a month. This was accomplished by shipping flour, grain, provisions, or petroleum to a European port, obtaining a foreign credit, and then waiting for its value to increase. The opportunity to speculate in foreign credits sometimes became a more important reason for engaging in foreign trade than the commercial reasons alone.[94]

Gold speculation became almost a national sport. Gold coins disappeared from circulation, making it difficult or impossible to make change. In 1864 R. H. Gallagher opened "Gallagher's Evening Exchange" so that speculators could indulge their fancies after regular exchanges had closed. Women were particularly affected by gold fever, and some pawned their jewelry in order to speculate. One disgruntled male complained that "some of our women are already infected with the prevailing passion of money-making as they have been long with that of spending it. 'What's the price of gold to-day, my dear?' escapes from the pretty mouth of your wife before she has impressed the habitual kiss of connubial welcome upon your expectant lips." Clergymen, whose fixed salaries were ravaged by wartime inflation, attempted to protect their purchasing power by investing their wages in the gold market. To some, this seemed most improper.

During the peak of the speculative mania, brokerage firms collected huge commissions, as much as $5,000 a day, and telegraph firms basked in the lucrative warmth of heavy wire traffic. Not to be left out, a significant number of telegraphic orders to trade gold originated in the South, as many as 800 per day. Since the price of gold fluctuated with the fortunes of the Union army, opportunities for cheating were frequent. One ring of speculators conspired to gain control of telegraph wires in order to delay reports of the outcome of the battle of Chancellorsville until they had invested heavily. Failing, they lost $100,000. In March 1865, rumors of peace sent gold and stock markets tumbling.[95]

Trading with the Enemy

A form of enterprise that was extremely profitable because of wartime disruptions and that was also extremely speculative involved trading with the enemy. As in the American Revolution and the War of 1812, trading with the enemy continuously plagued the Union cause. It is impossible to measure this illegal activity accurately, but there is little doubt that Abraham Lincoln's problems with illicit trading were several times more severe than George Washington's or James Madison's. As early as February 1860, fully a year before the fighting, Southern agents were in Springfield, Massachusetts, to contact manufacturers with thinly veiled proposals for buying arms in the event of war.[96] Despite the bitterness and the bloodiness of the Civil War, commercial relations between the North and South were never entirely suspended.

The most commonly traded commodity was cotton. The reason for the intense activity in cotton was the great difference in price on the two sides of the battle lines. Prices fluctuated, but cotton sold for three to ten times as much in the North as in the South. For example, in 1862 cotton bought at Helena, Arkansas, at fourteen cents per pound could be resold at St. Louis for 45-46 cents per pound. This may even understate the situation: other estimates state that cotton purchased in Arkansas at $25 to $50 per bale sold in St. Louis at $250. By 1863 the price of cotton reached $1.00 per pound in New Orleans and $1.50 in Boston, inviting the payment of bribes to Union officers who would allow it to be shipped through Union lines.[97]

There were many reasons for the extensiveness of the illicit cotton trade. Smuggling is an ancient theme in American history, and there are very few if any occasions in which a market as lucrative as the Civil War cotton trade has not been served. The cotton growing area was vast, and there were innumerable routes to the North, where as many as 400,000 bales of illegal cotton may have been sold, leading Allan Nevins and others to suggest that the Northern market may have been even more important to the Confederates than was the European market.[98] The smugglers were ingenious and often very difficult to identify—who could tell exactly which shipments originated with a Union sympathizer and which with an enemy? Certainly some of the cotton imported from England had been grown in the Confederacy and smuggled through the Yankee blockade. The policy of the Lincoln administration was inconsistent about trading with the enemy. Early in the war Lincoln was most reluctant to suspend trade with the border states, whose loyalty he was anxious to retain, even though contraband might enter the Confederacy. Kentuckians who lobbied for government business found a receptive audience in the president and his cabinet. Treasury Secretary Chase formulated a policy by which the government permitted trade with Kentucky, Delaware, Maryland, and Missouri, unless the "ultimate destination" of the articles was the rebel states. But how could the final destination be ensured? Chase was also quick to reopen trade once an area had been subdued, establishing a policy in 1861 "to let commerce follow the flag."[99]

Since military occupation was by no means the same as establishing loyalty, and since occupied areas were often recaptured, Chase's policy was a loose one. It may even have been the result of misguided humanitarianism. Treasury agents were ordered to allow the shipment of "family supplies" to the South, but under this guise much contraband was moved. No such reason could excuse Secretary of War Stanton's actions in covering up evidence that implicated Chase's son-in-law, Senator William Sprague, in a scheme to trade Northern guns for Texas cotton. This was carried out in order to prevent personal and political embarrassment.[100]

President Lincoln was himself halfhearted about suppressing trade with the enemy. On 13 July 1861, Congress authorized Lincoln to suppress all trade with the Confederacy or, at his option, to allow it to continue under license. Not until August 16, when the war was fully four months old, did Lincoln take action. Others had acted more swiftly. On 23 April 1861, police in New York City seized 2,000 uniforms packed for shipment to rebel troops in Alabama. On the following day, the U.S. district attorney in New York convened a grand jury to stop the furnishing of supplies, food, and clothing to the Confederate army. On the same day, Daniel Fish, a gun manufacturer, was arrested and charged with treason for selling guns to the rebels.[101]

Elsewhere, Committees of Safety reminiscent of the early days of the American Revolution were established to ferret out traders. These appeared in April and May 1861 in Pennsylvania, Ohio, and Indiana. In Pennsylvania, extra-legal committees confiscated contraband disguised as buggy whips, flour, and other nonmilitary items. In Cincinnati, a committee seized bacon destined for the Confederacy. In Indianapolis, a Committee of Vigilance organized about 1 May 1861; Governor Oliver P. Morton hired two detectives to investigate rumors of illegal trading, and this led to the interception of shipments at Vincennes, Evansville, and New Albany. In May 1861, the Indiana legislature, acting more rapidly than Congress, outlawed trading with the rebel states.[102] Lincoln was very liberal in granting licenses to trade for Confederate cotton. He seems to have treated these special exemptions to the "no trade with the enemy" policy as a form of patronage. By 1864 Lincoln had granted at least forty exemptions that allowed trading in cotton. Amazingly, these exemptions could be sold to another trader who would carry out the actual transfer. The known availability of these special licenses caused rumors of bribery and corruption to circulate throughout the capital. Lincoln also ordered the release of an illegal trader who had been apprehended and sentenced to prison for the duration of the war.[103]

Lincoln reasoned that it was better for the Union to buy cotton directly from the South rather than to buy it indirectly from England after it had been shipped there and its price enhanced six times. In 1863 General Nathaniel P. Banks authorized the purchase of cotton as long as it was paid for in U.S. currency alone. In effect, this was supposed to establish an exchange in which the North sent pieces of paper to the South and received cotton in return. Of course, cotton traders with Confederate sympathies were very reluctant to accept payment in greenbacks, but when they found out that the alternative to trade was confiscation, they were usually persuaded. Some planters, however, successfully demanded payment in gold or salt. In 1862 Treasury agents determined that in a single three-month period, $355,000 in gold passed through Cairo, Illinois, on its way south. Trading for salt was more acceptable, since it was neither lethal nor easily traded in world markets. As General William T. Sherman observed, however, it was necessary to the curing of bacon and salt beef and therefore was of military value. As in the American Revolution salt was scarce in the South, and at the Union army's insistence, trade in salt was prohibited in December 1862. Nevertheless, trade remained brisk throughout the war. In 1862 the exchange rate was established at one bale of cotton for ten sacks of salt. Jefferson Davis agreed to the trade, although he objected to the exchange rate as unfair to cotton.[104]

While the cotton-for-greenbacks trade may have been somewhat advantageous to the North, there is evidence that the Confederates then used the U.S. currency to purchase much-needed military equipment in Northern markets. In 1864 the administration attempted to take control of the huge trade by organizing a government-operated cartel that alone could trade with the belligerent states. Treasury agents who accompanied the Union forces paid 75 percent of the New York price for cotton less taxes, insurance, and freight. This reduced profits of the illegal traders considerably, but the system remained open to abuse. In January 1864, Chase appointed Ralph S. Hart to be special agent at Natchez, Mississippi. Before Hart was suspended five months later, records of the Adams Express Company showed that he sent his wife $19,000. His timely death saved him from prosecution.[105]

Union army officers who were not trading with the enemy were commonly exasperated at the illicit practice. Although less eloquent than George Washington, General Sherman was equally blunt. "War and commerce are inconsistent," Sherman declared. "We cannot have commerce until there is peace and security." But while salt sold for $1.25 per sack in Union-occupied New Orleans, a few miles across Lake Ponchartrain in Confederate territory it brought $100. When a price differential of that magnitude existed, it became nearly inevitable that Sherman's hopes of interdiction would not be realized. Nevertheless, he did his best to stop illegal trading, and in August 1862 he had several soldiers shot for the offense. A somewhat milder policy advocated by General U. S. Grant was to have illicit traders drafted.[106]

Lincoln's attempts to regulate the cotton trade were constantly thwarted by poor administrators, and by 1865 the trade was huge and almost unrestricted. Memphis was its center, but Brownsville, Texas, lagged not far behind. Most Northern traders hailed from the border states, but in imitation of colonial chases between their ancestors and the Royal Navy, Rhode Islanders were prominent in the ranks of smugglers.[107]

Plunder

In partial continuation of earlier practices, military regulations during the Civil War legalized the confiscation of cotton by the navy but not by the army. The customary naval practice (which continued in effect) was that the quest of naval prizes was a legitimate encouragement for vigorous efforts. Prize money remained a very effective incentive for Civil War sailors. In May 1862, Captain David D. Porter ruefully observed of his sometimes languorous senior officers, "It is astonishing how much better and stronger these old fellows get when there is prize money in view; it resuscitates them completely."[108]

An interpretation of maritime law allowed contraband seized by the navy on inland waters as well as the ocean to be classed as a legitimate prize. Because Civil War naval operations frequently took place on broad western rivers flowing through prime cotton land, this policy unlocked abundant opportunity for economic gain. The rule stated that half of the cotton confiscated was to go to the government and half was to be divided among the fleet. In the Red River campaign, Porter's sailors fanned out along the banks in search of cotton for resale, unfortunately spending more time looking for cotton than searching for Confederates. In order to get around the question of whether the cotton they captured was actually owned by rebels rather than Unionists, the sailors kept a stencil marked "C.S.A." When they found cotton, they simply marked it as enemy property and declared it contraband. To stop the sailors' abuses and to quiet disgruntled army officers, Congress in July 1864 made it illegal for the navy to seize cotton as a lawful prize. That the practice was lucrative while it lasted was made clear by Porter, who had yielded to the temptation he had earlier deplored. "When this is over I will come North," exulted Porter, by then an admiral. "Then I will sit down and rest under my own vine and fig tree and as my share of the prize money ought to be large I can live quietly and after the fashion I have desired all my life."[109]

According to army regulations, confiscation of cotton by the troops constituted an act of plunder, which if it were allowed to continue would contradict the theory that the army was fighting the war for lofty purposes. The federal government accordingly appointed civilians to handle confiscated cotton, paying them 25 percent of what they could capture (often with the assistance of the army). These civilian agents, unfortunately, sometimes confiscated cotton from Unionists, pilfered cotton and sold it themselves, or cheated the government by delivering low-grade cotton and keeping the best for themselves. The switch to civilian agents was but a qualified success.[110]

Sometimes plundering was carried out under the guise of foraging. As one young Yankee explained, "We are in Secessia and the meanest part of it, too, and anything the boys can forage they consider it theirs. A field of potatoes, five acres, was emptied of its contents in short order. . . . You ought to see them clean out the fences." In the bitter fighting along the Kansas-Missouri border, both sides plundered pitilessly. In their famous raid on Lawrence, Kansas, William Quantrill and his men openly declared their purpose as being "plunder." The Kansas militia, the Jayhawkers, swept into Missouri and plundered without restraint—horses, livestock, and slaves. In an echo of practices past, in 1862 Union Major General Ormsby M. Mitchel, a world famous astronomer, permitted his men to plunder cotton, watches, and whatever else they could find. He was charged, like his military ancestors, with illegally transporting the booty in government wagons.[111]

A more innovative scheme involved General Neal Dow. This officer, who served in Louisiana, confiscated many objects of fine furniture, as well as pianos, china, books, and knickknacks, which he shipped back to his home in Maine. After the war General Dow was sued for his seizure of cotton, but in 1880 a decision of the U.S. Supreme Court upheld his right to plunder. Not so fortunate were four enlisted men of the Thirteenth Connecticut. Apprehended in 1862 with $400 plundered from the home of a New Orleans woman, all four were hanged.[112]

Fraud

The great value of Confederate cotton in the North led to the corruption of at least some quartermasters. This office was continually scrutinized for fraud during the American Revolution, and while professionalism had greatly reduced the number of cheating quartermasters, a few continued to observe the hoary traditions of their trade. The chief quartermaster of the Department of the Gulf, Colonel Samuel B. Holabird, sold $205,000 worth of seized cotton and kept $102,000 for himself. His assistant, Captain Jacob Mahler, had a deficiency of $266,000 in his accounts. There also were a few scattered examples of corruption in the purchase of other commodities: a Detroit horse trader offered a $10,000 bribe to Major A. A. Selover of General Frémont's staff in order to get a horse contract; the Indiana quartermaster received a kickback of 50¢ on every saddle sold to the state, as well as 5-10 percent on other contracts (he was forced to resign); and R. B. Hatch, the quartermaster at Cairo, Illinois, received a $300 bribe from Chicago dealers for awarding a contract for the purchase of lumber for the construction of Camp Douglas. There were a few other frauds disclosed, but the striking characteristic is the overall honesty of the Quartermaster Corps.[113]

Manpower

The American Civil War was the last and greatest occasion in history in which military manpower procurement offered a large opportunity for profit. Numerous scholars, from Fred A. Shannon to Eugene C. Murdock, have illuminated unethical aspects of the procurement business. The most common offense was bounty jumping, which was even more lucrative and extensive than in the American Revolution. Although it is impossible to determine how extensive the practice was, there were 268,000 desertions from the Union army, and a substantial number were in order to reenlist and collect another bonus. The known record for successful bounty jumping is thirty-two times. Groups of men enlisted and jumped together; Fred A. Shannon reported a group of 3,000 to 5,000 such gangsters who lived on Manhattan Island. In one raid, 590 were captured. Men who successfully enlisted twenty times received as much as $8,000 each. But as in the American Revolution, the penalty for bounty jumping could be extreme. In December 1864, three Indiana men were shot for the offense, although President Lincoln reprieved a fourth who was just nineteen years old. A group of 150 bounty jumpers were paraded through the streets of Indianapolis with placards hung from their necks to identify their crime.[114]

A second enterprise that produced great controversy was substitute bro-kering. In 1862 Congress invoked a draft law that included a provision to allow drafted men to gain exemption by furnishing a substitute. Firms appeared whose business was to locate substitutes for reluctant warriors who were unable or unwilling to find substitutes for themselves and who were willing and able to pay for the service. Although there was nothing inherently unethi-cal in offering this service in return for compensation, the draft was so un-popular and so often abused that substitute brokers became lightning rods for criticism. Historians have argued that the draft was terribly unfair to the poor (a fact that has been sharply disputed), and substitute brokers—men who profited upon the sacrifices of patriots—became a kind of showcase exhibit of its essential unfairness.[115]

Substitute brokers were charged with encouraging bounty jumping, over-charging for the service, and furnishing defective recruits, among other faults. They had a clear incentive to promote bounty jumping, since they needed po-tential recruits badly, and it is entirely likely that they did so. They also charged heavily for their service. Fortunes of $10,000 to $15,000 and even as much as $100,000 were made in New York. Hawley D. Clapp, known as the "King of Bounty Brokers," was arrested in 1864 and imprisoned in Fort Lafayette. He was alleged to have swindled more than $400,000 from recruits by persuading them to enlist and then keeping their bounties for himself.[116]

Several spin-offs of the substitute brokering business were reminiscent of notorious practices of the past. As had their predecessors, Civil War substitute brokers sometimes passed off injured men and claimed them as healthy. This was so common that the New York Herald claimed there was a parallel to Shakespeare's Henry IV, in which Falstaff protests, "I have pressed me none but good householders sent to the Tower by a summary order of Lord Chief Justice."[117] Harper's Weekly claimed that substitute brokers furnished, besides criminals, cripples, underage boys, herniated men, the mentally retarded, and even men who were partially blind. There were also some escaped Confeder-ate prisoners of war furnished as substitutes. They could enlist in the Union army, collect their bounties, and then desert to their comrades with pockets full of cash and heads full of military intelligence. In a belated attempt to gain control of the situation, in March 1865, Congress made it illegal for brokers to enlist the insane, convicts, persons under indictment for felony, deserters, minors, or men "in condition of intoxication." In addition, it became a crime to deprive a recruit of his bounty. The penalty was set at two years in prison or a $1,000 fine.[118]

In a new twist, a kind of subbroker cheated on charges for housing recruits before the army could provide quarters for them. The army paid brokers for temporarily supplying room and board for enlistees, but it was slow to send the money. Not wanting to wait for payment, the brokers sold at a discount their claims against the government to a group of "special contractors." These "special contractors" then falsified the records before presenting them to the government. The case of Solomon Kohnstamm illustrates the scheme. In 1864 a broker named Louis Pfeffer of Albany, New York, subsisted 116 recruits for the 58th New York Regiment. His bill was for $100, which he sold to Solomon Kohnstamm of New York City. Kohnstamm altered the document so that the figure became $1,366, which he then submitted to the government for payment. He was apprehended and tried on forty-seven counts of fraud. After deliberating only fifteen minutes, the jury found him guilty. Before these brokers were discovered, they defrauded the government of an estimated $700,000.[119]

The recruitment of so-called mercenary troops aroused great controversy. Facing a domestic shortage of raw material for their business, the enterprising brokers decided to import from Europe. Since a large number of Union soldiers were German immigrants, Confederates and their sympathizers charged that Abraham Lincoln was following a cruel practice of the infamous King George III, who hired Hessian mercenaries. Although this was an exaggeration, a new disgrace to the already inglorious record of mercenary warfare occurred in 1864 when a shipload of 700 German immigrants were brought to the United States by bounty brokers. The brokers agreed to pay the Germans' passage if the Germans would enlist in the Union army. In return, the brokers were allowed to collect and keep the Germans' enlistment bonuses, thereby turning a gross profit of $700,000. A military investigation determined that the Germans were not deceived, despite many allegations to that effect. Although Confederate charges that the U.S. Army recruited mercenary soldiers abroad have been accepted by several historians, there is no evidence that this happened with great frequency.[120]

Censure and Control

The Civil War era was unusual in American history in many ways, not the least for its atmosphere of mistrust. Americans of the 1860s expressed doubt about the motives of their countrymen. Loose charges of malfeasance fell upon a receptive audience. The first Civil War Congress, as Allan G. Bogue has shown, was most notable for its willingness to launch investigations, and these investigations were rooted in suspicion. Contemporaries recognized this phenomenon. "The last Congress," wrote the *New York Herald* in 1863, "will be memorable for its inquiring disposition."[121] This milieu of suspicion gave credence to charges of profiteering, and it also prompted demands for change. In comparison to the great issues of the war—saving the Union and ending slavery—reducing profiteering was decidedly of less importance.[122] Nevertheless, the Civil War generation began to grapple with it.

Reports of fraud in contracting that began to circulate during the chaotic weeks after the opening of hostilities produced the first outraged calls for punishment. In August 1861, Senator Orville Browning of Illinois, reacting to reports of bribery in the award of verbal contracts, called for the imposition of the death penalty. "I would not stop to carry that penalty to the extremity of death against any officer of the Government who, in such a time as this, would be so lost to the principles by which we all ought to be governed, as to seek . . . his own pecuniary advancement at the sacrifice of the interests of the public," Browning told the Senate. "If we were to shoot one or two of the rogues, it would have a more salutary effect in putting an end to the pilferings that are alleged than all the others we could adopt." Browning was by no means alone. There were repeated calls for the death penalty in the press. "Every dishonest contractor and conniving inspector," demanded the *Scientific American,* should receive a "trial by a drum-head court-martial and instant military execution."[123]

Although the death penalty was not imposed (except for bounty jumping), the government experimented with other devices for controlling the profits of its suppliers. The government had no choice except to buy in a seller's market, but it was not wholly without the ability to negotiate. A case in point was its dealing with Jay Cooke, the general subscription agent appointed by Secretary Chase to sell government bonds to the public. In July 1861, Cooke informed Chase that "we could not be expected to leave our comfortable homes and positions here [in New York] without some great inducement and we state frankly that we would, if we succeeded expect a fair commission from the Treasury in some shape for our labor and talent." The "great inducement" or "fair commission," according to Cooke, ought to be a commission of three-eighths percent (.375 percent) of the price of the bonds. This would be divided three ways: one-eighth percent for subagents, one-eighth percent for expenses, and one-eighth percent for Cooke himself. Chase objected and successfully cut the commission to .25 percent, which was one-third less than Cooke had asked. As bond sales increased to enormous quantities, Chase lowered the commission further. For selling $362 million in "five-twenty" bonds, Cooke received $220,054.49—less than one-sixteenth percent. By 1865, a near panic on the stock market caused Chase's successor, William P. Fessenden, to meet Cooke's demand, even though Cooke would have settled for less. For five war years, Cooke's total income (including both government and private business) was about $844,000.[124]

The government could also use its great buying power to affect prices. Quartermaster General Montgomery Meigs learned that he could restrict prices by placing contracts carefully. When inventories in government arsenals declined, Meigs ordered that replacement should take place gradually, in order to prevent "a field day for the profiteers." Secretary of War Stanton was willing to threaten suppliers with confiscation if they sought what he considered to be an unfair price. In 1862 the War Department needed to buy boats for use on the Mississippi River. When he learned that the owners were asking "high figures," Stanton exploded: "The [War] Department will submit to no speculative prices. . . . I will authorize the quartermaster to seize such boats as may be needed as other property is taken for military purposes, leaving the parties to seek remuneration from Congress."[125] Of course, a claim against the government might take years to recover, and the amount would be problematical.

In 1863 the speculative crisis deepened sufficiently to prompt the War Department to resort to commandeering if necessary to obtain supplies. As the Confederate army advanced into Pennsylvania, Quartermaster General Meigs ordered the purchase of clothing for emergency troops. "Do not allow speculative prices," Meigs telegraphed the quartermaster in Harrisburg. "With the approval of the commanding general fix prices and compel supplies." Subsequently, on several occasions there were discussions in Congress and the press about the possibility of applying martial law to defense contractors. This would expedite the prosecution of fraud and probably increase the severity of the punishment, perhaps even allowing the death penalty. Only the doubtful constitutionality of the idea prevented its adoption.[126]

Instead of declaring martial law, Congress approved landmark legislation that made fraud in defense contracts a federal crime. Sponsored by the House Committee on Government Contracts, the Frauds Act of 1863 included an innovative concept that would spark much controversy in the twentieth century. Any person who informed on a fraudulent contractor (in modern parlance, a "whistle-blower") could claim half the monetary judgment obtained; in the twentieth century this could and did amount to tens of millions of dollars. John K. Stetler of Philadelphia became the first contractor convicted under the law. He received a five-year sentence for supplying the army with adulterated coffee.[127] Since the law was effective against quality degradation, but did not address the basic question of price gouging, it was of modest effectiveness during the Civil War. The nation preferred to rely on indirect or informal means of controlling extortionate pricing.[128]

One alternative that the government employed was to produce military supplies itself. Although the Union did not utilize government-owned defense facilities as much as did the Confederacy, nevertheless the war saw the expansion and creation of new plants. By 1863 rifle production at the Springfield armory reached 25,000 weapons per month, while all private armories combined produced 60,000 small arms. The government also constructed three new arsenals. In 1862 Navy Secretary Gideon Welles proposed the construction of government-owned shipyards on the grounds that private yards were too small and too slow. The idea died when opponents argued (perhaps with some prescience) that government yards would be filled with lackadaisical political appointees. The government did build the largest bakery in the world. Near Alexandria, Virginia, it was built next to a railroad track so that flour could be delivered directly. Consuming 400-500 barrels of flour each day, it employed

200 men in a building that covered an entire acre. Normal production was 90,000 loaves per day. There was a public demand for price controls on bread, but the government at least had little to fear from greedy bakers.[129]

In 1862 the Holt commission had forced the renegotiation of more than a hundred rifle contracts, establishing renegotiation as a principle of defense contracting. In a later case, the ordnance bureau learned that an arms dealer, acting as a middleman, had turned a windfall profit of $280,000 for work of "a few days at most." Confronted with the evidence, the dealer proposed to surrender 40 percent of his proceeds. Unfazed by this offer, the War Department unilaterally rescinded 90 percent of his gain. In 1862 a governmental commission examined claims for monetary compensation filed at Cairo, Illinois, by 1,696 hopeful petitioners. It cut the claims by 25 percent (from $599,000 to $451,000). Agents of the Navy Department also used renegotiation as a way of reducing prices they deemed exorbitant. At least one attempt to appeal these reductions to the U.S. Supreme Court met with failure.[130]

As the crisis deepened, Northerners began to see the war as a struggle against twin evils, the rebels at the front and the laggards at home. The latter, who would be known as "slackers" during the Great War, were often identified as greedy defense contractors. "When a nation goes to war from a high motive," declared the *North American Review*, "the motives which influence individuals are tested and disclosed. . . . Thus war develops the immoral, no less than the moral, elements in a society, and the worth of that society depends on the relative power which each of these elements secures to itself. . . . The soldier who gives, not only his life, but his heart to his country; the contractor who cheats the government and abuses the soldiers with his shoddy . . . such are two among the contrasts which the test of war displays."[131] Northerners were determined not to fail that test.

Although the government possessed a few methods of limiting the profits of contractors, the Northern people preferred to trust voluntary action combined with moral suasion to thwart profiteering. Most Northerners had little doubt that the war was being fought for worthwhile principles, and they denounced vehemently anyone who would dishonor the cause by using the occasion to seek personal fortune. "Vultures that prey upon the hearts of the dead on the battlefield," thundered the *New York Tribune* in 1861, "are human compared with monsters who furnish rotten blankets and rotten meat to the living in camp." "The self-styled loyalist, who puts money in his purse at the expense of soldiers who go to fight rebels," added the *New York Times*, "is worthy only of unqualified detestation, and no fate can be too severe for him. Our prison doors gape for such knaves. Hustle them in."[132]

The outrage expressed toward defense contractors intensified as the war brought new social strains. Some actually welcomed the hardship they expected the war to bring. "For the past twenty years we have been getting rich, lazy, and luxurious," insisted the *New York Herald* in 1861. "Our dresses and our diamonds were as rich and costly as any aristocrat's." But when the hardship of war set in, the *Herald* optimistically promised, the nation would experience a "regeneration." "We will be more American and less foreign. We shall have cheap houses. . . . Our ladies will wear gingham dresses, instead of silks and satins, and for ornaments they will have red roses . . . and no pasty, glassy diamonds." But after 1861 the war brought only more prosperity for much of the civilian population, not the anticipated cathartic suffering. Many worried that the war was becoming deeply caustic to the virtue of a republican people. The *Herald*'s prediction that "wealth will no longer be the criterion of what a man is worth. . . . Merit will rise to its proper position in the social scale" seemed sadly premature.[133]

A favorite means of moral suasion was the antiluxury campaign. After the first year of depressed production, the war-stimulated economy brought a marked increase in the demand for luxury goods. Many items were targeted, but the most commonly denounced were silk dresses, laces, diamonds, fancy carriages, expensive theater tickets, and yachts. Wartime prosperity extended from Leavenworth, Kansas, where the population tripled and opulent homes were built on the bluffs overlooking the Missouri River, to New York City, the center of the nation's financial markets. "Never was New York so brilliant, so captivating," remarked one resident.[134]

The antiluxury campaign was led by women and the press. In the capital, women led by wives of cabinet officers entered into a "Ladies National Covenant" in which they pledged not to purchase luxury goods. Members identified themselves by wearing a badge showing a black bee upon a tricolored ribbon. In New York, a large meeting of women voted against the purchase of "unnecessary foreign luxuries." Some women, led by Susan B. Anthony, objected to the action on the ground that the pledge was not strong enough. In Boston, the antiluxury movement was led by the most socially prominent local women. Elsewhere, women pledged to cut expenses by raising hemlines by three inches and by replacing silk with mousseline de laine.[135]

The press, led by the *New York Herald*, heaped scorn on extravagant expenditure. In an editorial on "The Striking Contrasts of the War," written only a week before the battle of Gettysburg, the *Herald* described with a pen dipped in vitriol the conspicuous display of luxury in Central Park. "If the weather be pleasant this afternoon, the Park will be crowded with rich equipages," the *Herald* wrote. "Our shoddy aristocrats will be found enjoying themselves in state and style. To them the war has brought only riches and luxury. Their prancing horses, their elegant carriages, their silks and laces and diamonds, are the results of the war. The breeze, elsewhere laden with the shrieks of the wounded and the groans of the dying, brings to these wealthy contractors and their families the sweet scents of the flowers, the voluptuous strains of the music, and the delicious coolness of the hills and the woods."[136] Similar editorials on such topics as "Luxury against Patriotism," "Immorality and Political

Corruption," "Democracy on Trial," and "The Age of Extortion" could be found in such diverse publications as the *New York Times, North American Review, Christian Examiner,* and *Scientific American.*[137]

The public outcry against war profits became institutionalized in the demand for a personal income tax. The Civil War income tax was imposed by Congress in response to western objections to a federal property tax imposed in 1861. Under a plan advocated by Secretary Chase at the beginning of the war, a direct tax on real estate apportioned by state would be adopted. Western states protested their people were not as wealthy as Easterners, and they thus would bear an unfair share of the burden. Western congressmen demanded redress in the form of an income tax that would bear heavily on Eastern manufacturers and highly salaried public officials.[138]

The income tax worked approximately as designed. In 1863 New York alone contributed fully one-third of the total revenues, and other important manufacturing states such as Pennsylvania, Connecticut, and Massachusetts also paid disproportionately large sums—from 10 to 15 percent each. The income tax was not the principal producer of revenue, however. It furnished only about one-fifth of the revenues collected, rising to 29 percent in 1865. Overall, the revenue system was not as regionally discriminatory as these figures imply.[139]

The income tax was clearly designed to be progressive and particularly to strike prosperous manufacturers. Theories of public finance were not well advanced in the 1860s, but the idea of progressivity was certainly grasped. The *New York Times* declared that the "correct principle" was to tax luxuries but not necessities, and the *New York Herald* advocated taxing "the richest and most able to pay." Manufacturers argued that imposing both corporate and personal income taxes constituted double taxation, but merchants replied that manufacturers were given monopolies through tariff protection and that the income tax was a fair quid pro quo.[140]

The income tax rate structure was progressive by the standards of the times, if not of the twentieth century. The lowest bracket was 3 percent on incomes from $600 to $10,000 per year, but since only one American in seventy-two earned $600 or more per year the tax clearly struck the affluent. Incomes over $10,000 paid 5 percent. In 1864 the rates were raised to 5 percent on incomes from $600 to $5,000, 7.5 percent on incomes from $5,000 to $10,000, and 10 percent on incomes over $10,000. Only 31,000 Americans (out of 36 million) acknowledged that they had incomes in excess of $5,000—the richest .09 percent. The income tax was a modest antidote to a system of war finance that was burdensome to low-income wage earners because it relied heavily on inflation as a means of transferring resources to the government.[141]

The war tax system had several other progressive features. Congress discussed but rejected a plan to impose a dollar per year direct tax on gold watches and pianos. Instead, the government adopted an annual charge of 50¢ per ounce for gold plate and 3¢ an ounce for silver. (In an unprecedented display of federal power, tax collectors entered private residences and weighed the precious metal to establish the fee.) Yachts were assessed an annual fee ranging from five dollars upward, and carriages paid from one to ten dollars each year. Billiard tables were charged at ten dollars, and beer and liquor at 2-5¢ per gallon. Predictably, indignant brewers, billiard table manufacturers, and carriage makers descended on Washington to protest the taxes, but their efforts were in vain. To get at railroads and their passengers, a 3 percent federal tax was imposed on dividends on railroad bonds and on sales of passenger tickets.[142]

There were many contemporary charges that income and other taxes were flagrantly evaded, and some of these have been accepted by historians. As with many illegal activities, tax evasion is difficult or impossible to measure. New York gas utilities were frequently cited as evaders, and the city fathers threatened to take over the firms. In order to force tax payment, some newspapers published the names of taxpayers and the amounts of taxes paid. The tax law also contained an innovative provision that allowed a person informing the government of tax evasion to claim half of the fine imposed. Although there was inevitably some cheating, George Boutwell, Lincoln's able commissioner of revenue, reported that income from the taxes far exceeded expectations. Noah Brooks, an astute journalist, remarked that "it is indicative of the spirit and temper of our people [that] people everywhere are more ready to comply with the far-reaching requirements of the law than the most sanguine had supposed to be the case. Our people talk a great deal, but they usually do about right."[143]

Contributions

In 1985 a respected scholar described the effort on the Northern home front as an "endless tale of fraud, corruption, profiteering, and near-treason" that belied the conspicuous valor of Yankee troops at the front.[144] Yet despite all the evidence assembled thus far, the record of the Northern people during the Civil War was by no means a litany of unmitigated greed. There was a great outpouring of support from businessmen and other civilians who were willing to give much more than lip service to the cause. The history of the war is filled with instances of businessmen who charged less than the market would bear, contributed their services or their possessions without charge, raised money to help the troops, or bought war bonds at charitable prices.

Although some railroads charged handsome fees, others carried troops for less than cost or even for nothing. From April through June 1861, the Michigan Southern and some other lines carried troops without charge. Erastus Corning, president of the New York Central, proposed in 1862 to transport troops to Washington at a discount of 40 percent off the regular fare. Volunteers who were rejected by the army for medical reasons would receive a free ride home.[145] The record of defense contracting would be incomplete if these sacrifices were excluded.

Shipowners also sometimes generously loaned their vessels to the navy without charge. In 1862 James Gordon Bennett, the editor of the *New York Herald,* loaned his yacht, the *Henrietta,* to the navy for a year at no cost. When the Confederate navy threatened to break the Union blockade of Hampton Roads with its ironclad warship CSS *Virginia,* President Lincoln called Cornelius Vanderbilt to the White House and requested the assistance of Vanderbilt's fleet, asking him to name his price. Vanderbilt refused compensation, but committed to the navy as a gift a new steamship built at a cost of $800,000 and christened the USS *Vanderbilt.* Vanderbilt even put his life on the line. He sailed his namesake vessel to Hampton Roads with the intention of ramming the *Virginia,* but when the Confederate ship did not accept battle (probably fortunately for the amateur admiral), Vanderbilt turned over command to the navy. The USS *Vanderbilt* served throughout the war, and after the Confederate surrender a grateful Congress passed a resolution of gratitude and ordered that a gold medal be struck to signify the nation's appreciation. The commodore reportedly was miffed because the navy did not return the ship, but he accepted the medal graciously. "Should our government be again imperilled," Vanderbilt declared, "no pecuniary sacrifice is too large to make in its behalf, and no inducement sufficiently great to attempt to profit by its necessities."[146]

Some businessmen patriotically sold their products to the government below the market price. Just one example was John Baldwin of Chicago, who in 1861 rejected price enhancement on lumber he sold to the army for the construction of barracks. In explanation, Baldwin showed a deep sense of wartime ethics. "We were bound to put the lumber as low as we could sell it to the government," he said simply. There were many other examples.[147]

Merchants with thorough knowledge of overseas markets also rendered important service to the Union cause at little or no cost. The international arms dealer George L. Schuyler of Schuyler, Hartley, and Graham was given authority to purchase $1 million in European arms. Working for only $10 per day plus expenses, Schuyler rejected as insults all offers of bribes intended to induce him to accept inferior weapons. Henry Shelton Sanford, the Union spymaster on the Continent, spent $15,000 of his own money on bribes in 1863, winning a great but secret victory in preventing the Confederates from obtaining warships in France. In 1863 William H. Aspinwall and John M. Forbes served as secret agents seeking to buy ironclads and commerce raiders. They accepted compensation for their expenses but not for their services.[148]

The war effort received millions of dollars in voluntary contributions from businessmen. On 20 April 1861, amid the excitement and outrage of the post-

Sumter period, in twenty minutes members of the New York Chamber of Commerce contributed $22,450 to the Union cause, with the pro-secession editor James Gordon Bennett making the largest donation ($3,000). By 26 April contributions reached $87,790, aided by $10,000 from the famous merchant A. T. Stewart. On the following day, William B. Astor topped Stewart and Bennett with a $15,000 donation, and the fund grew to $118,890. Solomon Sturgis, a merchant from Chicago, contributed over $120,000, even equipping at his own expense an entire company of soldiers with Sharps repeating rifles. Francis Loomis of New London, Connecticut, a wealthy clothing manufacturer, offered to replace the entire garrison guarding New London at his own expense. President Lincoln declined, although the plan would have cost Loomis nearly a million dollars.[149]

Businessmen institutionalized their contributions in such organizations as the Union League Club and the U.S. Sanitary Commissions. Both were pro-Lincoln, quasi-political institutions that selflessly raised large sums of money and tirelessly distributed assistance for the Northern war effort. Nevertheless, the Union League was erroneously denounced as merely a group of *nouveaux riches* who lacked sufficient status to gain admission to the more exclusive gentlemen's clubs. Allegedly, they were "debarred by their lack of education, culture and refinement from associating with the really intellectual portion of the community."[150]

In fact, the Union League included old wealth (the shipping merchant R. B. Minturn, the tea merchant George Griswold, the department store king A. T. Stewart), distinguished scholars (the scientists Francis Lieber and Wolcott Gibbs), accomplished surgeons, and prominent men of affairs (Franklin H. Delano, grandfather of Franklin Delano Roosevelt, and Hamilton Fish, the future secretary of state). The rival Loyal Leagues were made up of men who were not admissible to the Union League because of membership in the Democratic Party, failure to support President Lincoln, or insufficient prominence. In any case, the Union League and the U.S. Sanitary Commission rendered invaluable aid to the war effort, and the troops knew it. "Where'er we went—in the pine lands of Georgia or the swamps of Mississippi," wrote General William T. Sherman, "letters came to us from your members telling us not to heed the cost or expense; 'You shall have whatever is necessary; there shall be plenty of money and plenty of men; go and do your duty like soldiers, and you shall be backed.' And we knew we would be, and we *were*."[151]

Although there was confusion and fraud in the Civil War, the full story, in Allan Nevins's deft words, was of an organized war replacing an improvised war. By 1865 the term *shoddy* had fallen into disuse as the troops recognized the growing efficiency and basic selflessness behind the war effort.[152] Although the "shoddy" legend would be sustained into the late twentieth century,[153] the real Civil War was a mixed record of waste and efficiency, of profit and loss, of fraud and sacrifice.

5

Toward the Great War

Those who scent from afar the cadaverous odor of lucre have for the most part furnished war's dominant motive.

DAVID STARR JORDAN (1914)

For half a century after the Civil War, the United States was at peace, interrupted periodically by frontier skirmishes, colonial incursions, and a short war
with Spain. Arms production declined abruptly after 1865, and American suppliers faced a major shakeout. For a few years, lively foreign sales staved off
bankruptcy. By 1872 nearly 1.4 million weapons had been cast off to other nations' armies. Obsolescent guns were shipped all over the world—to Russia,
Spain, Denmark, Sweden, Egypt, Cuba, Greece, China, Japan, and even the
Vatican City. Turkey bought 350,000 of the best and cheapest rifles, while a
desperate France, threatened by Prussian attack, paid a dear price for 706,000
lower-grade arms, along with 54 million rounds of ammunition. Nevertheless,
the inevitable reduction in the weapons business took its toll. Of the forty-
eight major arms contractors at the close of the Civil War, only eleven remained in business after 1870.[1]

Foreign arms sales kept the profiteering controversy smoldering. Some
Americans harbored grave doubts as to whether the United States should be
the arms merchant to the world, particularly the arms merchant to France. The
French had spent lavishly on Confederate bonds during the Civil War, and, although the Union army was distracted, French troops had invaded Mexico, in
defiance of the Monroe Doctrine. Conversely, German-Americans had loyally
supported the North, and their Prussian cousins were sympathetic to the
cause. After the outbreak between France and Prussia in 1870, France placed a
large order for rifles with the Army Ordnance Department. When Secretary of
War W. W. Belknap learned of the impending sale, he withdrew authorization
immediately in order to maintain American neutrality. Belknap agreed, however, to sell guns and ammunition to the American agent of France, the firm of

E. Remington and Sons of Ilion, New York. This decision brought a protest from the German government and opposition from Secretary of State Hamilton Fish.[2]

Because of the bitterness toward France and the dubious legality of the sale, a controversy developed that anticipated the larger and more bitter disputes of the early twentieth century. Secretary Belknap's leading critics were Senators Charles Sumner of Massachusetts and Carl Schurz of Missouri, two former abolitionists who were now leaders of the Liberal Republican faction that had broken with President Ulysses S. Grant. Schurz was also the recognized leader of the German-American community, an exiled combat veteran of the Revolution of 1848.[3] The coming dispute with arms merchants would be heavily indebted to abolitionists and their descendants, and it would often draw upon the ideas of German liberalism. The Massachusetts-Missouri alliance also suggested the future cooperation between New England progressives and western agrarians.

Early in 1872 Senators Sumner and Schurz demanded a probe of arms sales to France. A discrepancy had been discovered in the French accounts: although the French government had paid $10 million for the American arms, the U.S. Treasury Department reported that it had received only $8,286,000. The senators suspected that the missing funds had been used to bribe American officials to approve the sale. The Remington firm had ample motive for chicanery, thought Sumner, as it was to receive a handsome commission of 5 percent on the order.[4]

The Grant administration was reluctant to support a legislative investigation, but after extended debate the Senate approved a resolution to establish a committee of inquiry. Owing to Sumner's illness, Schurz spoke in favor of organizing the committee, and he was assisted by women who packed the galleries and filled the Senate chamber to overflowing. The ladies applauded wildly to express their support.[5] (This demonstration foreshadowed feminist backing for the investigation of the munitions industry conducted by Senators Gerald Nye and Arthur Vandenberg in 1935.)

The Senate committee could find no evidence of corruption on the part of any American official, although it did uncover a letter from Samuel Remington in which he claimed that he had the "strongest influences working for us." Upon investigation, this turned out to be an exaggeration of the political influence of Remington's friends. The theft of funds was most likely committed by French agents.[6] Nevertheless, neither Schurz nor Sumner was persuaded that the truth had been found. Following their lead, the historian Allan Nevins declared (without substantiation) that "it was more than suspected that some [War] Departmental officers made a pretty penny from [the transaction]."[7] A minor incident in itself, the arms "scandal" featured abolitionism, feminism, German liberalism, pacifism, politically inspired charges, and unsuccessful investigation. All these elements were conspicuous in the larger controversy half a century later.

In the 1870s the United States relinquished leadership in weapons production to Britain and Germany, where American designers of machine guns and armor plate were forced to seek business.[8] Few Americans noticed or mourned the decline in military arms production, and concern about war profits was nonexistent. Meanwhile, the industrial revolution was steadily but inexorably transforming military technology. Like other technologies of the period, weaponry became greatly more sophisticated. In no area was this more evident than in sea power, where the world's leading navies were shifting to steam power, metal hulls, screw propellers, armor plate, and breech-loading cannon. The American fleet, smugly oblivious to these changes, rotted and rusted away.

Profits and the New Navy

By 1880 the U.S. Navy was so decrepit that even Chile's tiny fleet seemed menacing to some, and a consensus formed that the ironclads and wooden-hulled frigates of Civil War vintage must be replaced with more modern vessels. During the 1880s and 1890s, the American fleet shifted from what James L. Abrahamson has termed a "peace navy," suited only to coastal and harbor protection and limited commerce raiding, to a "war navy," which featured a powerful battle fleet capable of challenging any great power for supremacy at sea.[9] Modern warships required a thick and expensive belt of armor, and it was in the furnishing of this steel plate that the war profits question returned. It appeared in the two most time-honored forms of profiteering: first, that prices charged by manufacturers of armor plate were exorbitant and, second, that armor plate was supplied in criminally degraded quality.

Following the battle of the ironclads USS *Monitor* and CSS *Virginia* in 1862, European naval architects had raced to implement the revolutionary technology. By 1885 armor plates had thickened dramatically, but even so they could still be penetrated by rapidly improving naval cannon. To catch up with these developments, naval officers pleaded with American steelmakers for help. Although the interest of the steel industry was initially lethargic, the mid-1880s weakness in the primary market for steel rails finally prompted some interest in diversification. Steelmakers told Congress, however, that prices of at least $500 per ton would be required to induce them to enter the armor plate business. When in 1887 the Navy Department finally accepted a bid from the Bethlehem Iron Company for 6,703 tons of armor, the average price was $538.67 (expected to rise to $632.61 after testing losses). This was within 20 percent of established European prices.[10]

Bethlehem's entry into the military market, though it would make the firm a central figure in the coming arms controversies, was a shrewd business decision. Located in eastern Pennsylvania, Bethlehem Iron (later Bethlehem Steel) was at a disadvantage to its western competitors in serving the rail market. Bethlehem also depended on Cuban mines for iron ore, which burdened its products with heavier transportation costs. In serving the eastern naval shipbuilding industry, these disadvantages disappeared. Bethlehem's location became an advantage, and Cuban ore was better suited for making the higher-grade steel required in military contracts than it was for rails. Bethlehem was nevertheless unable to complete its mill and deliver steel quickly enough to satisfy the naval appetite. In 1890 Secretary of the Navy Benjamin F. Tracy, with the assistance of President Benjamin Harrison, persuaded a reluctant Andrew Carnegie—the leading business pacifist—to enter the military market. Tracy promised Carnegie, Phipps and Company the same prices as were given to Bethlehem three years earlier. Both Bethlehem and Carnegie received assurances of obtaining steady business, and the armor plate industry settled down as a two-firm oligopoly, with the U.S. Navy as the only customer.[11]

This was a prescription for discord. Steel rails made by the open hearth process could be produced profitably at the turn of the century for $30 a ton.[12] When it was reported that open hearth armor plate brought more than $500 a ton, a throng of critics in and out of Congress gathered. There were many critics who doubted the need for a modern "war navy" in the first place, and numerous others explicitly disliked paying unbid and allegedly exorbitant prices for its armor. Bethlehem and Carnegie (and later Midvale Steel, a third invited entry into the armor plate business) became known collectively in the Populist-Progressive argot as the "Armor Trust" or the "Armor Ring."[13]

In 1893-94 opposition to the naval rearmament program spread into a controversy that became known as the "armor plate scandal." Four discontented employees of Carnegie Steel—they would be known today as "whistleblowers"—presented evidence to Grover Cleveland's secretary of the navy, Hilary A. Herbert, that Carnegie Steel had fraudulently sold defective armor plate to the navy. Herbert investigated, agreed with the informants, and recommended that Carnegie be fined. President Cleveland set the fine at $140,489.[14]

Carnegie complained, and a lengthy congressional investigation ensued. This investigation examined several charges: that Carnegie Steel had failed to follow specified procedures in making armor plate; that it falsified results of tests of armor plate; that it had concealed or disguised defects in plate (allegedly the steel had been weakened by blowholes that were secretly and improperly patched); and that it had secretly re-treated samples of armor plate scheduled for ballistic tests. The committee disregarded the explanations of the steel firm and reported that frauds of a near criminal nature had transpired.[15]

The most careful student of the episode, Robert Hessen, has determined that the facts of the matter belie the committee's report. No defective armor was sold to the navy, no sailor's life was endangered, and the retreated plates submitted for testing were rejected by the navy. Even so, the behavior of Charles M. Schwab, Carnegie's supervisor of armor plate production, was seriously improper and deceitful.[16] Because of Schwab's misconduct, the brief and simplistic report of the investigating committee became accepted as the standard account of the incident. It became a centerpiece in the broad but ingenuous indictment of defense contractors presented by at least six major books.[17] The overstated charge of Carnegie's perfidy joined the erroneous claim that J. P. Morgan's fortune had been founded on ill-gotten gains from the Civil War. Together these charges formed the "historical basis" for the progressive indictment of munitions makers.[18]

Not only was the particular lot of steel that was held in question during the armor plate scandal not defective (except cosmetically), but American plate had become the world's strongest. In 1889 Hayward Augustus Harvey, an American inventor, had devised a process for case-hardening steel that gave American plate a technological superiority. In 1892 firing trials in the United States and at the Vickers steel works in Sheffield, England, showed that Harveyized steel was significantly better than English types. Vickers hastened to catch up, in the process infringing on Harvey's patents. Meanwhile, the Krupp firm in Germany developed a similar, patented process that could be combined with the Harvey method. In 1894 the various patent holders and steel firms formed a pool to divide the business—and fix the prices—among the ten principal armor plate manufacturers of Britain, Germany, France, and the United States. This was a true armor trust, but the agreement applied only to sales of armor plate to third-party countries; the agreement did not apply to steel sold to the British, German, French, and American navies. Nevertheless, it gave substance to congressional and other critics who alleged collusion in bidding on armor plate.[19]

The armor plate scandal and the existence of the patent pool raised serious questions in respect to the appropriateness of profits on armor plate. In 1895 Secretary Herbert initiated a review of armor prices with the intention of eliminating excessive profits. He demanded that the prevailing price for armor steel ($600 a ton) be cut, on the ground that the initial investment in plant and equipment had been retired. The steelmakers offered a 10 percent reduction, which Herbert rejected as inadequate. He undertook an investigation of world armor prices, even making a secret trip to England and using naval attachés to help uncover the truth. But the Carnegie firm learned of Secretary Herbert's plan, and it successfully undermined his investigation. Herbert's inquiry discovered wide variation in foreign armor prices, with some similar to and some lower than the American price.[20]

Because of the secretiveness of the steelmakers and the absence of records, it is impossible to determine today the level of profit they were obtaining. Influential congressmen, however, wanted a price reduction. In 1897 the populistic Democratic senator from South Carolina, Benjamin R. Tillman, demanded that the maximum price for armor be fixed at $300 a ton or, if this was not accepted, that a government-owned armor plate plant be constructed. Tillman, in what his biographer would describe as a "heroic last stand," declared that any senator who would pay more than $300 was willing to grovel "in the mire of corruption and rottenness." The $300 figure originated from a revelation in 1896 that Bethlehem Iron had agreed to sell armor to the Russian Navy for $250 a ton—less than half the price it charged the U.S. Navy! The lower figure for foreign sales exemplified a practice common in the armor plate and elsewhere in the international arms industry—selling steel at a price below cost of production in order to maintain an ability to service the domestic market. In an election year, the commonality of this practice (known today as "dumping") was disregarded by senatorial critics, and they also ignored the fact that the Russian price was later raised to $524. Secretary Herbert, who was anxious not to impede rearmament, was willing to pay $400, and the House of Representatives approved $425. When the Navy Department asked for bids at $300, Carnegie Steel demurred, and it also rejected bids at $400. Instead, both Carnegie and Bethlehem offered to sell their armor plants to the government and to abandon the armor plate business entirely.[21]

For Carnegie Steel, the armor plate business was a minor specialty—its plant represented just 2 percent of its total investment, and profit margins were declining. In 1893-94, according to company records, the armor plant earned a rate of profit of 30 percent on the investment. At that time, the price of armor was $671.15 a ton. By 1898, after congressional critics had conducted their price reduction campaign, the profit margin fell to 15 percent, which was lower than that obtained on other steel investments. In 1898, after prolonged negotiations with the new Republican administration, the price was fixed at $460. This was undoubtedly profitable, but it was probably less so than the margin that pertained in Germany, where the House of Krupp was obtaining a profit on armor plate of 100 percent of the cost of production. (Krupp prices were generally higher than those obtained in the United States.) Nevertheless, the profits obtained by Carnegie and Bethlehem were far too high to satisfy those largely agrarian skeptics who doubted the need for the rearmament program in general and its growing cost in particular.[22]

The Spanish War

As the 1890s drew to a close, the rearmament program had obtained some success, although certainly not without generating considerable controversy. By 1896 the U.S. Navy had become the world's sixth largest. Although still far from formidable, it was respectable. The peacetime army, whose duties were formerly restricted to Indian control, riot duty, coastal defense, militia training, and river and harbor improvement, was becoming a "war army," whose basic mission was to fight a small war with a foreign power.[23] Yet the war with Spain revealed that America's organizational capacity had lagged behind its military aspirations.

In 1898, when the United States declared war, the regular army had not been enlarged in many years, and its complement remained fixed at 25,000 men.[24] The tenfold increase in strength would have to be made up, in the American tradition, of volunteer troops who were assembled, trained, equipped, and transported at nearly a moment's notice. This haste would necessarily entail a considerable financial burden, and this would offer another opportunity for charges of mismanagement and war profiteering.

The Spanish War was too brief, too cheap, and too popular to produce severe dispute, but it did foreshadow what was in store. There were numerous charges of malfeasance, some well grounded, but most accusations of profiteering were really matters of haste and inefficiency rather than avarice. The best-known of these incidents was the "Embalmed Beef" controversy, which materialized because provisions supplied to the troops for once fully justified the soldier's eternal gripe about army food.

The basic ration issued to the troops in 1898 was canned roast beef. The army had little experience in fighting in the tropics, and its supply officers failed to anticipate that canned beef would spoil quickly in the heat. There were also some cans of beef that contained gristle, pieces of rope, and even some dead maggots. These problems formed a good example of the degraded quality form of profiteering. Charges of malfeasance reached the press, which in classic muckraking fashion exaggerated the evidence, attributed sinister motives, and assumed corruption in government. The army's commissary general, Charles P. Eagan, was forced to resign, terminating a long career and staining a distinguished record. In 1899 a court of inquiry found that Eagan had made a mistake in judgment, but had done so honestly. Nevertheless, the myth of meatpackers corruptly providing the troops with adulterated beef became deeply engraved on the social memory of the Spanish War.[25]

There were a few other allegations of profiteering related to uniforms, arms, and training camps. As in the Civil War, there was not enough blue cloth of military grade available to supply uniforms for the multitudes of adventuresome volunteers. Inferior grades were again employed, this time knowingly. These uniforms faded or wore out quickly, allowing press critics to recycle old charges of corruption and to bring back the timeworn accusation of "shoddy" quality. Due to congressional stinginess, supplies of modern, bolt-action Krag-Jorgensen rifles were insufficient. These weapons went to the regular army, while volunteers were armed with obsolescent single-shot Springfields of Civil War vintage. The new Krags used smokeless powder, but the older models used black powder that left a cloud of smoke to give away the troops' position. Powder manufacturers (principally the DuPont Company) hastened to produce greater quantities of smokeless powder, but they were charged with improperly hiking their prices as well. (The original price, a dollar per pound, was reduced to 80¢, which one congressman termed "extortion.") James A. Frear, a veteran of 1898 who served for many years in Congress, maintained that contractors had overcharged the army for the construction of training camps. Lumber that allegedly should have cost thirty dollars per thousand board feet was sold to the government at sixty dollars, and laborers who supposedly should have received $4 a day were paid $7.[26]

Brief though it was, the War of 1898 was not too short to produce its share of disgruntled veterans. Sanitary conditions in the thoughtlessly constructed camps were poor, and 20 percent of the recruits contracted typhoid fever. One in seventy died of disease, and doubts about this sacrifice appeared. The war offered an excellent opportunity for the era's leading dissenters, the Populists, to oppose its cost, but most Populists were slow to seize it.[27] The war was generally popular in the South and West, but Tom Watson, the Populist presidential candidate from Georgia, contended that foreign involvement only benefited the "privileged" classes. Watson lamented that imperialism diverted attention from "the unjust system they have put upon us." Congressman Frear, who would cast his vote against American entry into the war with Germany in 1917, said that his predecessors in Congress had been "swept off their feet" by war hysteria in 1898. His colleague in the Senate, James K. Vardaman of Mississippi, a combat trooper in Cuba, believed that the worthy goals of the war had been corrupted into an imperialistic venture by greedy businessmen, particularly bankers. Though the war lasted but three months, one volunteer uttered sentiments as old as war itself: "We have been ignored, neglected, dishonored, belittled, and more or less forgotten! . . . The enlisted man received a pittance of $15.60 while in the field."[28] The charges of expensive munitions, adulterated food, costly cantonments, and underpaid soldiers were all revived in the bitter disputes of the Great War.

The Progressives and Defense Profits

From 1898 through 1913, the United States embarked on a program of accelerated military expansion. In 1899 bitter fighting resumed as American forces sought to consolidate control of the Philippines. Congress voted to triple the prewar army and to double the naval construction program. Between 1898 and 1907 nineteen battleships and ten armored cruisers were authorized, and the naval buildup produced results. The new warships were twice as large and twice as heavily armed as their predecessors. Between 1879 and 1921, the number of sailors increased by ten times, and naval appropriations rose by forty times. The U.S. Navy rose from sixth largest in the world to second place, behind only Great Britain, and even that was not enough to satisfy some navalists. The army increased in manpower by nine times and in its budget by thirty times. As this unceasing growth continued, progressive critics of expansionist diplomacy asked a series of simple, reasonable, and searching questions: Why did the United States need these forces? Would they be used to good purpose? When would the expenditures stop rising? And perhaps most important, would great military strength have harmful effects on the nation?[29]

These troublesome questions foreshadowed the larger and increasingly more divisive debates of the following decades. Progressive critics of military force and expansionist diplomacy—who might be called "isolationist progressives"—added two more considerations to the list: what part in American defense policy was being played by private profit, and was this private profit serving the national interest? Their answer to these new questions—that profit played an inordinate and improper role—would shape their answers to nearly all other military questions.

The most fundamental defense issue of the Progressive era was whether the United States faced any serious danger at all. In 1837 Abraham Lincoln himself had eloquently articulated a traditional belief in American invulnerability to attack. "All the armies of Europe, Asia, and Africa combined," Lincoln declared, ". . . could not by force take a drink from the Ohio, or make a track on the Blue Ridge, in the trial of a thousand years." In 1880 the illustrious Civil War general William T. Sherman said that the chance of invasion was "simply preposterous."[30] For many Americans, this safety from foreign attack was as complete in the twentieth century as it had been in the nineteenth. "Three thousand miles of salt water separate the hell over there from God's country," wrote one Oklahoman. "Surely, surely there [is] no reason why the USA should get mixed up in the bloody mess on the other side of that blessedly deep and wide ocean."[31]

Military critics specifically rejected the assertion that the principal potential enemies identified by champions of defense spending—Germany, Japan, and Britain—represented true threats. David Starr Jordan, the pacifist president of Stanford University, dismissed the notion that Japan had any design on the Philippine Islands or upon gaining territory in the Western Hemisphere. "We cannot conceive of a war between Japan and the United States," Jordan confidently announced in 1914.[32] Lucia Ames Mead, a pacifist who was one of three women on the board of directors of the Anti-Imperialist League, believed that American control of the Philippines was a matter of reversed priorities. She believed that navalists desired the Philippines in order to justify a larger navy, rather than that the United States needed a strong navy in order to protect the islands. Progressive critics further believed that to continue to increase defense appropriations would divert funds from social programs, would violate traditional American opposition to standing armies (thus undercutting American exceptionalism), and would even invite battlefield carnage by making available the weapons it required. Opposing the Spanish War, David Starr Jordan expressed views that were common among American pacifists throughout the progressive era. He told the Stanford Class of 1898 that it was "not what we shall do with Cuba, Porto Rico, and the Philippines [sic]. It is what these prizes will do to us." Jordan believed that the only practical uses for the United States Navy were to conduct ceremonies and to render assistance to American citizens in trouble abroad."[33]

Forcefully rejecting the logic in support of defense spending, Progressive critics began to speculate upon what other motivation might explain the advocacy of defense increases. Their suspicions centered on the corrupting influence of money. Their conclusion became the American version of what became known internationally as the "Merchants of Death" theory, namely, that defense contractors aided and abetted the outbreak of war in search of profit. Its origins came partly from within the United States and partly from Europe. But American conditions—particularly the heritage of the Civil War—ensured that the Merchants of Death idea would gain a large American following.

In 1909, Oswald Garrison Villard, a leading Progressive journalist and a grandson of the abolitionist William Lloyd Garrison, declared in a speech in New York City that support for increased naval spending came from "a combination of very wicked persons who stand to profit from a big navy." (Villard referred to Colonel Robert M. Thompson, a Naval Academy graduate who had become chairman of the board of directors of International Nickel. Nickel was an important component of armor plate, and Villard believed that avarice explained Thompson's support for navalism.)[34] Villard was not only among the first to link the rearmament program directly to greedy profit seeking, but he also remained the most outspoken advocate of that position. Villard displayed some of the suspiciousness and contentiousness for which his grandfather was famous, and the antipreparedness movement (and later the antiwar movement) exhibited some of the zealousness and crusading self-righteousness of the abolitionist movement.

Others soon picked up Villard's cry. In 1910 Nicholas Murray Butler, president of Columbia University, offered an early and candidly simplistic statement of this idea. Butler declared, "I am one of those who look for the simplest motives in explanation of action or of conduct. My impression is that somebody makes something by reason of the huge expenditures in preparation for war." In 1913 a leading Progressive journalist, Hamilton Holt, charged "the agents of ordnance manufacturers and shipbuilders" with working against friendship between the United States and Japan. By 1914 Congressman Clyde H. Tavenner was writing in the Progressive journal *La Follette's Weekly Magazine* that millionaire munitions executives were "agitating" for a larger defense in search of profit. Tavenner went on to allege the existence of a "World Wide War Trust" that endangered the peace. In early 1914 David Starr Jordan added confidently, "About war scares and war equipment, matters inherent in the War System, centre the grossest exhibitions of human greed. Those who scent from afar the cadaverous odor of lucre have for the most part furnished war's dominant motive." Thus the allegation that munitions makers in search of fat profit promoted war preceded the debate over the Great War by several years.[35]

The Great War was profoundly disturbing to Progressives who blithely entertained a romantic, nineteenth-century belief that humanity had marched so far down the road of progress that war had become an anachronism, even an impossibility. "Except as a result of accidental clash in uncontrollable war machinery," proclaimed David Starr Jordan early in 1914, "war is already impossible." Since war was inconceivable, progressives reasoned, preparing for it was senseless. "There is, in fact, something primitive, outworn and unprogressive in the spectacle of a civilized nation composed of millions of clever people trusting for its defense in forts and ships," Jordan proclaimed. "Europe recoils and will recoil [from war]," he predicted with supreme overconfidence.[36] When the events of August 1914 smashed forever this dreamy innocence, the reaction among Progressives was despair and anger. "It is impossible now," Jane Addams recalled later, "to reproduce the basic sense of desolation, of suicide, of anachronism, which that first news of war brought to thousands of men and women who had come to consider war as a throwback in the scientific sense." But David Starr Jordan knew immediately where to place the blame: "The mailed fist has crashed through the delicate far flung fabric of our civilization, sweeping away as cobwebs all that we have cherished. . . . THE WAR SYSTEM MUST GO."[37]

Jordan's intellectual journey to a belief that war was caused by profit seeking was illustrative. As a youth, Jordan was a militant republican, hating every respect of royalism, and he maintained that faith throughout his life. Jordan spent the summer of 1912 studying the American Civil War, and he later dedicated one of his early pacifist tracts to the memory of his brother, who was killed in it. He concluded from his study that while civilization was progressing, war, as an anachronistic remnant of monarchism, was the worst enemy of progress. Jordan also reached the conclusion that threats to peace in Europe were being conjured up by greedy weapons merchants in England and Germany, whose baneful alliance with militarists was a wretched carryover from feudal days. "It is the gigantic and remorseless naval military lobby, not the menace of Germany, which keeps up the British naval budget," Jordan wrote to Charles W. Eliot, the president of Harvard University, in September 1912. "In the United States," Jordan added, "with no possible fear of attack of any kind from any quarter we are spending nearly a million dollars a day, (1) to follow the fashion of Europe, (2) to create an ideal perfection of defense where no defense is needed, (3) to gratify our own armor plate lobby, and (4) to keep national defenses so high as to discourage tariff reduction." Eliot demurred. "I cannot believe that all these war preparations are caused by the machinations of a few manufacturers interested in the making of war materials," he wrote, arguing that defense contractors were too few to cause a great war. This exchange between the leaders of Stanford and Harvard portended a debate that would absorb the nation two decades later.[38]

Antipreparedness

The year 1915 marked the fiftieth anniversary of the end of the Civil War. In one way or another, its bittersweet memory affected the views of most participants in the preparedness controversy. The Civil War reminded many Americans not only of gore and glory but also of swollen fortunes, an unwanted outcome that had never been reconciled. Those who lamented the failure of the Northern cause to achieve its idealistic goals—and thus to justify its sacrifice—found that the charge of war profiteering resonated. For a few the memory was vivid and direct. At eighty, Isaac R. Sherwood of Ohio, who had served as a brigadier general in the Union army, was one of the most venerated members of Congress. In January 1916, Sherwood echoed his Civil War heritage by denouncing "that powerful group of war exploiters in Gotham who value blood-coined dollars as more vital than orderly self-government." Senator John D. Works of California, who had served in the Union cavalry, was a leading opponent of preparedness. Works displayed his long (but not infallible) memory by declaring that the call for preparedness came from "the influence of plutocracy, wealth, big business. . . . Plutocracy and militarism are concordant and congenial evils."[39]

Others formed their views of Civil War profiteering by hearsay. In 1922 Warren G. Harding, by then president of the United States, recollected how a chance meeting a decade earlier had left a deep impression: "I remember," wrote Harding, "in 1913, I attended the fiftieth anniversary of the Battle of Gettysburg. Quite by accident I ran on to a small camp-fire group of Union and Confederate veterans. . . . The chief theme of discussion, in this reunion of former foes, was that the Civil War was fought at the command of the capitalists of the North and South. . . . Among these grim old veterans, rollicking in the enjoyment of their reunion, there was that inevitable resentment of those who had made fortunes out of the war or those whose fortunes had something to do with bringing on the war."[40]

Many of those who were outspoken against preparedness and profits were profoundly affected by the social memory of the war as related by historians of the time. Gustavus Myers's depiction of *The History of the Great American Fortunes* was one of the most widely read and influential tracts of the Progressive era. It made the oft-repeated charge that the foundation of J. Pierpont Morgan's fortune was gun-trading during the Civil War.[41] Since Morgan's firm, under the leadership of his son, was the leading supporter and supplier of Britain before the American war declaration, Myers's charges focused the intervention question sharply: Should the United States participate in another gory conflict if the probable result was to intensify the inequality of the distribution of wealth?[42] Other writers cited James Ford Rhodes on Civil War profiteering, conveniently overlooking Rhodes's opinion that the charges of corruption were "much exaggerated."[43]

Some remembered the economic cost of the Civil War from Confederate perspectives. The Southern historian Frank L. Owsley blamed the Confederacy's defeat on the failure of Britain to intervene, and he believed that Britain made this decision for reasons of profit. Secretary of the Navy Josephus Daniels, a North Carolinian, was the son of a Confederate soldier who was killed in the war. His mother taught her sons to hate war, and Josephus remained suspicious of military ways. Daniels became arguably the most aggressive opponent of war profits in the Wilson administration.[44]

Perhaps the most important legacy of the Civil War was creating the conviction among antipreparedness Progressives that if the United States entered the Great War it would mean the end of the Progressive movement. Recalling the vitality of the antebellum reform spirit, as exemplified by abolitionists, pacifists, feminists, and 48'ers, they sadly also recalled how the Civil War had drained away the zest for reform. From this perspective, the Civil War became a profoundly reactionary event.[45] Since progressives regarded businessmen as the sworn enemies of reform, they suspected that businessmen would seek to enter the war not only in order to gain a blood profit but also to kill off the vigor of the reform movement. Jane Addams pessimistically summarized this view: "Everything that we have gained in the way of social legislation will be destroyed."[46] Thus, progressives argued, war profits had a dual role in bringing on American participation in the Great War.[47]

In a curious way, the battle over preparedness and war profits pitted descendants of both sides of the Civil War profiteering quarrel against each other. Those in favor of preparedness included J. P. Morgan, whose father was at the center of the Civil War dispute, Marcellus Hartley Dodge, and Charles Schwab. The descendant of Marcellus Hartley of the Civil War era, Dodge owned all the shares of the Remington Arms–Union Metallic Cartridge Company.[48] Unquestionably, American entry would greatly benefit Dodge's ammunition business. Charles Schwab of Bethlehem Steel, the leader of the "Armor Trust," was the grandson of a mill owner who had held a contract to make blankets and overcoats for the Union army. The du Pont family was another case in point. The family firm had supplied powder for the nation's arms for a century, and the latest generation of du Ponts was expected to gain from American participation.[49]

Thus while defense contractors were often heirs of family businesses that had long been engaged in supplying military needs, the chief critics of defense spending were often descendants of abolitionists, like Oswald and Fanny Villard. Congressman Daniel R. Anthony Jr. of Kansas was another preparedness foe of staunch abolitionist lineage. Anthony's father, a fiery Quaker editor, was an abolitionist of considerable local standing who later worked for temperance and woman suffrage. The younger Anthony's better-known relative was his illustrious aunt, Susan B. Anthony, the renowned abolitionist and feminist.[50]

In charging that profit was a major purpose of preparedness, the antipreparedness camp was taking up a cry that had resounded in England and Germany. In the decade before the war, British defense contractors had built nineteen plants in Russia, making everything from cannon shells to battleships. These investments were highly profitable, and critics at home claimed that British foreign policy was designed to protect the investment. A serious controversy developed in 1906 concerning the Coventry Ordnance Company, a major contractor for the Royal Navy. Allegedly, H. H. Mulliner, Coventry's managing director, supplied Parliament with counterfeit documents purporting to show a substantial increase in German naval spending in hopes of increasing British naval appropriations. This episode included a frightened outcry from the English public for the construction of more dreadnoughts: "We Want Eight and We Won't Wait!"[51] This allegedly contrived danger led to the desired increase in spending and contributed to the Anglo-German naval arms race, which was one of the causes of the Great War. In 1906 the famous English dramatist, George Bernard Shaw, presented *Major Barbara,* which featured as a principal character Sir Andrew Undershaft, an amoral defense contractor who cheerfully sold arms to both sides in any war. (Undershaft was loosely modeled on the Krupp family, the cannon kings of Germany.)[52] First produced in the United States in 1915, *Major Barbara* attracted considerable attention because of its author's literary prominence.

There was a substantial production of British nonfiction which was also very critical of defense spending. An illustrious galaxy of writers including John A. Hobson, Norman Angell, Henry Noel Brailsford, George H. Perris, and H. G. Wells attacked the military from different perspectives.[53] Besides the publication of English books and journals in America, these ideas crossed the Atlantic in several other ways. Norman Angell, the best-known British pacifist, made a speaking tour of the United States, and several American pacifists visited their British comrades.[54]

In Germany, the principal prewar critic of military spending was Karl Liebknecht, the leader of the socialists in the Reichstag. In 1913 Liebknecht gained international prominence by charging that members of the Krupp steel enterprise were warmongers and by presenting evidence to the Reichstag that confirmed his claim. A Krupp agent was eventually imprisoned and several naval officers were cashiered for bribery and theft of documents. Krupp agents used this secret intelligence to plant a war scare in a French newspaper. From the Krupp point of view this had the happy and profitable effect of winning the Reichstag's approval of an expanded military budget. But it also brought wide attention to Liebknecht's book, *Militarism,* the American edition of which appeared in 1917. In *Militarism* Liebknecht denounced "profiteers" and "the bloody international of the merchants of death." (Both the terms "the bloody international" and "merchants of death" would later become titles of book-length exposés of war profiteering.) Liebknecht also charged that German navalists demanded a fleet that would be "highly profitable to themselves."[55]

The German Left had many admirers in the United States, and to them Liebknecht's claims were highly plausible. Americans who sympathized with German military critics included pacifists Oswald Garrison Villard and David Starr Jordan, the socialist Scott Nearing, Senator Robert M. La Follette of Wisconsin, and Congressmen Richard Bartholdt of Missouri and Henry Vollmer of Iowa, who were widely identified as spokesmen of the large German-American voting minority in Congress. Together, they were the intellectual descendants of Carl Schurz.[56]

Although a fledgling antipreparedness, antiprofiteering movement had appeared by the highwater mark of the Progressive era, the outbreak of war in Europe in August 1914 forced Americans to think more deeply about defense questions. Isolationist Progressives of both parties united against American participation. Devoted to the Jeffersonian heritage, they deeply distrusted professional soldiers, whom they regarded as unproductive and parasitical. The war itself was European, therefore wicked; its principal value was merely to demonstrate anew the decadence of monarchy and imperialism. War was a logical outcome of the uneven and unfair distribution of wealth, which monarchy made justifiable and which agrarian progressives found deeply reprehensible. War meant killing, and by connecting maldistribution of wealth to war, an economic problem was transformed into a profound moral question.[57]

By 1914 a broadly based, pacifist, antipreparedness phalanx existed. The antipreparedness persuasion produced several organizational expressions, the most prominent of which was the American Union against Militarism. Other such groups included the League for Peace and the Anti-Preparedness Committee. The militantly pacifist American Union against Militarism counted among its leaders such illustrious journalists as Oswald Garrison Villard and Frederic C. Howe; Crystal Eastman, a well-known socialist who organized the Woman's Peace Party of New York; and Amos Pinchot, a wealthy New York lawyer whose brother Gifford was the Progressive era's leading conservationist. At the suggestion of a number of congressmen, in 1916 the Union against Militarism organized a public speaking tour to protest military buildup.[58]

The members of the Union against Militarism, like the Anti-Imperialist Committee, were mostly New Englanders. The Union against Militarism opposed secret diplomacy, atrocious weapons, and war toys. It demanded a national referendum on a war declaration and the outlawing of the private manufacture of war matériel. The position of the Union against Militarism was that businessmen were looking forward to making profits out of war. "Neither the United States nor any other country can carry on a war which will make the world safe for democracy and for plutocracy at the same time," proclaimed Pinchot. "If the war is to save God, it cannot save Mammon." It followed logically that by taking the profits out of war businessmen would work against war. If people had a financial incentive to maintain peace, war would be ended as a human institution.[59]

The antipreparedness movement had strong ties to the feminist movement as well as to progressivism in general. As early as 1872 the abolitionist Julia Ward Howe, famous as the composer of the "Battle Hymn of the Republic," went to England to work for peace, although her mission failed when she was denied permission to speak at a convention of the English Peace Society. Returning to the United States, Howe held peace meetings in eighteen American cities, denouncing the American role as arms purveyor.[60] Jane Addams, the famous settlement house worker, became actively involved in pacifist work in 1907. Her prominence and indefatigable zeal gained her the leadership of the women's peace movement. Other prominent feminists who became active in the peace movement included Lucia Ames Mead, of the Anti-Imperialist League, Crystal Eastman, settlement worker Lillian Wald, and the writer Charlotte Perkins Gilman. (Gilman did not attribute warfare to greed, but believed that war had masculine causation: men had a naturally combative instinct, and they loved glorious ostentation and noise.) In August 1914, Fanny Garrison Villard, the daughter of William Lloyd Garrison and mother of Oswald Garrison Villard, gave the peace movement further prestige. She identified herself as a vigorous promoter of a permanent women's peace organization by chairing a Woman's Peace Parade in New York City. She believed that men were too willing to compromise with militarism. What the peace movement needed, in the family tradition, was militancy.

In April 1915, at the call of Jane Addams, a group of militant pacifists meeting at The Hague organized the Women's International League for Peace and Freedom. The delegates denounced war profits as a cause of the Great War: "The private profits accruing from the great arms factories [are] a powerful hindrance to the abolition of war." They approved and vigorously supported a resolution to make the manufacture and sale of arms a state monopoly. Three of eleven points in the resulting platform of the Women's Peace Party concerned the economic causes of war. In November 1916, the Women's Peace Party proposed an investigation of the munitions industry, a goal that stretched back to the Franco-Prussian War and was realized two decades later.[61]

Although the opposition to war profits had its roots in the Civil War legacy, in religious pacifism,[62] in progressivism, in feminism, in British anti-imperialism, and in German liberalism, its geographical strength was in the West and South. During the prewar decade, "little navy" men in Congress usually represented the rural South and Midwest.[63] Very early, the antiprepared-ness, antiprofiteering campaign displayed some aspects of midwestern parochialism. Far from the seacoast, little navy men felt no significant danger of invasion. Shipbuilding was entirely an Atlantic seaboard industry, which in the midwestern view produced little benefit but brought great cost. Often believing that prices of agricultural commodities were manipulated to their benefit by Wall Street brokers, westerners suspected that unnecessary naval spending was another instance of devious easterners fleecing the pockets of helpless farmers. Small-town newspaper editors were frequent proponents of this analysis.[64]

Opposition to defense spending was most intense in places like Nebraska and North Dakota. Besides the occupational affinity of grain farming for paci-fist, antimilitary ideas, many midwesterners maintained ethnic affections for Germany or Scandinavia (the latter being neutralist, socialist, or both), as well as other non-English regions of Europe. Economic links to the military were likewise weaker in the Midwest and South. North Dakota ranked last among the states in defense expenditures, receiving only 1/2500th that of New York. Thus in October 1914, only a few weeks after the guns of August had first thun-dered, Congressman Henry Helgesen of North Dakota was already charging that Woodrow Wilson's love of big businessmen would involve the United States in the European war. In the presidential election of 1916, the Prohibition Party, which had much of its support in the Midwest, included in its platform a plank calling for the removal of private profit from the manufacture of war matériel. "The feeling of New York is not the feeling of America," wrote the Californian David Starr Jordan. "The big business of New York and [New York's] big voting population look on the world with very different eyes."[65]

The war profiteering issue, in the sense that large wartime gains were im-proper amid sacrifice, was secondary in the quarrel over preparedness. The core of the dispute was the appropriateness of American sales of war matériel to belligerent powers, which might lead the nation into military as well as eco-nomic engagement. The principal questions consisted of the legality of arms sales, the amount of credit that should be extended to support them, and the degree to which the United States should prepare to defend this commerce. Inevitably, though, the debate confronted the question of who benefited from, and who paid for, American mobilization.

6

Warhogs and Warsows

War brings prosperity to the stock gamblers on Wall Street. The stock brokers
would not, of course, go to war, because the very object they have in bringing on
war is profit. . . . They will be concealing in their palatial offices on Wall Street, sitting behind mahogany desks, covered with clipped coupons—coupons dyed in the
lifeblood of their fellow man.

SENATOR GEORGE W. NORRIS (1917)

The demise of peace in 1914 startled and frightened Americans, and two and
a half years would pass before the United States reluctantly entered the Great
War. During the interval between Congressman Clyde H. Tavenner's sweeping
arraignment of the munitions industry and the American war declaration,
profits on military products climbed to unknown levels, the market for
military-related stocks reached record highs, and the isolationist progressive
bloc dueled steadily with a growing coalition of supporters of preparedness.

The harshest words were exchanged over the war declaration itself, but the
acrimony of the dispute flared much earlier. In 1916 the opponents of preparedness persuaded Congress to enact a special tax on profits gained from
making munitions. The munitions tax was yet another example of progressivism's evangelical campaign to master monopoly capitalism, but it became
the harbinger of modern attempts to control war profiteering. The campaign
against monopolism was much older than the debate over the Great War, and
the longing to obliterate such particularly odious and allegedly dangerous examples of monopolistic profiteering as the "Armor Trust," the "Powder Trust,"
and the "Aircraft Trust" would continue for decades. Yet as dividends and
prices of military stocks swelled in 1915 and 1916, the campaign against arms
makers took on increasing urgency, and the war profits issue surfaced in various forms.

By 10 August 1914, the Wilson administration was already considering the
American economic role in the conflict. Secretary of State William Jennings

Bryan informed President Wilson that the firm of J. P. Morgan and Company had asked on August 9 whether the government objected to loans to belligerent powers. Bryan recommended that Wilson disapprove the Morgan Company's request on the grounds that loaning money would be an unneutral act (although not illegal). Perhaps somewhat inconsistently, Bryan did not oppose *sales* of munitions to the Allies by private firms, but that had been the established American policy since the Franco-Prussian War of 1870-71. Wilson first agreed with Bryan's view (that sales should be allowed but not loans), but he changed his mind and overruled his aide, deciding to permit both sales and loans on March 31 1915. Wilson did follow Bryan's advice in part, however. As Bryan had recommended, Wilson took the first action designed to limit war profits; by purchasing fifty-four German and Austrian merchant ships trapped in U.S. ports, the government attempted to counteract what it considered to be extortionate rates charged by private shipping firms. These rates had risen by 700 percent during the first six months of the war.[1]

In part to protest Wilson's decision to approve the Morgan loans, Bryan resigned his post and commenced a speaking tour. This decision by such a highly visible figure focused national attention on war profits. On 20 June 1915, Bryan addressed a crowd of 70,000 at Madison Square Garden in New York City. The Great Commoner denounced preparedness and loans to belligerents and called for government ownership of munitions plants in order to take the profits out of war. Disregarding his earlier position, by 1916 Bryan also contested sales to belligerents as well as loans.[2]

Champions of a strong defense roared back at Bryan. Between December 1914 and June 1915, advocates of preparedness formed two organizations, the National Security League and the Navy League, to advance their cause. Since both organizations were well financed, progressives suspected that the funds were coming from profit-seeking defense contractors who stood to gain handsomely from a defense buildup.[3] In 1915 Hudson Maxim, the brother of one of the principal inventors of the machine gun, published an intemperate argument for preparedness titled *Defenseless America*.[4] This was followed by the first of many motion pictures that would figure in the war profits debate. *A Battle Cry for Peace,* the film version of *Defenseless America,* luridly depicted the awful consequences of an America conquered by Germans. The antipreparedness circle regarded the film as reckless demagoguery (it received the assistance of the large and powerful Hearst newspaper chain), and in 1916 they alleged that it was deliberately shown in theaters located in districts represented by antipreparedness congressmen seeking reelection. The wealthy automaker Henry Ford, a business pacifist, took out full-page advertisements in 250 newspapers in the hope of countering the effect of the film. Ford charged that Maxim was a war profiteer.[5]

The antiwar progressives noted that Maxim's brother was earning substantial profits from supplying rapid-fire weapons to the Allies. They inferred from this relationship that the preparedness campaign was backed by wealthy businessmen who hoped to profit from the war and who would stop at nothing to gain their ends, including the use of the new motion picture medium as a pernicious means of rousing people's fears. In suspecting that the preparedness campaign was profit motivated, the progressives were substantially wrong. Advocacy of the cause of American preparedness fell mainly to businessmen with little or no previous involvement in military contracting, and the largest defense contractors generally shied away from the preparedness debate.

Executives of the Du Pont Company, for example, assumed that the war would be brief: a conflict perhaps on the scale of the Franco-Prussian War. Like other American business leaders, the du Ponts were wary of losses they could suffer if they tooled up for a war that might terminate abruptly. Du Pont executives had distinct and bitter memories of the Spanish-American War when their firm had swiftly expanded its facilities to produce brown prismatic cannon powder. After Spain's surrender, the firm was left with expanded facilities, a large powder inventory, canceled contracts, and political charges of extortionate pricing. The U.S. Navy afterward attempted to soothe its disgruntled supplier, but Du Pont president Eugene du Pont declared a preference for accepting the loss rather than remaining in the military powder business.[6]

In 1915 the Du Pont company's new president, Pierre S. du Pont, specifically requested that no officer or employee of the firm become active in the preparedness movement, although he broke his own rule by personally subsidizing distribution of *Defenseless America*.[7] The position of the J. P. Morgan Company early in the war was that it would be foolish to think that America would not be harmed by a general European conflict. Since Morgan's brother-in-law, Herbert L. Satterlee, was a founder of the National Security League, the League's views were contaminated in the eyes of isolationist progressives, who saw it as a mouthpiece of the Morgan interests. Morgan did become a contributor to the NSL by 1918, as did T. Coleman du Pont (Pierre's brother), but investigations of the National Security League and the Navy League were unable to show that major funding came from large defense contractors.

Most of the money for the preparedness organizations actually came from businessmen whose wealth derived from enterprises that served the civilian economy, and 94 percent of the money behind the National Security League came from New York City.[8] In 1918, after the United States entered the conflict, the National Security League assembled a $400,000 war chest to defeat congressmen who had voted against preparedness or against the war. The League was thus a latter-day equivalent of the Union League of the Civil War era, even to the extent of holding some of its meetings in Union League clubrooms. The ample resources of the National Security League, as well as its willingness to question the loyalty of its opponents, quite naturally inspired rancorous reactions, as had the Union League.[9]

The sinking by Germany of the British liner *Lusitania* in May 1915 staggered the neutralist position, and two days later President Wilson asked Congress for increased appropriations for the army. A clamor for preparedness developed, and although it came principally from outside progressive ranks, the preparedness impulse drove a wedge into the progressive cohort. By 1916 Wilson, backed by growing consensus of American opinion, advocated further enlargement of the American fleet. This phase of naval expansion appealed to some who had been previously uncertain or opposed—southern and western farmers who had become resentful of the British blockade. As the president put it succinctly to his key adviser, Colonel House, "Let us build a navy bigger than hers and do what we please."[10]

Former president Theodore Roosevelt, who needed no new excuse for championing preparedness, reacted to unfolding war news by extending his reasoning in respect to arms sales. Although the United States had always defended its *legal* right to sell arms to belligerents (just as a local gun dealer could sell arms to private citizens), Roosevelt argued that German atrocities in Belgium meant that the United States now had a *moral* duty to sell arms to those who were attempting to resist the war criminals. "[The] issues at stake are elemental," wrote the Rough Rider. "The free peoples of the world have banded together against tyrannous militarism and government by caste. It is not too much to say that the outcome will largely determine, for daring and liberty-loving souls, whether or not life is worth living." Roosevelt's rather romantic view of the war issues was not atypical. Just as pacifists entertained the romantic notion that civilization had progressed so far that war had become impossible, supporters of preparedness and intervention gave romantic, even poetic,[11] reasons for participating in the conflict. It was not uncommon to describe the great issue of the time as simply whether to allow the continuation of "Prussianism" or "Kruppism." These terms lumped together Germany's allegedly evil qualities: atrocities, militarism, attacks on peaceful shipping, and the "sordid enormous trade in the instruments of death." Herbert Hoover, writing in 1917 on "German Practices in Belgium," said: "I have neither the desire nor the adequate pen to picture the scenes which have heated my blood. . . . I myself believe that if we do not fight and fight now, all these things are possible to us—but even should the broad Atlantic prove our present defender, there is still Belgium. Is it worth while for us to live in a world where this free and unoffending people is to be trampled into the earth and to raise no sword in protest?"[12] The Great War was presented to the American public in glowing, romantic terms by its advocates, and when the clamor for the crusade stilled, an angry reaction struck those who profited during the hostilities.

In 1914-15 there was ample reason to be concerned about national defense. Despite twenty years of rearmament, the nation remained hopelessly unprepared for a major war. Unlike European nations, the United States maintained no agency to coordinate defense mobilization. President Wilson, obedient to the isolationist progressive wing of the Democratic Party and distrustful of business influence in government, opposed serious efforts to make plans for converting the American economy to a wartime basis, although he moved closer to clear support for a defense buildup.[13]

Wilson's preparedness program appealed to a new branch of progressivism—what John M. Cooper Jr. has termed the "Liberal Internationalist" wing.[14] Most progressive luminaries feared American entry into the war substantially because they believed that war would herald the end of the reform spirit as it had in the Civil War era. The nascent internationalist wing took a different view. This dissenting group, which included such progressives as Herbert Croly, Walter Lippmann, and Ida Tarbell, welcomed preparedness as an opportunity to advance the liberal agenda. They believed that preparedness would bring a flow of power to the national capital. They were joined, within the Wilson administration, by Secretary of War Lindley Garrison, Assistant Secretary of the Navy Franklin D. Roosevelt, and chief foreign policy adviser Colonel Edward M. House. With Woodrow Wilson in office, progressives controlled the White House, so the liberal internationalists believed that the expected flow of power could be channeled into support for progressive programs. In particular, they believed (correctly as it turned out) that the funds required for preparedness could be obtained only from the expansion of the federal income tax. This would afford an opportunity to increase tax rates on persons in the upper income brackets significantly. War would thus be used to accomplish their leading goal of wealth redistribution through taxation, a dream which had not been realized in peacetime. (The Spanish-American War experience offered a promising lesson. It was financed by a progressive inheritance tax, although this lapsed in 1902.) Indeed, by 1916 the distribution of wealth in the United States was more unequal than in any other year in recent history.[15]

The most active advocate of the redistributionist position was Amos Pinchot, who, while remaining prominent in the American Union against Militarism, now formed and bankrolled the New York-based American Committee on War Finance. Other groups involved included the Association for an Equitable Federal Income Tax and the Public Ownership League of America. As Pinchot put it, "If we ever get a big income tax on in war time, some of it—a lot of it—is going to stick."[16]

Even more supremely optimistic about the possible benefits of preparedness were certain members of the clergy, who anticipated that preparedness would create a renewed sense of community in the nation. One even went so far as to predict that by revitalizing community spirit preparedness would lead to a reawakening of Christianity in America, and this might even bring the Second Coming![17]

The thunder of 1914 presented new and inviting opportunities for American businessmen with a keen eye for the main chance. Most businessmen who aspired to take advantage of the new circumstances did so first by seeking to capture markets in neutral or Allied countries abandoned by their British,

French, and German competitors. In prewar years Germany had nearly monopolized the European market for dyestuffs, potash, toys, and film. Britain's naval blockade made it possible for American firms to compete for this trade for the first time. American toys now sold briskly in the British Empire, and American motion pictures displaced British productions as the dominant attractions in both the British and the world markets. Machine tool sales blossomed overseas, and American firms replaced the German concerns as the largest exporters of civilian merchandise to Russia. German torpedoes sank hundreds of ships, and American shipbuilders busily laid down new keels to replenish the world's supply. Meatpackers did well, and even American sealskins found eager new customers. All these were examples of a classical form of profiteering: gaining windfall profits as a consequence of sectoral shifts in demand brought about by the disruptive propensity of war.[19]

Serving these flourishing new markets was a practical approach to profit. As long as the United States followed the neutralist policy declared by President Wilson, disruptions in the world economic structure would offer Americans trade advantages that would fatten dividend statements. The stalemate on the western front was immensely profitable to American manufacturers, who commenced production of rifles, cartridges, cannon, uniforms, buttons, canteens, shells, fuses, and every other accoutrement a British or French soldier required. "Workmen and engineers who never before had seen blueprints of these death dealing instruments of war quickly familiarized themselves with their manufacture," remembered a steel executive. There was stiff competition for Allied war contracts, and a sharp broker who could secure a lucrative military contract for a manufacturer acquired (for obscure reason) the sobriquet "liceman."[19]

These markets were a far more certain and appealing path to prosperity than was promoting American military engagement, which would likely lead to governmental controls and elevated tax rates. Nevertheless, the antipreparedness camp continually attempted to find evidence to show that military contractors had sinister and greedy intentions of stirring up a war declaration. In 1935 Senator Gerald Nye of North Dakota sought to show that the president of the Colt's Patent Fire Arms Company, Samuel M. Stone, had endeavored to bring about American entry into the European war. To the contrary, the evidence disclosed that the real interest of the Colt company in the prewar period was in selling automatic pistols in Latin America, a market from which European competition had become excluded.[20] There was no need to incite a war with Germany, as there were ample business opportunities in serving the former customers of European producers and in producing war matériel for the Allies. In 1916 a business journalist estimated that while no more than 3,000 workmen had been employed in only six firms in the munitions business in 1914, by the end of two years of war that number had risen to 500,000 workers in more than a thousand shops. The greatest gains, moreover, had come in manufacturing machinery rather than in manufacturing munitions.[21]

The fight over the income tax of 1916 became the first direct engagement in the long quarrel over corporate profits obtained during the First World War. On 24 November 1915, Secretary of the Treasury William C. McAdoo announced the Wilson administration's version of an income tax bill. The Treasury had included a provision to impose a special tax of two cents per pound on the production of dynamite, gunpowder, and nitroglycerine, but President Wilson ordered its deletion.[22]

The attempt to impose a special munitions tax was a direct strike at the E. I. du Pont de Nemours Company, which would henceforth figure prominently in the profiteering controversy. The Du Pont Company was controversial even before the war. It was one of several combinations that had been successfully prosecuted during the progressive era trust-busting campaign. In 1911 the "Powder Trust" (as Du Pont was known colloquially) was divided into three separate firms, and the U.S. court enjoined it from carrying on further anticompetitive commercial practices. In the interest of protecting national security, however, Du Pont was allowed to maintain its monopoly on the military powder business.[23]

By 1915 Du Pont was again a focus of controversy. This was partly because the wealthy du Pont family, which had relatives in France, was very active in the French relief program, thereby revealing its sympathy for the Allied cause. But mostly the controversy developed because the company was heavily engaged in selling munitions to Great Britain.[24] The British, like their American cousins, were inadequately prepared for war. British ammunition supplies were too small, and when Winston Churchill became First Lord of the Admiralty he was shocked to find that these vital but vulnerable supplies were not even guarded against saboteurs. As the war intensified, Britain became dependent upon foreign (which meant American) suppliers of powder (which meant Du Pont). The reason for this developing dependency was the German submarine.

In order to produce one ton of military-grade high explosives, eight to twelve tons of raw materials were required. In order to produce one ton of smokeless powder, as used in rifle and cannon ammunition, fifteen to twenty tons of raw materials were needed. Because of this unfavorable ratio of raw materials to finished product, which reached twenty to one, it was impossible to produce the required munitions in Britain. There were simply too few vessels in the British fleet to transport the huge quantity of raw materials, and even if the ships had been available, the German submarine menace was a decisive complication. Most of the nitrates came from mines in Chile, so the ships would be in jeopardy to the U-boats during a long voyage. Britain had no choice but to ask Du Pont for help.[25]

Pierre S. du Pont drove a hard bargain. Contrary to his company's reputation, before the war military sales were only a small part of the firm's business—only 5-10 percent. Commercial sales of explosives were far more important to the profit stream. Du Pont was very reluctant to convert or expand his commercial production facilities to serve the British and French military market. If the war terminated abruptly, the firm's investment would be lost, since a smokeless powder plant, because of the corrosiveness of the reagents employed, would deteriorate very rapidly if taken out of production. Pierre du Pont therefore insisted that the price of powder be sufficiently high to pay off the cost of plant expansion in a single contract. In practice, this meant that the Du Pont firm demanded that the British pay a dollar per pound for powder. Since powder could be produced profitably at fifty cents per pound, this meant that the company was allocating fifty cents per pound to pay for the enlarged factory. Du Pont even insisted that the British deposit the funds for expansion in advance, so that if the war ended quickly, Du Pont would lose nothing. If the conflict raged for years, profits would be immense. The British government, whose only alternative—surrender—was no alternative at all, was compelled to accept the terms. Du Pont later defended his demands as a cold business calculation.[26]

As the war disaster lengthened, Du Pont's profits became, in the words of Pierre's biographers, "truly staggering." Du Pont ultimately supplied about 40 percent of the propellant powder used by the Allies. The firm's net profits averaged 11.57 percent of invested capital for the last three years before the Great War commenced, or about $5,263,000 per year. By 1915, profits zoomed tenfold to $57,399,000. The following year they reached $82,107,000.[27] These earnings formed an inviting target for the opponents of preparedness.

Although Wilson had canceled McAdoo's scheme for a two cent per pound tax on munitions, this provision was restored by the Ways and Means Committee of the House of Representatives. The powerful chairman of that committee was Representative Claude Kitchin, the longtime little navy progressive who doubled as majority leader. He remained a staunch (but not an entirely consistent)[28] foe of preparedness, and his determination was bolstered in an election year by heavy pressure from his North Carolina constituents and by the hometown press that opposed American involvement. Despite his prominent position in Congress, Kitchin faced a challenge for the Democratic nomination for his seat, and he became the architect of the special munitions tax.[29]

The Kitchin committee proposed a graduated tax on gross sales of munitions. This would vary somewhat by industry. Manufacturers of gunpowder would pay a tax of 8 percent of sales, while makers of cartridges and firearms paid 5 percent and suppliers of copper only 3 percent. The reason for this discrimination is obscure, but the effect was to single out the Du Pont Company. A conciliatory provision specified that if profits fell below 10 percent of invested capital, the special tax would be forgiven.[30] Since the Du Pont Company did better than that even in peacetime, this provision offered small consolation.

Pierre S. du Pont was outraged. He pointed out, somewhat reasonably, that the special surcharges on makers of gunpowder, guns, and copper left other vendors of war matériel untouched. Items like tanks, trucks, passenger automobiles, aircraft, and woolen goods had military value, and those selling them to the Allies were also enjoying elevated profits. Why should he be singled out? The Senate Finance Committee listened receptively to this complaint, but the Joint House-Senate conference committee retained the special munitions tax. (The rates, however, were changed from 8 percent of *gross* sales of powder to 12.5 percent of *net* income of powder manufacturers.) Du Pont's umbrage was certainly partly due to affronts by backers of the tax. "All this . . . is to be done," wrote the Progressive Senator Robert M. La Follette, "because wealth is not loyal enough to assume the burden it ought to bear."[31]

The special munitions tax of 1916 thus became the first excess profits tax in American history. The Du Pont Company paid 90 percent of the money the government collected from this source. With some justice, particularly since the tax was imposed retroactively, the du Pont family and others resented this tax as sectional and class legislation. It was intended largely to please farmers, laborers, the lower middle classes, and isolationist Progressives in an election year. In the case of North Carolina congressman Claude Kitchin, the tax was partially a form of revenge against supporters of preparedness. However, the American decision to impose an excess profits tax on the beneficiaries of the Great War directly followed that of Britain, which had recently imposed steeply graduated war levies.[32]

The Revenue Act of 1916 was on the whole a victory for the preparedness forces, as the government now had the funds to commence the military buildup. Kitchin and the isolationist Progressives had won only a consolation prize, the special tax on munitions makers. Nevertheless, the preparedness faction faced daunting problems, as in 1916 the nation was still far from ready to consider seriously the implications of converting the American economy to a wartime basis.[33]

Rising commodity prices were a major source of concern. The price of basic materials in military products rose well above peacetime levels long before the U.S. declaration of war. By 1916 the price of copper, one-third of which went into artillery ammunition, had more than doubled to twenty-eight cents per pound (from thirteen cents in 1915). Between January 1915 and July 1917, steel billets rose in price fivefold. Steel plates that sold for $33.60 in 1913 brought $200.00 in 1917. Zinc prices tripled, and mercury went from $40 per flask to $225. Before the war Germans controlled the world tungsten supply. Tungsten prices advanced so rapidly that everyone who had a tungsten supply thought he would eventually become a millionaire. Sales of canned meat broke all records, driving up the price of pork and beef, and demand for cotton, wool, and wheat was also brisk. The war encouraged new production of benzol, toluol, ammonia, and napthaline, much of which went into explosives. This level of demand produced tumultuous days on Wall Street and on regional commodity exchanges as fortunes were made and occasionally lost.

Successful stock and commodity speculators became known, if male, as "warhogs" and, if female, as "warsows."[34]

A few examples serve to illustrate. The price of Bethlehem Steel stock stood at 33 3/4 on 30 July 1914; it rose to a price of 600 during the war, a seventeenfold increase. In 1917 the firm paid its shareholders a 200 percent dividend. During 1915 U.S. Steel stock rose from 48 to 120 5/8, and General Motors went from 78 to 750. At Atlantic Steel, the return on stockholders' equity, which had averaged 4.9 percent for the three years before the war, rose to 45.6 percent in 1916. These returns were not approached again for twenty years. An index of nine ordnance stocks rose 311 percent in eighteen months. This was in sharp contrast to stocks that served the civilian market—between 1914 and 1918 the stock market as a whole dropped by 60 percent in real value. In these circumstances of inflated prices in some industries, depression in others, and rampant speculation, there was ample room for resentment, for suspicion of overcharging, and for charges of profiteering. Progressives were particularly bitter, as they felt that the war was rejuvenating the wealth of corporate monopolists, whose power they had been vainly hoping to destroy. Financial gains obtained by appreciation in stock and commodity prices—a more incidental form of profiteering than windfall profits obtained directly from market shifts—were no less contemptible to Progressives. Fighting Bob La Follette wrote bitterly to his son, "The war-hogs never get enough."[35]

Besides the enmity toward the Du Pont firm and toward the commodity speculators, a wellspring of the profiteering controversy as well as a major obstacle to preparedness was the broad reservoir of bitterness toward the banking firm of J. P. Morgan and Company. As in the case of Du Pont, animosity toward the Morgan firm was well developed long before the war. In 1912 a congressional investigation had charged that the Morgan firm monopolized financial markets, and it became known in the progressive idiom as the "Money Trust." Interchangeable terms like "The Morgans" or the "Money Trust" served Populist and Progressive orators well when they wanted to provoke anger in the grain belt or the cotton belt. As part of their general indictment of Morgan's allegedly ill-gotten gains, Progressive critics constantly repeated the myth that the foundation of the Morgan fortune was illicit trading during the Civil War. Once the European war began, it seemed logical that the Morgans would attempt to repeat their previous success by capitalizing on the opportunities the new war presented.

Perhaps more so even than Du Pont, the Morgan firm attracted the wrath of the opponents of preparedness and war profits.[36] In January 1915, barely four months after hostilities opened, the Morgan banking firm was selected by the British government as its official purchasing agent in the United States. The Morgan Company thus landed one of the most lucrative contracts in American history: exclusive control over all British military purchases in the United States during the First World War. The terms of the purchasing agreement were quite attractive: for the first $10 million worth of contracts that Morgan placed, the company would receive a 2 percent commission; for everything above that figure, the fee would revert to 1 percent. The plan was also very attractive to Britain, as the British not only gained Morgan's expertise but also cut their rates considerably. They had been paying as much as 10 percent to American commission agents. Because British purchases reached astonishing amounts, Morgan's profits were enormous. At its peak in 1916 Britain was spending $83 million per week in the United States. Morgan later testified under oath that British purchases for the entire war amounted to about $3 billion.[37] This gave his firm gross earnings of at least $30 million from that source alone.

That quantity of money alone would have attracted scrutiny, but several other aspects magnified the squabble. It was Morgan money, banking money, money gained from providing financial services rather than from producing tangible assets. The Morgan firm had an exclusive contract, thus raising anew the issue of how much control a single person or firm should wield over the American economy. The Morgan firm gave most of the British contracts to firms located in the industrial heartland, the North and East, but this of course won it few friends in the South and West. Successful bidders were often firms with strong previous links to the Morgans, raising charges of cronyism.

Critics hotly charged that this trade was the cause of growing American involvement in the war. In January 1915, immediately after Morgan's appointment as Britain's agent, a group of congressmen from Iowa, Indiana, and Nebraska sought by law to prohibit the export of munitions. This bid was unsuccessful, as public opinion narrowly supported war profits. In January 1915 as well, the *Literary Digest* polled representatives of 1,000 American newspapers on whether they favored stopping the exportation of war matériel to belligerents. Of those who replied, supporters of continuing the sales were in the majority by 244 to 167. This support was by no means uniformly distributed, however. Residents of small towns in the South and West preferred an embargo, whereas the big cities of the North, East, and South wanted business. The *Omaha World-Herald* denounced the trade as "blood-money," and the Nebraska legislature approved a resolution calling for an embargo. The German-language press also overwhelmingly opposed sales to Britain. Some Americans were, however, evidently swayed by economic interest. A writer from the copper country of Michigan, where business was booming due to war sales, declared, "We are torn with grief at the desperate state of affairs in Europe, [but] we can not refuse to supply them with whatever goods they need." The *Nashville Banner* added bluntly, "Let 'em shoot! It makes good business for us."[38]

The dispute over sales to Britain was complicated by the issue of making loans to Britain, which Wilson had approved at Morgan's request. In 1914 most businessmen sided with Bryan and the agrarian progressives in opposition to lending the Europeans money because the financiers considered the Europeans to be poor risks. By 1915, however, military sales were so lucrative that most businessmen wanted the trade to continue, even if loans were necessary to facilitate it. Sales were brisk not only to the British and French. Greece was buying uniforms, rifles, gunpowder, and tents, as well as 500,000 canteens, 300,000 sheepskin overcoats, and five million cans of corned beef. Other large orders came from Belgium, Bulgaria, Rumania, Serbia, and Russia. Fast selling items included horses, "caterpillar" tractors, railroad ties, absorbent cotton, and even Cossack boots.[39]

The opponents of sales and loans to the Allies complained that economic commitment would inevitably lead to military commitment, but their advice was not persuasive. Besides the growing value of American military sales abroad, the Germans used submarine warfare in a bloody attempt to interdict the flow of trade. Although the advisability of this trade was acrimoniously disputed, however, the *legality* of these loans was not seriously contended. Under the Hague Convention of 1907, neutral powers could legally sell arms to belligerents. Germany could hardly protest, as it had sold arms to Britain during the Boer War and to both sides during the Russo-Japanese War. (German-Americans were not so constrained, however.) Austria-Hungary did file protests against American military sales, but Secretary of State Robert Lansing flatly rejected them. Lansing garnered hearty applause from the industrial Northeast.[40]

The Morgan loans became a source of controversy in one other way. By 1917, when the United States declared war, Britain and France had run out their line of credit. Private investors could no longer safely loan money to these insolvent governments, so the Wilson administration faced an unwelcome decision. The United States had either to guarantee the repayment of the Allied loans, or shipments of war matériel must cease. Since the latter was totally unacceptable—it would mean defeat—there really was no choice in the matter.

Wilson's inevitable pledge of repayment thus guaranteed that Morgan and other bankers would be reimbursed in case the Allies defaulted. This was exactly what happened, and Midwestern Progressives were livid. "Shall we without organized resistance," beseeched Congressman John Nelson of Wisconsin, "permit plutocracy to put the Forty Billion war debt with interest on our backs and they to enjoy in comfort and luxury the partly untaxed interest on liberty bonds bought at a discount out of swollen incomes from excess war profits?"[41]

When the German government announced the resumption of unrestricted submarine attacks against neutral vessels, the probability of an American declaration of war became evident to everyone. This touched off a new and even angrier phase of the dispute over the interrelated issues of intervention and war profits. The United States went to war in 1917 in defense of neutral rights. This was a time-honored American principle, and it was truly and deliberately violated by the German submarine campaign of 1917.[42] But the dissenters raised legitimate questions. Were these rights vital to the security of all Americans or only to the interests of a few well-heeled shipowners and ocean travelers? Was securing these rights worth the price of fighting a major war with fifty thousand deaths? To those who answered either of these questions negatively, another explanation seemed necessary. Some reasoned that the war was fought for profit. When German submarines actually sank some American ships, President Wilson asked Congress for a war declaration, and the moment of truth on the great question of war or peace had arrived.

Fifty-six congressmen, representing a substantial minority of the American people, voted against American participation in the war. Most were Progressives, most were Republicans, and most were from the Midwest. They had long abhorred the "Money Trust," the "Powder Trust," and "Wall Street." Memories of the Civil War lingered and mixed with these new enmities. Together, they raised the temperature of the debate to a new level of acrimony.

The question of war profits always figured prominently in the thinking of the war critics. One Progressive congressman, John M. Nelson of Wisconsin, simply reasoned about war that the "effects are evil so [the] causes must be evil." Since the source of all evil was money, Nelson deduced that the war must be the fault of "Big Business, The Interests, the System, The Corporations, The Monopolies, The Trusts, Special Privileges, Industrial Autocracy, Plutocracy, the Money Power, Wall Street, or the Rockefellers and the Morgans."[43] Arthur C. Townley of the Nonpartisan League, which was rooted in the upper Midwest, charged angrily that "hundreds of thousands of parasites, the gamblers in the necessities of life, use the war only for the purpose of exacting exorbitant profits. We are working, not to beat the enemy, but to make more multimillionaires."[44] Townley added, "It is apparent that munitions, armor, and steel plants would be the gainers by a conflict. It is generally believed that the munitions plants are responsible for a propaganda to involve this nation in the European conflict."[45]

A leading war opponent was Senator Robert M. La Follette Sr. of Wisconsin. He made it clear that his opposition rested substantially on the war profits aspect. "Fellow citizens," warned La Follette in 1917, "it behooves a nation to consider well before it enters upon a war . . . how much it has got at stake. If all it has got at stake is the loans of the house of Morgan made to foreign governments and the profits that the munitions makers will earn in shipping their products to foreign countries, then I think it ought to be weighed not in a common hay scale but in an apothecaries' scale."[46] La Follette even suggested rather preposterously that big business had intentionally placed innocent American passengers on the munitions ships in a vain attempt to protect their highly profitable cargoes from German torpedoes. These words seemed sufficiently treasonable to some of La Follette's colleagues to warrant his expulsion from the Senate. La Follette replied that the campaign to expel him was just another attempt by wealth to keep its war profits, and he prevailed.[47]

No war opponent was more embittered than Nebraska's Progressive senator George W. Norris. "War brings prosperity to the stock gamblers on Wall Street," accused Norris. "The stock brokers would not, of course, go to war, because the very object they have in bringing on war is profit. . . . They will be concealing in their palatial offices on Wall Street, sitting behind mahogany desks, covered with clipped coupons—coupons soiled with the sweat of mothers' tears, coupons dyed in the lifeblood of their fellow man."[48]

By April 1917, when the United States finally drew its sword against Imperial Germany, the allegation of appalling, even criminal, war profiteering was well entrenched. Rooted in the rearmament debate of the first years of the century, the charges of profiteering thrived in the dispute over participation in the European conflict. Until a new war subdued the contention two decades later, the profiteering dispute would linger as a fundamental memory of the Great War. In 1935 the popular magazine *American Mercury* portrayed the Great War as "No. 4" in its series entitled "Thieveries of the Republic."[49]

7

Supplying the Doughboys

Who then is willing to consecrate his service this day unto the Lord?

1 Chronicles 24:5

Despite fierce opposition in some quarters, most Americans supported the declaration of war with Germany with determined enthusiasm. The Spirit of '17 was perhaps not so deep-seated as was the Spirit of '76 or the Spirit of '61, but it was earnest and resolute. Nevertheless, a definite sense of apprehension was present even among interventionists, who sensed very early the peculiar but marked ability of the Great War to modify or destroy old verities. "If war is declared," commented an editorialist in the *Commercial and Financial Journal*, "it is needless to say we shall support the government. But may we not ask, one to another, before that fateful final word is spoken, are we not by this act transforming the glorious Republic that was, into the powerful Republic that is, and is to be? . . . Must not we admit that we are bringing into existence a new republic that is unlike the old?" Some were convinced that the venerable "Glorious Republic," with its familiar Victorian fashions like refined manners, modest government, and isolation from Europe's problems, was to be preferred above the emerging "Powerful Republic," with its enormously expensive military, its intrusive government, and its troublesome foreign engagements.

Nagging doubts about the purpose and long-term effects of the Great War formed the basis for continuing attacks on war profits, but there were several other reasons which caused the criticism of war profits during this conflict to be particularly heated. The Progressive mentality sustained a fundamental trust in rationality, a deep passion for efficiency, and a simplistic belief in conspiracy. The war seemed manifestly irrational, grossly inefficient, and plausibly conspiratorial. These qualities were a jarring introduction to the twentieth century, and "Munitions Makers" became the rod that attracted mighty bolts of Progressive anger.

Even had the Great War not arrived as a most unwelcome intrusion that threatened the heady but incomplete success of Progressivism, mobilization issues would have provoked vehement disagreement. When the United States declared war on Germany, it formalized a military relationship with Britain and France that had begun in 1914. Thus the period of de facto economic alliance preceded the period of direct military participation by nearly three years. This led eventually to a blurring of the distinction between *war* profits and *prewar* profits. War profits were limited by ethics, tradition, and law, whereas prewar profits (which were much greater in this instance) were not so clearly constrained.

Before formal entry, the heavy burdens imposed on American firms by British and French purchasing had brought them nearly to full production. Accordingly, there was little slack in the American economy, and when orders for the United States military effort were added to those of the Allies, heavy demand drove prices upward. Of all American mobilizations, that for the First World War produced more windfall profits in the form of increases in the value of inventory (the classical definition of profiteering) than did any other. Surprisingly, the single most profitable commodity to own before the war disrupted normal trade patterns was machinery suitable for war production—despite years of use the price of such equipment as lathes and screw machinery appreciated.[1]

It was unfortunate that these unanticipated increases in inventory values occurred when they did, because the United States by 1916 was at a peak in its history of income inequality.[2] Thus the profits gained during the prewar and wartime periods tended to exacerbate a national problem of income inequality—regional and sectoral as well as individual—which had been a major matter of controversy throughout the Progressive era.

The Great War was also America's first European conflict, which meant that American forces were deployed at much greater distance than ever before. This placed a premium on the price of transportation (both land and sea), which was unexpected and could be reduced only with considerable difficulty. Never in recent memory had the United States been in a position like that of 1917, when it found itself, ill-prepared, in a life-and-death struggle against an enemy as dangerous, as fully mobilized, and as distant as Germany. If victory was to be achieved, mobilization would require a brisker pace and a conversion more complete than even in the Civil War.

Another defining quality of the Great War was its frightening new weapons: submarines, tanks, poison gas, and, in particular, aviation. Aviation was a very new field of military endeavor, and it possessed an unmatched ability to excite the imagination. Despite America's pioneering role in civilian aviation, the development of military capabilities lagged well behind Europe's leaders.[3] Since aviation technology was so new and unfamiliar, it was unusually difficult to measure the degree of quality degradation (a secondary form of profiteering) that took place.

Finally, mobilization for the Great War was initially managed by Wilsonian Democrats, who were the leadership of the minority party, and by military commanders. The war managers were not generally drawn from the highest rank of American business management. Wilson doubted the need for emergency measures to force conversion to war production, relying instead on the goodwill and virtue of the American citizenry.[4] In this respect, Wilson hearkened back to the strategies of Washington and Lincoln. Both of them had found difficulty with the voluntary approach, and so too would Wilson. When mobilization was managed by Democrats and generals, criticism by Republicans and businessmen was inescapable.

As the United States pondered its formal entry into the hostilities in the fateful spring of 1917, the elements that deepened the profiteering controversy and made it permanent were concealed or disregarded: apprehension about the purposes and probable effects of the war on the nation, the problems of conducting a grand military effort amid a reform milieu, the likely inflationary effects of massive military spending on an overheated economy, the difficulty of managing the effects of large sectoral shifts on an economy that was beset by wealth inequality, the disruptive effects of a distant war fought with frightening weapons, and the obstacles to harmony associated with management of the war effort by a minority party. Successful management of these elements would require a very sophisticated administrative capability, and the American system of military procurement was long neglected.

Procurement

The American system of procurement was little changed since the Civil War. The fundamental assumption underlying the old custom was that in war as in peace the U.S. Army Quartermaster Corps, Ordnance Corps, Signal Corps, Corps of Engineers, and Medical Corps would independently purchase the supplies each agency needed. Each bureau had its own purchasing agents, warehouses, financial system, and transportation management. Military reformers during the Roosevelt and Taft administrations made some halfhearted efforts to modernize the system, but their initiatives were brushed aside by entrenched interests in the various bureaus. If there were competing claims among the bureaus for the same resources (a possibility seldom if ever considered), the general staff would presumably balance them. The civilian economy was believed to be so large as to be able to meet any conceivable military need without breaking stride. If war came, it was expected to expand to meet the new demand automatically.[5]

In peacetime this system was satisfactory, but in 1917 four different army corps were competing to buy clothing, trucks, and automobiles, and the general staff was too small and too inexperienced to manage the competition among branches. The Corps of Engineers was building wharves, but so too was the navy. The Quartermaster Corps and the navy were bidding against each other for the same food supplies. The navy's supply system was much better prepared for war than the army's system, so army officers discovered that their naval counterparts had already cornered key markets before the army had even determined its priorities. There was no agency to coordinate or control army-navy rivalry. None could prevent a profit-minded and unpatriotic supplier from playing off one service against the other.[6]

At peacetime levels the amount of military purchasing would not strain the American industrial colossus—appropriations for the ordnance department were only about $10 million per year before 1914. Furthermore, the military bureaucracy was sufficiently large and had the competence to let the contracts and to supervise delivery. But military purchasing officers, accustomed to spending at comparatively low peacetime levels, lacked the comprehensive knowledge of the American economy needed to manage massive wartime spending proficiently.[7]

There was ample evidence of the elementary character of prewar American military planning. The normal complement of the Ordnance Corps was 85 men. The plan was that in case of war this agency would increase to 142 men in five one-year increments. In December 1916, five months before American entry, the chief of ordnance, Brigadier General William Crozier, asked that the entire wartime complement be appointed. Secretary of War Newton D. Baker, who was deeply suspicious of growing state power, rejected Crozier's request on the grounds that no wartime emergency existed. When war was declared, the plan automatically increased the ordnance staff to a mere 96 men. When the real needs of the army were discovered, however, the complement of the Ordnance Corps reached 5,000 men. Amid the confusion that was inevitable during such a swift expansion, there was little time or inclination to consider how to limit war profits.[8]

This haphazard approach was not only inefficient, which the progressives abhorred, but also demonstrably obsolete. In April-May 1915, British war production had fallen dangerously short of demand. This forced Britain to create a Ministry of Munitions, an agency which was intended to rationalize and plan procurement and to prevent wasteful and unfair advantage gained from the war effort. In practice, this step proved to be more than just an emergency defense measure. It was a major leap toward a modern economy: it amounted to the replacement of a primarily free enterprise system of war production by a primarily statist system.[9]

Not everyone understood that if the United States entered the war a similar transformation would have to take place. In the months before the declaration of war, some members of the Wilson administration, particularly Secretary of the Navy Josephus Daniels and the Bryan faction, resisted comprehensive planning for conversion, although European governments maintained agencies for this purpose. Part of the military bureaucracy also feared the creation of a ministry of munitions that might interfere with its prerogatives.

The chief of ordnance, General Crozier, described such a reorganization as "dangerous radicalism." The quartermaster general, General George W. Goethals, did advocate a ministry of munitions, but President Wilson disapproved.[10]

Perhaps the chief intellectual force advocating the need for prewar planning for the control of profiteering was Howard E. Coffin. Coffin, a vice president of the Hudson Motor Car Company, was the highly respected president of the Society of Automotive Engineers. A leading member of the National Security League, Coffin spent $20,000 of his own money to promote the cause of preparedness. Appointed by Wilson to the Advisory Commission of the Council of National Defense, Coffin urged the commission to inaugurate discussion of a standard form for war contracts and also of the appropriate means for regulating profit on these contracts. The question of how to limit profiteering was thus being discussed by civilian executives months before the American declaration of war. Nevertheless, the inability of the government to make the changes necessary to manage war profit satisfactorily was evidenced by the experiences with constructing training camps, building airplanes, and managing the copper and steel supplies.[11]

The difficulty was in managing the shift of resources from civilian to military markets efficiently. A major enigma was trying to predict the duration of the war as well as the amounts of men and materials required in order for Allied arms to prevail. All of these quantities are indispensable requirements for rational planning. The accurate prediction of the length and price of a war is, however, among the most intractable of all social problems.

In retrospect, the estimations of the Army War College were reasonably accurate, if somewhat belated. After initially refusing to provide an approximation of future needs, in February 1917 the War College, in collaboration with the navy, submitted an estimate that a force of four million men would be needed. This proved to be about the number of troops that were eventually mobilized. The prewar appraisal was that a year would be required to train, equip, and transport a major portion of this force to the western front. The weight of the American effort would thus begin to be felt during the campaigns of 1918, and this estimate proved approximately correct. Late in 1917, however, army planners were forced to revise their estimate because of defeats in Russia, Italy, and Flanders. Early German victories in the spring offensive of 1918 added to the consternation. An additional million men would be required, and victory would not come until 1919.[12]

Despite the general accuracy of the army's estimate, there was little real understanding in 1917 of what this would mean for the civilian economy. Wilson's Council of National Defense, for example, sought to measure the effect of mobilization by using an elaborate chart of the supplies that would be needed for an army of *one* million men, rather than the four million men actually required. As army officers in the various branches prepared their individual shopping lists, none were based on an accurate estimation of the productive capacities of American industry. Prewar industrial surveys were of little use.[13]

In the early months, mobilization was hasty and disordered, if not chaotic. The American economy was quickly strained by the massive demands of military spending. Because of heavy Allied purchasing prior to American entry, there was little slack in many sectors of the economy. Of prime importance was an acute shortage of machine tools that remained evident despite soaring prewar prices. As in the early months of the Civil War (and later during World War II), the first step was to construct the machines that were needed to manufacture the weapons that would fight the war. The shortage of machine tools was particularly troublesome in the aircraft industry, which was still in its infancy. As one beleaguered machine tool executive (who understood the fundamental problem very well) lamented, "We have no automatic way . . . of going to war. We declare war and depend upon the patriotism and genius of our people to supply, to any degree that may be necessary, what we as a people are unwilling to maintain . . . a fully equipped military organization." Other industries in which the strain was initially severe were motor trucks (the war had demonstrated their military value), gunpowder, steel, and electrical and mechanical equipment. On the other hand, residential construction, advertising, and printing suffered immediate cutbacks.[14]

The task of mobilizing an army of four million soldiers meant that there was little time for careful consideration of possible excess profits. The sheer size of the buildup dwarfed all past experience. "None foresaw the gigantic size of those orders," recalled one contractor. "Accustomed to millions [of dollars], we were confronted with hundreds of millions, with the billion mark actually in sight." Even the total number of war contracts remains uncertain. There were at least 100,000 contracts let by the War Department during World War I, with at least 25,000 of them in amounts exceeding $100,000. Charles Eisenman, a member of the Committee of Supplies of the Council of National Defense, was given the responsibility of approving contracts in the first days of the war. He later testified that he reviewed about 200 contracts per day, making a total of about 45,000 contracts in the first 200 business days of the war.[15]

Expenditures by the Ordnance Corps went from $10 million per year before the war to $4 billion in the nineteen months of the war—an increase in the rate of purchasing of 250 times. Some of the contracts were so large that they were difficult to comprehend, let alone oversee: for example, the army ordered 41 million pairs of shoes and took receipt of 32 million pairs. Who could administer contracts of this magnitude capably? The construction of an ammonia plant at Muscle Shoals, Alabama, employed, at its peak, 24,000 workers. There were allegations of widespread and fraudulent waste of time and materials at Muscle Shoals, causing the Department of Justice to investigate. As one humbled agent reported, "I am frank to say that I started out with the theory that this crowd at Muscle Shoals was a gang of crooks, and I was primed for that, and blinded by that same opinion until I met them and had occasion to investigate. . . . I saw no evidence of waste that was not incidental to an enterprise of that character."[16]

Careful supervision of contracts of this number and of this size was very nearly impossible in a crisis situation. Even obtaining the required office space was a difficult task. Congress was slow to approve funds for the expansion of the Washington bureaucracy. Some clerks employed by the Ordnance Corps placed their typewriters on window sills, some worked at home, and one group of army officers rented a loft above a garage with their own funds and then divided it into makeshift offices. Even if space had been available, trained manpower was not. The Ordnance Corps hired Lester W. Blyth, an accountant from Cleveland, to head its finance division, which audited war contracts. When this division finally became fully staffed, which was not until 1918, it employed 1,200 accountants, plus numerous military officers, enlisted personnel, and clerical workers. There were simply not enough trained accountants in the United States (perhaps even in the world) to review all the war contracts thoroughly. Even if there had been more accountants available, it is doubtful that the state of the accountant's art was sufficiently developed to meet the national need.[17]

Heavy military purchasing soon altered the relationship between the government and its vendors. General Crozier's Ordnance Corps was among the largest purchasers. He recalled that "the state of affairs very soon got to be such that they knew they would get the business, and they knew I could not withhold the business from them on account of dissatisfaction with the price, because they knew I could not escape the most destructive criticism if I left the army without munitions because of price." Crozier was not entirely at the mercy of the contractors; he had authority to commandeer any factory and to force it to cease civilian production, under penalty of a fine of $50,000.[18] These were awkward tools, nevertheless, and a wise administrator would be most hesitant to use them. This reluctance did not escape the notice of the vendors.

The army had a long-standing institutional lack of interest in the careful review of war profits. This was evident in American history at least as early as 1711, during Queen Anne's War, and on numerous occasions thereafter. Bernard Baruch, who chaired the War Industries Board during the Great War, described the twentieth-century version of the military concept of expediency bluntly if inelegantly: "You must not forget this about the army: They want to get this stuff as quickly as possible." When victory or defeat hung in the balance (or was thought to hang in the balance), soldiers cared nothing about price or profits. As General H. M. Lord, the director of finance of the War Department, candidly explained: "If a cable was received from General Pershing that they needed shoes, and to get them right away, the only thing to do was to get them as quickly as possible—inside the law if possible, outside the law if necessary."[19]

Even if the army officers had been deeply concerned about war profits (which they were not), the problem of defining and identifying profiteering in a modern economy had become increasingly vexing. By 1917 the ascent in prices had become widespread, although not universal. It was extremely difficult to distinguish between increases that were cost-driven and increases that were greed-driven. There was also a thorny question of distinguishing between products that were munitions of war—and that might therefore deserve price control—and products that were essentially civilian in nature. As the Progressive *Nation* observed (with some exaggeration), in order to be rigorously fair in controlling profiteering, it would be necessary to distinguish between each bushel of grain sold by each farmer (as well as all beef cattle, all sheep, all fruit, and all vegetables) to determine which went for war purposes (such as feeding soldiers or war workers) and which did not. "We must think this thing through," the *Nation* warned. "Where shall we draw the line in the process of making munitions of war? How far back shall we go in the demand that no individual citizen be allowed to coin money out of the miseries of war?"[20]

Cantonments

Plagued by the troublesome obstacles of haste, inexperience, and intricacy, the effort to mobilize American resources without excessive gain plunged forth. The first requirement was to induct men into the army and to train them as rapidly as possible. In 1917 the British and French armies were faltering on the western front; indeed, the German submarine campaign that provoked the American belligerency was a gamble that the Allies would collapse before American might could be brought to bear.

In a very real sense, training and equipping the troops was a race against time: delay would mean defeat. Training the inductees required barracks and training grounds, and these must be built as quickly as possible. In later years, when the victory had been won (and its value doubted), and when the fear of defeat had disappeared (and its memory forgotten), the cost of these camps or cantonments became an object of partisan controversy.

The army was woefully unprepared to build training camps both speedily and efficiently. In peacetime the construction division of the Quartermaster Corps normally supervised the building of barracks. In April 1917, this agency numbered just one colonel and four assistants. The war required, however, the expenditure of $150 million on construction in the first six months; in comparison, the Panama Canal, the largest previous construction project, had a budget of only $46 million in its largest year. Wilson timed registration for the draft for June; if the camps were not ready to shelter the 1.2 million inductees before the September frosts, the boys would freeze, the war effort would falter, and Wilson would sustain a major political liability which he did not relish.[21]

The army's original plan was to shelter the future doughboys in tents, but the impossibility of this scheme quickly became evident. The army's estimated requirement was for 87 million square yards of tent-grade cotton duck cloth. The largest total production in the United States before the war was only 12-13 million yards, or about one-seventh of the demand. The army considered seizing tents used by American circuses, but this proved impractical. The camps would have to be built of wood.[22]

The size of the cantonment project was truly remarkable. There were to be sixteen camps, and they would require in total four billion board feet of lumber, which was approximately equal to the total output of American mills in a normal year. The barracks would need 177,000 doors, 46,000 water closets, 38,000 shower heads, 38 million feet of wall board, 200,000 kegs of nails, and 5,000 refrigerators. Delivery of these materials would require about 5,000 railroad cars for each camp. It would be necessary to unload an average of 50 cars per day, and at the peak, 150 cars per day. These materials had to be manufactured, delivered, and installed in only two months.[23]

In May 1917, the civilian-dominated General Munitions Board of the Council of National Defense decided that this job was far too immense for the Quartermaster Corps. Since the Corps of Engineers was occupied by construction problems in France, and since its expertise in handling such a task was doubtful in any case, it became clear that the only available alternative was civilian management. "We commenced to see the great need for the finest men we could get," recalled a veteran of the Munitions Board. "We got hold of big firms all over the country and we asked them to send big men. . . . Twenty-five men who were earning $15,000 to $25,000 a year." These civilian construction managers seized control of military construction projects, displacing army officers in all but name. "It was like a camel that got its head into the tent, and then pushed all the way in," remembered one construction executive. "We built a great big organization . . . and got its [the army's] tacit consent."[24]

The civilian managers moved swiftly. The locations for the camps were chosen by a panel of army officers in order to limit political interference. At Camp Grant, built near Rockford, Illinois, the construction contract was let on 21 June 1917, the same day that the site was selected. Work commenced on June 24, and the camp was sufficiently complete to receive its first 27,800 trainees by September 17.[25]

The construction contracts became highly controversial. The terms of the sixteen contracts, each of which was awarded to a large, established firm, specified that each general contractor would be compensated at the rate of the cost of construction of the camp plus a fixed fee of $250,000. These contracts, then, went to large firms, were in undefined but certainly large amounts, and had quite unusual conditions. The contracts were let without competitive bidding, and the "cost-plus" feature was peculiar and open-ended. This was enough to raise the suspicions of bypassed bureaucrats, unlucky competitors, business critics, and Wilson's political opponents. The saga of the wasteful and improper cost-plus contracts was beginning.

Although much criticized, the cost-plus contract of the First World War was not entirely novel. This type of agreement was occasionally employed in private industry and in military construction contracts before the war. Its use was preferred in instances when costs could not be determined before a project was begun—for example, when a building was to be repaired and the extent of damage could not be known until the foundation was exposed, as happened occasionally to post offices and courthouses. Cost-plus contracts were also the mode for factory construction in New Hampshire, Massachusetts, and Ohio, and some large firms, notably the tire manufacturers U.S. Rubber and Goodyear Tire and Rubber, even preferred them. At least one experienced construction firm worked exclusively on a cost-plus basis, and the U.S. Shipping Board utilized them for a time in ship construction. The cost-plus contracts used for wartime army contracts were based on models first employed by the navy for repair work, though the navy preferred not to use the cost-plus device except in unusual circumstances. The navy's experience was that significant amounts of unnecessary material would be ordered and that workmen tended to shirk on the job. Nevertheless, the navy did utilize the cost-plus feature for its new training camps and naval air stations during the war.[26]

There were several reasons for the use of cost-plus contracts in the construction of the cantonments, all of which became matters of dispute. The first consideration was that the process of advertising sixteen contracts, preparing bids, and selecting the winners was time-consuming. A second obstacle was the unknown cost of materials. The heavy demands for lumber (a whole year's supply would be needed) and other materials would presumably drive up prices. Labor costs were also uncertain. How much wages and prices might increase, as always, was anybody's guess. Finally, there was the unknown cost of site preparation. When the contracts were first written, the actual location of the camps was undecided. Accordingly, a contractor could not reasonably estimate the cost of clearing trees, leveling roads, and grading soil for drainage.[27]

Critics contended later that none of these reasons should have been compelling. The process of bidding the contracts would have taken about six weeks, which, they maintained, would have been justified by the savings. In almost any business proposition, furthermore, the costs of labor and materials are somewhat variable, and to remove this variation as a factor in pricing removes much of the risk, which is the justification for the profit. The type of building under construction as a barracks was architecturally very simple and very standard. Having been built many times before, it was simple to approximate its cost, since the army had prepared both a standard set of blueprints and a standard list of materials. Finally, there was really no need to know the exact position of the camps, because in practice the cost of site preparation is seldom highly variable.[28]

An assessment of these charges necessitates some speculation. Indisputably, advertising the cantonment contracts would have delayed the decisions, perhaps by as much as six weeks. Although under wartime conditions the paperwork would have been expedited, another period of time would have been required to survey the sites properly.[29] Due to these delays, American troops would have arrived in France either somewhat later and/or somewhat less well trained. The carnage on the western front would therefore have been prolonged, and American casualties might have risen. Thus the economic cost of having unadvertised contracts at the outset of the war must be weighed against the economic and human cost of continuing the war for a period of unknown length.

Besides unadvertised contracts, cost-plus contracts were allegedly a source of excessive waste. After the Armistice, a special congressional committee was charged with investigating the war's cost. It was chaired by Representative Sylvester Graham, a Republican from Illinois, and hence became known as the Graham committee. One of its subcommittees studied the construction of the cantonments and reported that the total cost of the sixteen camps was $206.6 million, or about $12.9 million each. The majority of the committee asserted that the "proper cost" was only $128.1 million, so that the "loss to the taxpayers" due to waste and excess profit was $78.5 million.[30] The majority report, however, was signed strictly by Republicans, who had an incentive to criticize a Democratic administration; the Democratic minority unanimously rejected this verdict. The straight party-line vote raises suspicions about accuracy of the conclusions.

Although no one could foretell the exact cost of the camps, the government's estimate was that each camp would cost between $3.5 million and $5 million, or about one-third of the final expenditure. In keeping with this approximation, the Munitions Board specified that each general contractor's fee would be fixed at $250,000. This fee was calculated to be 6 percent of the assumed cost. These contracts, then, were of the type known as "cost plus fixed fee." The fee of 6 percent of the estimated cost was slightly lower than, but in general accord with, the customary fee structure of the prewar construction industry, in which a successful project typically yielded the contractor a return of about 7 percent of the cost. As costs escalated, of course, the fees fell as a percentage of the total expenditure. They averaged just 2.84 percent of the cost, which seemed eminently reasonable to the contractors, who pointed out that Canada had paid 10-15 percent. Since risk was almost totally eliminated, however, the quarter-million dollar fee for four months' work seemed excessive to some.[31]

The real problem with the cost-plus contract was not excessive profit but rather excessive waste. Each contractor found it necessary to assemble a large labor force, and each used wage incentives to attract workers to relocate. This was unquestionably necessary, as local labor was completely inadequate. The construction of Camp Sherman near Chillicothe, Ohio, required 11,000 workers, but the entire population of the city numbered only 15,000.[32] Yet even if all the skilled carpenters in the country had been diverted to the camps, the number would still have been insufficient. Therefore, contractors had to hire and train unskilled workers to become carpenters, plumbers, electricians, and other tradesmen. This circumstance became the source of much of the controversy.

At Camp Grant near Rockford, Illinois, at least half of the 4,000 carpenters employed on the project had no previous training. This workforce was made up of a hodgepodge of grocery clerks, farmers, tailors, machinists, cooks, and bartenders. They came, an army general pronounced, "from the flotsam and jetsam of the laboring world." Despite their inexperience, these green workers received what was then an attractive wage: 62.5¢ per hour, or $6.25 for a ten-hour day. Neophyte electricians made 70-75¢, plumbers, 75-80¢, and bricklayers, 75-85¢. To no one's surprise, the few bona fide craftsmen did not applaud this invasion of their vocations by the well-paid and the untrained.[33] Although the new artisans could not meet the customary definition of "journeyman," or be recognized as such by the craft guilds, the buildings did get built and were fitted out, so that the titles "carpenter," "plumber," and "electrician," although somewhat embellished, were not wholly inappropriate terms.

Much of the subsequent bitterness felt by returning doughboys toward civilian workers was rooted in the contrast between the comparatively handsome wages paid to these instant craftsmen and the pittance paid to army draftees. Even those who were not able to proclaim themselves to be skilled tradesmen still received 35¢ per hour as common labor, or $3.50 a day.[34] A new private, by contrast, received just $30 per month plus room and board. As one disgruntled doughboy complained, "The man at the front must not be made to feel that while he risks his life and his legs for a dollar a day, his exempted neighbor in the comparative safety of his own home town is earning more money in a day than he used to earn in a week."[35] The historic relationship in which soldiers' wages ranked between the wages of unskilled and skilled workers had been dangerously violated.

The labor shortage combined with the frenzied supply situation to create further disorder. Under heavy wartime pressure to reach completion, it was impossible to synchronize the arrival of men and materials at the worksite. This meant that gangs of workers all too frequently were assembled before sufficient materials had arrived. With little else to do, these employees were often observed simply lazing around the project, or worse. As one disgruntled worker hired at Camp Sherman colorfully recalled, "When they first started construction there, the wheat was still in those fields; why, you could go out in those wheat fields any time and see a poker game and a crap game—I don't mean one but several of them. . . . Every day more or less the same thing from the time they got there in the morning until the time they quit; they wouldn't even stop to eat dinner; didn't have time; the game was too big. . . . There is some of the best crap shooters in the country right here in Camp Sherman now." "How about the saloons?" he was asked. "Oh, they were filled," was the reply. "There wasn't standing room in those two up by the camp."[36]

Since disclosure of this situation could obviously prove embarrassing, some foremen simply ordered idle workers to stay out of sight or to hide from probing reporters. Inevitably, some of these idlers were discovered, with multitudinous expressions of outrage and demands for dismissal resulting. This produced a startling revelation: at most of the cantonments, general superintendents had ordered that no worker be dismissed. Labor was so scarce that managers ordered miscreant workers to be demoted or reassigned before dismissal was permissible. Critics suspected foul play; contractors must be seeking to drive up costs so as to fatten profits in a cost-plus situation.[37]

Contractors defended themselves by pointing out that, since their fees were fixed at $250,000, they had no incentive to inflate costs by padding the payroll. On the contrary, if there was gross waste and inefficiency at a job site, their professional reputations would suffer and their hopes of obtaining new contracts would be jeopardized. The real source of payroll padding was labor. Well-paid construction workers had a strong incentive to make the job last as long as possible. As a result, there were numerous reports that workmen had slowed down or "slackened up" as soon as they heard that the open-ended, cost-plus contract was in vogue. Reports of shirking in order to stretch out the job were most common among plumbers.[38]

The cost of the cantonments spiraled upward for related reasons. Supplies of lumber arrived erratically, and this caused foremen to order extra sawing to modify the lumber that was on hand rather than waiting for proper sizes and lengths. Much of the lumber was low-grade, with many bad sections that had to be cut out. Bad roads in the camps destroyed some lumber in transit as well. As a result, there was substantially greater waste of materials—about 10 percent more scrap—than would have occurred in peacetime work.[39] The scrap wood was destroyed by fire, and this reinforced the view that the camps were constructed carelessly and wastefully. The version that circulated was that contractors deliberately burned good lumber in order to inflate costs and profits.

Reports of reckless waste, idling, and fraud at the projects prompted the War Department to take action. Government auditors swarmed to the sites, accompanied in some cases by undercover agents of the Bureau of Investigations of the Department of Justice, the forerunner of the Federal Bureau of Investigation. Keeping tabs on the contractors proved to be difficult, as the federal auditors often had little knowledge of construction methods. Friction between the contractors and auditors developed quickly, particularly when government auditors attempted to influence the order in which buildings would be erected.[40] Nevertheless, the auditors and undercover men ferreted out a few cases of fraud.

Several workers were apprehended at Camp Grant for signing on to two construction crews at the same time. They were jailed for ninety days before trial, but their juries found them not guilty. At Camp Sherman, a secret agent unearthed a conspiracy in which a contractor apparently leased about two hundred superfluous trucks from a subcontractor and collected a $5-10 a day kickback on each vehicle for doing so. Also at Camp Sherman, the general superintendent, Frank J. McGrath, received a salary of $4,362 for supervising construction. When auditors discovered that McGrath also owned 25 percent of the stock in the contracting firm of D. W. McGrath and Sons, they ruled that he was engaged in double-billing and denied him his $4,362. There were other instances of minor fraud and graft, but what was really notable was the general absence of dishonesty amid boundless opportunity for corruption. Although at each camp under construction the Bureau of Investigations commonly recruited as an informant the stenographer to the general accountant, only one prosecution resulted. There were a few other convictions, but no general chicanery. The presence of undercover agents to inspect the performance of defense contractors, nonetheless, signified a heightened level of sophistication in the antiprofiteering effort.[41]

These instances of minor crime or misconduct are insufficient to prove, one way or the other, the degree of general fraud or waste on the construction of the camps. The Graham committee, which investigated the cantonment question, made several attempts to evaluate the broader issue. Their method of evaluating performance was to invite a well-qualified construction engineer to review in detail an individual contract. A professor of civil engineering at the University of Michigan, selected as a disinterested observer, reviewed the performance of contractors who erected 1,528 buildings in ninety days, or one every forty-two minutes. He concluded that cost and waste were not excessive.[42] An experienced estimator from Chicago reviewed the performance of the general contractor at Camp Grant, Bates and Rogers. He testified that in his professional opinion, Camp Grant should have been built for $8,820,000, which was much less than the actual cost of $12,851,000. He further concluded that the total profit collected by the general contractor and all subcontractors, $532,000, was also excessive in the amount of $112,000. He calculated the "correct" profit at the rate of 5 percent of the "correct" cost. Yet when Bates and Rogers reviewed the outside evaluator's work, they discovered that he had neglected several important expenses, including the costs of building a 1,000-foot bridge, sixty miles of underground pipe, and a rifle range. His expertise was successfully impeached, if not destroyed.[43]

An even more general standard was applied to the cost of building Camp Sherman. This camp was the third most expensive in terms of the price per bed, and it was much vilified as the source of excessive waste. In defense, the general contractor, A. Bentley and Sons, measured the total volume of all the buildings they erected and determined that the price of the camp was 14.75¢ per cubic foot of building. Since the labor cost at Camp Sherman was 6.25¢ per cubic foot, the ratio of labor cost to total cost was 42.4 percent. In peacetime, the industry expected that a well-managed construction project would have a ratio of labor cost to total cost of 40 percent.[44] The price of Camp Sherman was therefore slightly high, but not excessively so when the speed of construction is considered.

Summarizing the price of the construction of the cantonments, it is reasonable to conclude that the Republican progressive critics of these projects seriously underestimated the cost of delaying completion by putting out the contracts to sealed bid. The charges of excessive, unwarranted profits gained by the contractors from the terms of these contracts were similarly overstated. The actual cost far exceeded the estimated cost, but the extra money went to hire labor and to pay for materials, not to fatten the pocketbooks of greedy contractors. The Democratic defenders of the Wilson administration were much closer to the mark when they argued that while the rushed construction schedule carried a heavy price, the money was well spent because it shortened the war. The cantonment issue, however, was just one facet of the progressive indictment of American business during the war. The progressives were also angry at the aviation industry.

Aircraft

Next to the controversy over the Morgan loans, which addressed the very purpose of American participation in the Great War, the quarrel over the performance of the American aircraft industry was the most furious. This topic served as a kind of surrogate for the entire conduct of the war effort. The aircraft industry was investigated by the Department of Justice (Hughes commission, 1918),[45] by the Senate (Thomas committee, 1918),[46] by the House of Representatives (Graham committee, 1920),[47] and again by the Senate (Nye committee, 1935-36).[48]

The fundamental problem of the American aircraft effort, as with other aspects of defense mobilization, was a late start. Although the United States was the first nation to fly an airplane, its early technological leadership had languished as far as military applications of the invention were concerned. By 1916 the United States certainly lagged behind Britain, Canada, France, Germany, and Italy, and perhaps behind others as well. This dawdling proved to be costly, both in terms of wounded pride and wounded sons.

The first appropriations by the United States to explore military applications of the airplane were made in 1908. For the first eight years of lethargic experimentation, the total appropriations were only $1.5 million, and total number of airplanes shipped to the U.S. Army Signal Corps was just fifty-nine. In 1916, eighty-three more planes belatedly arrived. As the military value of the airplane on the western front became increasingly apparent, Congress finally raised the appropriation to $13 million per year. During the war, expenditures on warplanes reached $1.21 billion, a hundredfold increase, but even this sum was too little and too late.[49]

The United States, once the proud originator of the flying ship, only fifteen years later was incapable of producing a state-of-the-art combat aircraft. The few planes it did build—213 observation planes—were of negligible value. Meanwhile, American soldiers huddled in the French trenches with inadequate protection from enemy air attack. The American Expeditionary Force was forced to purchase its warplanes from Britain and France, and in 1918 the AEF had only 20-25 percent as many planes protecting the doughboys as did comparable units of the French army. American pilots, moreover, were killed in battle about three times more frequently than British or Belgian fliers. The American people demanded to know who was to blame and where the money went. Had the United States received a fair value for its many dollars? Or was degraded quality—a perennial form of profiteering—rampant in the aircraft industry?[50]

When the United States declared war, Raynal C. Bolling was the U.S. Army's leading technical expert on aircraft production. In December 1917, Bolling led a group of high-level officials to London to confer with Allied leaders in order to determine the appropriate role for the United States in the air war. The Bolling commission, which included Wilson's key foreign policy adviser, Colonel Edward M. House, met with British and French leaders, including Winston S. Churchill, who was then the minister of munitions. The three Allied powers penned an agreement which stipulated that the United States would not attempt to manufacture its own pursuit (fighter) plane but would instead purchase these types from the British and French. This was a recognition, Bolling admitted two years later, that the United States lacked the technical expertise needed to design and produce an advanced warplane. This view was confirmed by Major General Mason M. Patrick, the chief of the Air Service, American Expeditionary Force. Patrick said that it was wise not to attempt to build a pursuit plane because the design was "changing rapidly," which meant too rapidly for American designers.[51]

Secretary of War Newton D. Baker later defended this decision as rational and deliberate. The assumption, according to Baker, was that the United States possessed the vital raw materials for aircraft production, while Britain and France had sufficient productive capacity to meet the needs of all three powers. This version of the story was true as far as it went: the United States did have the materials and the Europeans did have factories. Their productive capacity, nonetheless, did not fully meet American needs. The United States' late start, furthermore, was partially the result of careless planning. An American aviation designer, Dr. W. W. Christmas, had met with Baker well before the American war declaration. Dr. Christmas told Baker that the United States would soon be in war and implored him immediately to commence a pursuit plane program, while there was still time. Christmas even offered to surrender all royalties on his patents if Baker would agree. Baker, nevertheless, rejected both Christmas's premise and his conclusion.[52]

Meanwhile, the Advisory Commission of the Council of National Defense was beginning to consider a possible role for the United States in producing warplanes. The farsighted Howard Coffin, vice president of the Hudson Motor Car Company, perceived accurately how far the United States was behind. In March 1917, some two weeks before the United States declared war, Coffin reported that no American warplanes could be produced and sent abroad until 1918. He thought that 2,000 machines might be built in 1918 with an increase to 4,000 in 1919. Coffin predicted, correctly, that there would be a serious shortage of wood and cloth for the fuselages but that the "controlling factor is and will continue to be the production of aeroplane engines." Since the estimated need for an army of one million men, which was less than half of what was ultimately required, was 4,500 to 5,000 planes, Coffin knew in 1917 that the doughboys could not be adequately protected until 1919.[53]

Once a decision was reached not to attempt to build an American pursuit plane, American leaders turned their attention to other possible contributions to Allied air power that the nation might make. John D. Ryan, president of the Anaconda Copper Company, became director of the Air Service, and his appointment immediately raised eyebrows in the Progressive camp. Anaconda, as one of the largest copper producers, was the leader of a highly oligopolistic industry with frequent labor problems. The copper business featured several anticompetitive practices, including interlocking directorates and a single firm whose purpose was to sell the entire production of the leading copper producers. This agent was the United Metals Selling Company, and its presidency was held by Anaconda's John D. Ryan. Progressives lumped these practices and these firms together and denounced them as the "Copper Trust." They regarded profits earned by the copper producers before the war as highly improper.[54]

Ryan selected Colonel Edward A. Deeds as head of aircraft production. Deeds was also suspect on antitrust grounds. He was a former executive of the National Cash Register Company of Dayton, Ohio. In 1913 he was convicted of having violated the Sherman Anti-Trust Act and was sentenced to a year in prison. His conviction was overturned on appeal, but this did little to restore his reputation among Progressive business critics. Deeds owned 30,000 shares of the United Motors Corporation, a holding company whose assets included a valuable property known as the Delco ignition system. When Deeds was appointed head of aircraft production, he transferred his stock certificates to his wife's name, later claiming that this constituted a bona fide divestment. During the war, Deeds awarded a large contract for the purchase of airplane engines to the Packard Motor Car Company, but the contract specified that the ignition systems must be made by Delco. Several other valuable contracts also went to Deeds's friends in Dayton. After the war the special investigating commission headed by Charles Evans Hughes recommended prosecution for these infractions, but Ryan intervened. Ryan commended Deeds's contribution to the war effort and maintained that this "outweighed any technical violations that may have occurred."[55] Deeds, who before the war had been earning $85,000 per year, had originally declined to serve owing to his concern that his various interests might create an impression of a conflict of interest. He ultimately reconsidered and agreed to come to Washington and serve for $1 per year only after Coffin assured him that his investments would not be questioned. Because Secretary of War Baker distrusted the Hughes inquiry, Deeds escaped prosecution.[56]

Ryan and Deeds determined that the principal focus of the American air effort must be on the gathering of raw materials for the factories of Britain and France, which would make pursuit planes. A secondary role was to produce a powerful new aircraft engine, the Liberty motor, which would power the next generation of fighter planes. Of third importance was the fabrication of a warplane of auxiliary value, the combination observation-bombing aircraft known as the de Havilland DH-4. Each aspect of this endeavor became part of the profiteering controversy in respect to the price, quality, and rewards obtained from the materials supplied.

In the long and inglorious record of American unreadiness for war, there are few instances of such woeful inadequacy as the absence of aircraft materials in 1917. The state of the aeronautical art in 1917 required the use of exotic substances that were extremely difficult and expensive to obtain on short notice. The only suitable lubricating oil then available was castor oil, which the airplane engines of the era consumed with a thirst that would embarrass a drunkard—sometimes six quarts an hour. The United States not only did not grow sufficient castor beans but it even lacked the seed to do so. A shipload of seeds had to be imported from India, and American growers planted 110,000 acres to match the need. Propellers were made of mahogany grown only in the tropics, and the harvesting of this wood depended on heavy rainfall to float the logs to coastal ports. There were also serious shortages of linen to cover the wings, as not enough flax was grown, and of acetone for the lacquer that sealed them. Huge amounts of acetone had to be distilled from wood, of which there was also a shortage.[57] These glaring shortages led to unusual haste and concordant waste, which inflamed the profiteering controversy. None of these problems, however, approached the difficulty of obtaining enough spruce trees to build the aircraft.

There was only one material that was strong enough and light enough for the framework of World War I fighter planes. This was first-growth clear spruce, the kind found only in stands of virgin timber in extremely remote and mountainous regions of the Pacific Northwest. Virgin trees had very thin growth rings, producing lumber that was denser and stronger than that which came from second-growth trees. Even this wood was seldom satisfactory; only 10-15 percent of the spruce logs produced a grade of lumber suitable for fighter planes. Lower-grade spruce as well as some fir was strong enough for trainers, but only clear spruce could withstand the strains of combat aerobatics. To build the planes necessary to defeat Germany, about 100 million board feet of clear spruce were required. To obtain this yield, American lumberjacks would have to cut approximately one billion feet of first-growth spruce.[58]

Before the war, spruce had little commercial value, so no one had bothered to build roads to the vital areas. To further complicate matters, spruce trees did not grow in large clusters, but were intermixed with fir and scattered throughout the Pacific slope. Surveys showed that the richest stand of virgin spruce in the world was on 11,000 acres of the Olympic Peninsula in western Washington, estimated at holding between 250 and 330 million board feet of lumber.[59] The declaration of war instantaneously changed this property (known as the Blodgett tract) from one of uncertain value to one that was crucial to the war effort. How could a fair price be set for timberland that was vital to the nation's security?

The land was owned by William Blodgett, a Michigan lumberman who was holding it for speculative purposes. Blodgett believed that oil might be found on his property, and he further anticipated that in the next ten years the value of the lumber would appreciate, perhaps to as much as $2 million. Despite these hopes, in January 1918 he offered to sell the tract to the government for $635,000. It was later alleged but not proved that before the war he had offered the land to a private purchaser for $450,000, but the government decided to accept his offer because of its perceived need to build a great fleet of bombers for a massive air offensive expected to take place in 1919.[60] But to gain access to this resource, it would be necessary to construct a new railroad line. The cost, necessity, and location of this line became significant aspects of the war profiteering dispute.

When the pressing need for spruce became evident, a consortium of patriotic Washington lumbermen had offered to harvest the vital lumber from nearby trees. Twenty logging companies were involved, and each agreed to accept only a nominal fee of one dollar for its services. After surveying the easily accessible timber, the army concluded that even if the commercial loggers increased their production to the maximum extent possible, only one-third of the required yield would be provided. It was therefore necessary to draw upon the Blodgett stand, and the army awarded a construction contract to an eastern firm, the Seims-Carey-Kerbaugh Company, to build a railroad into the tract. This decision markedly annoyed the local loggers. Some felt that this expansion was unnecessary, creating needless capacity that would depress prices after the war. Others resented the presence of an allegedly inexperienced eastern firm, which they regarded as a slight to the Pacific Northwest. The contract was let on a cost-plus basis, which was automatically suspicious, and it seemed even more dubious because the forty-mile route through mountainous terrain appeared to be exceedingly long.[61]

When the cost of building the railroad was made public, the westerners became convinced that their suspicions were well founded. To meet the war schedule, the railroad had to be built in the midst of a rainy Pacific Coast winter. As a consequence, the road would cost $4 million, or about $100,000 per mile. This was roughly six times the customary prewar price, when roads were built more deliberately and in good weather. The westerners complained that the road's eastern terminus connected all-too-conveniently to the Chicago, Milwaukee, and St. Paul Rail Road. When the war ended, the Milwaukee Road would be the only conceivable purchaser and therefore the likely beneficiary of a give-away price. The track seemed to be of high-quality construction, as if it were meant to be a permanent line. The westerners noted that John D. Ryan, the director of the Air Service, was simultaneously president of Anaconda and a director of the Milwaukee Road, and they believed they had detected a monumentally fraudulent scheme to enrich the railroad.[62]

Although this explanation was plausible, under careful examination the charges proved groundless. A railroad construction engineer employed by the Union Pacific Railroad was invited to examine the site, and he concluded that the shortest feasible route had been selected. All parties agreed that the winter construction costs were dear, but none could prove them exorbitant. John D. Ryan, moreover, took no part in choosing the contractor, as that decision had been made before he entered government service. He also studiously declined to participate in the railroad site decision on the grounds of a possible conflict of interest.[63] Despite these facts and despite the inability of critics to prove their case, many doubters remained unwavering, and that is the essence upon which the war profiteering legends were based.

The effort to retrieve the riches of the great forests of the Pacific slope left behind another festering wound. The demand for spruce was so enormous that there were not enough lumberjacks in the entire Northwest to fell all the trees. Also, the northwestern lumber industry was plagued by perennial labor strife. The War Department feared that production might lag and that the heavy demands might touch off a potentially disastrous strike. To avert these difficulties, President Wilson took the unusual step of ordering the army into the forests. A special "Spruce Division" was formed and sent to the Northwest under the command of Colonel Brice Disque. These 30,000 soldiers worked alongside civilian loggers to augment their output. The question of an appropriate wage for these soldier-lumberjacks promptly appeared: should these men earn the same wage as their comrades in the trenches, or should they earn the same wage as their coworkers in the forests?

Privates in combat in France earned a dollar a day, plus subsistence. This was far beneath the earnings of northwestern lumberjacks, where the daily wage was $5.00-6.00. To bridge this gap, the army gave the troops in the Spruce Division a daily bonus of $2.50-$7.00. The reason offered to excuse this inequity was (as customary in military matters) expediency. The army claimed that it simply could not draft men and get them to work in the forest next to civilians who were paid much better. Therefore, an anomalous and basically unfair situation developed. Young men were conscripted either to face hard work in the forest at home or death in the trenches abroad. Although neither was an appealing prospect, the wages and safety of the Washington timberland were clearly preferable to the misery and danger of the Argonne forest. That the compensation was better in the safer endeavor was an arrangement that the trench fighters of France would not soon forget. To further compound their indignation, there were numerous reports and a widely shared belief that the army sent the better men to France, at a dollar a day, while the lesser soldiers went to the spruce forests at six dollars. It was also noteworthy that before the war Colonel Disque, who commanded the Spruce Division, was only a captain with a salary of $3,000. He was eventually promoted to general, and after the war he found employment with one of the civilian operators (with whom he had negotiated contracts) at a salary of $30,000.[64]

While doughboys and lumberjacks were hewing the spruce that would form the fuselages, others were considering how to manufacture the motors that would lift the planes. Here the nation was the beneficiary of good foresight, although this did not come from the military. Correctly anticipating the need for a powerful new aircraft engine, the Packard Motor Car Company commenced designing an advanced power plant that became known as the Liberty motor. As a Senate investigating committee reported, Packard saw the need for a stronger engine "prior to the time that anybody else in the country, including the Government, saw any need of it." The Packard engine was also well ahead of the rest of the world. The normal power plants in World War I pursuit planes of British and French design were in the range of 200-225 horsepower; the Liberty motor, at 400 horsepower, nearly doubled the standard. When the United States entered the war, Packard donated the design to the government free of charge, although its contribution was not altogether selfless sacrifice.[65]

When the Bolling commission met with British leaders to decide against building an American pursuit plane, the group also resolved that the United States should undertake a determined effort to manufacture the Liberty motor as quickly as possible. Airplanes equipped with this engine would be able to carry large bombloads that would allow the Allies to break through the German lines in 1919. The Liberty motor was also thought to be adaptable enough so that it could power the next generation of pursuit planes.[66]

The Packard Motor Car Company was rewarded for its foresight by obtaining an initial contract for the manufacture of 6,000 Liberty motors. This contract was of a new and unusual type, a more sophisticated form of incentive pricing than was used in the Civil War. The new arrangement was designed to be used in an instance in which the actual cost of producing a complicated new mechanism could not be accurately predicted beforehand. A contract of this kind was supposed to be an improvement on the open-ended, cost-plus form in which costs could soar upward unchecked. Engineers first estimated the cost of making a device. This approximation was expected to be near the highest figure that might be anticipated, and thus it would serve as an upper limit on the price. This uppermost or target price was known in wartime parlance as the "bogie" price. When the production run was finished, the actual cost of manufacture would be determined. If everything went well, the actual cost would be less than the bogie price. The difference between them would decide the final sum that the government would pay, and the supplier would get the actual cost plus 25 percent of the difference. This latter feature was to serve as an incentive to keep costs low. If unexpected difficulties caused costs to escalate, the government would not be forced to pay a huge price as could occur in a cost-plus contract. This innovation worked better on paper than in practice, as the Packard contract revealed.

The predicted maximum or bogie price on Packard's Liberty motor contract was $5,000 per engine. This included a fixed profit of $625, or 12.5 percent of $5,000. The actual cost of manufacture was determined to be only $3,200. Under the terms of target pricing, Packard was also entitled to 25 percent of the difference, which amounted to a bonus of $450 per engine. When the fixed profit and the bonus were combined, the total profit reached $1,075, a healthy 33.6 percent markup. This was considerably more advantageous to the contractor than was likely with a cost-plus contract. To make the situation even more questionable, the facilities for production of the Liberty motor were paid for by the U.S. Army Signal Corps at a price of $4,470,000. After the Armistice, Packard was able to purchase the facilities at a price of $849,556, or nineteen cents on the dollar. After World War II, the famous Truman committee would denounce this kind of sale as "legal profiteering."[67]

Despite Packard's technological leadership, and despite the great expense involved, the Liberty motor contributed very little to Allied victory. Although 22,500 motors were ordered, by the time Germany surrendered only 264 had been completed. With considerable understatement, the Senate described this output as "gravely disappointing." The Liberty powerplant also proved to be too heavy for most of the existing airframes. An attempt was made to fit the Liberty motor to the English-designed Bristol fighter, which would increase the plane's horsepower from 225 to 400. Vibrations caused by the great speed of the ship and its general lack of airworthiness caused fabric to tear away and wings to fall off. All the test planes crashed with the loss of their pilots. There had been many earlier examples of quality degradation in the history of American defense contracting, but none had had results so spectacular and so deadly.[68]

As an interim solution, Allied commanders decided to add the Liberty motor to a heavier plane, the English-designed de Havilland DH-4, an observation plane that could be modified for use as a bomber. A total of 213 DH-4s were built in the United States and shipped to France before the Armistice. Despite this fact, an oft-repeated postwar charge was that "not a single American-made fighting-plane of any description ever reached the front." Still, considering the funds expended ($640 million), this production level was at the very best an indifferent performance, even considering that that sum also purchased 640 British and French-built planes. The DH-4 was also not an advanced design, and both its fighting utility and its safety were questionable. Production of this airplane became the most contentious instance of allegedly degraded quality of the entire American war effort.

As a combination observation and bombing plane, the DH-4 was not a sound design for either purpose. Even by World War I standards, the plane was slow and clumsy, its range was short, and the pilot and his observer were too far apart to communicate without effort. Nobody could reasonably claim, therefore, that the DH-4 was a state-of-the-art design. Some declared it to be obsolete, but a more judicious assessment would describe it as a contemporary but not advanced design. It was a slow and ungainly flier, but not excessively so.

The chief deficiency of the DH-4 was its dangerousness to flyers. The cockpit was located behind the engine, with the fuel tank behind the pilot. The horizontal members of the fuselage were wooden longerons. In an emergency landing these longerons tended to splinter, driving the fuel tank forward to crush the pilot against the engine and/or to ignite a deadly and inescapable fire. Since the exhaust pipe was located close to the fuel tank, and since the fuel tank was not armored, the DH-4 was prone to cause an even greater horror— fire in the air. Captain Eddie Rickenbacker, America's best-known pilot of the Great War, described this fate as his comrades' greatest dread.[69]

American pilots of the First World War were not outfitted with parachutes. (This was itself a sad commentary on the state of American preparedness, since the "barbaric Hun" pilots were equipped with such gear.) If a fire broke out at high altitude, there were only two alternatives. These were described, with fatalistic understatement, by one veteran of the Lafayette Escadrille as not "very pleasant to contemplate. Either stick with the ship and suffer the agonies of a slow roasting death or to jump into quick and merciful oblivion, a crushed and broken mass on the hard, unyielding earth." The great American ace Raoul Lufbery died this way (he chose to jump).[70] There were enough instances in which the de Havilland DH-4 caught fire that the plane acquired the grisly sobriquet "Flaming Coffin," a term which often appeared in the glossary of postwar antiprofiteering writers.

The "Flaming Coffin" notion was too harsh. The placement of the fuel tank to the rear of the pilot was a common design in the era. Some contemporary warplanes did have fireproofed fuel tanks, although the DH-4 did not. There were no statistics to demonstrate that the DH-4 was abnormally risky; all wartime flying was dangerous, and the demolition record of the DH-4 was not unusually poor.[71] Nevertheless, its reputation contributed to the legend of profiteering on aircraft: that vast sums of money were spent, that only a scattering of obsolete and useless planes were built, and that those few that arrived belatedly were criminally hazardous to fly.

The confusing and disappointing record of American aviation during the war was an illustration of an important characteristic of the war effort that contributed greatly to the developing profiteering controversy. Unlike any other major American war, the American role in the Great War ended sooner than American leaders originally expected. In the past, Americans declared war (or opened hostilities) and then commenced mobilization. Whatever minimal planning took place usually underestimated the length of the hostilities by several years and the cost by many millions of dollars. In the case of the First World War, however, early estimates overstated the probable duration of German resistance. This wisdom held that the war would last into 1919, at which time massive shipments of American men and matériel would supposedly inundate the Central Powers. With the American economy in full production, the war ended abruptly and rather unexpectedly in 1918. Great expenditures had been made on warplanes and other weapons in anticipation of the next year's campaign. These expenses appeared (suspiciously to some) to have produced scant benefit but large profit.

Copper, Steel, and Shipping

The increasing complexity of the American industrial giant was another chief contributing factor to the war profiteering saga. There is no better manifestation of this intricacy than the performance of the copper industry. This business was highly concentrated among a few interlinked producers. This cartel was dominated by Anaconda, which was owned by the Guggenheim family and administered by John D. Ryan. Before American entry, copper prices rose dramatically due to brisk sales to the Triple Entente countries. In the ten years before the war, copper sold on the average for about seventeen cents per pound. Since the average cost of production was around eleven and one-half cents in that period, the copper business was solidly but not spectacularly profitable in peacetime. Heavy war demand, however, caused the price of copper to move upward markedly. In 1916 copper rose to thirty-two cents a pound, and on the eve of American entry, it reached thirty-seven cents. At these prices, copper profits became robust; they were three or four times prewar levels. In 1917 copper firms were annually returning in profits a range of 70 to 700 percent of invested capital.[72] As the copper men fully realized, this was too good to last.

The catalyst for bringing about a reduction to more acceptable wartime levels was Bernard M. Baruch, then a member of the General Munitions Board and later the chairman of the War Industries Board. As a prominent and very successful speculator in metals, Baruch had entree to the Guggenheim family. Baruch, along with Eugene Meyer Jr. met with Daniel Guggenheim and asked him to cut prices to a level that would convince the public that the war was not being fought for profit. John D. Ryan also participated in bringing about Guggenheim's consent to a reduction, and Guggenheim agreed to discount the price to 16.6739¢, which was the ten-year prewar average. Since this reduction affected an order of 45 million pounds, the savings were nearly $7 million. The willingness of the copper industry to cooperate was partly patriotism, partly acceptance of the inevitable, and partly a desire to avoid antitrust prosecution. Nevertheless, this event signified the evolution of a new role for the government, as the United States moved away from a weakly regulated economy with significant concentration in key industries toward a more managed system. The copper agreement exemplified the Wilson administration's management of the war, which emphasized patriotism, voluntarism, and cooperation, rather than coercion. The utilization of voluntary agreements and self-regulation were early examples of what became known as associationalism.[73]

Unfortunately, establishing a wartime price on the basis of ten-year prewar average was not a permanent resolution of the copper problem. The wages of copper miners were tied by union contract to the price of copper, and to reduce the price abruptly invited strikes in the copper mines. This circumstance alone would have been reason enough to allow the price to climb, but there were other grounds as well. Other costs borne by the operators rose as a result of wartime inflation, so that small inefficient mines could not operate profitably at a market price of sixteen and two-third cents. Their cost of production exceeded twenty cents per pound. If the government was to keep these mines in production, prices would have to be higher, but how much higher? To raise them significantly would produce a windfall for the Guggenheim interests, but to fail to raise them would mean an unacceptable loss of production. Baruch met with leaders of the copper industry, who asked for a price of twenty-five cents per pound. This was acceptable to the General Munitions Board, but not to Baruch, who recommended twenty-two and a half cents. Secretary of War Baker sided with Baruch, vetoing twenty-five cents, and a compromise was finally reached at twenty-three and a half cents.[74] At this rate the Guggenheim mines earned handsome profits, and smaller mines stayed in production. It was a classic compromise between the twin goals of restricting profits and maintaining production.

Baruch hoped to repeat his feat by negotiating a price reduction in the steel industry. Steel, however, proved tougher to bend than copper. Judge Elbert H. Gary, the chairman of the United States Steel Corporation, had no love for the Wilson administration, having been heavily criticized by Secretary of the Navy Josephus Daniels. The steel industry had been a favorite target of business critics for twenty years, and the steelmen resented it. Baruch also lacked the close ties with Gary that he had with Guggenheim.[75]

Nevertheless, as in copper, the steel industry could not hope to continue to post the record profits of the prewar period. In 1914 U.S. Steel earned $46 million, but this increased to $333 million in 1916 and $585 million in 1917. Bethlehem Steel was even more fortunate. This company was poorly located to serve the midwestern automobile market or to obtain raw materials efficiently. Although poorly sited for the American domestic market, it was ideally situated for the English military market, and it received an order for 8,000 field guns while also producing 12,000 cannon shells a day. From 1904 to 1915 Bethlehem paid no dividends, but in 1916, owing to hefty war contracts, it was able to return 22.5 percent of its invested capital. Bethlehem profits were $9.4 million in 1914 and rose to $43.6 million in 1916. Before the war a share of Bethlehem customarily sold for about $25, but in just nine months the price soared to $275. By 1916 it had zoomed further to $700. This could not continue indefinitely. As Price McKinney of the McKinney Steel Company of Cleveland admitted, "We are all making more money out of this war than the average human being ought to."[76]

Steel executives were notably unenthusiastic about giving up their lucrative overseas profits and were not pleased when the Wilson administration, in an expression of Allied solidarity, demanded that American steel firms sell to Britain and France at the same price as at home. Secretary Daniels recorded in his diary that upon hearing this the steelmen "gagged." Nevertheless, the government forced price reductions. U.S. Steel asked for a price of four and one-quarter cents per pound for steel plate, but government accountants declared that two and a half cents was a fair price. The final compromise was three and one-quarter cents for all producers. This process of price adjustment ensured that high-cost producers stayed in business, but only at the cost of maintaining near-record profits for U.S. Steel and other firms. Before taxes, total earnings of all 131 American steel producers quintupled between 1915 and 1917. The average pretax rate of return on investment soared from 7.4 percent in 1915 to 28.7 percent in 1917, falling back to 20.0 percent in 1918.[77]

The advantages enjoyed by large corporations during the war emergency did not go unnoticed by business critics. U.S. Steel recognized the quandary its efficiency forced upon the government. In March 1918, its chairman, Elbert H. Gary, advocated a high excess profits tax to soak up the profits of low-cost producers such as U.S. Steel while guaranteeing a fair profit to high-cost enterprises. Gary said this "would be fair and reasonable. . . . That would be satisfactory to all of us and certainly ought to be satisfactory to the Government."[78] This also seemed better than dividing the contracts among the various producers and assigning each a different price. That might have violated the Sherman Act, leading to an unwanted prosecution. A different approach was followed in the cotton textile industry, where a composite price was calculated. This figure was based on an average of one high-cost, one low-cost, and several middle-range manufacturers. It included a markup of 10 percent for profit, which Charles Eisenman of the War Industries Board testified was a "reasonable profit."[79] (Thus yet another practical definition of the meaning of "reasonable" entered into war profiteering discourse.) All efforts to spread the work had only limited effect, however, as the war effort consistently favored large corporations. Nevertheless, war taxes did soak up some of the profit. In 1915-16 steel firms paid less than 1 percent of their net investment in federal taxes, but this figure rose to 12 percent in 1917, receding to 8 percent in 1918.[80]

A related enterprise in which the bothersome question of "reasonable profit" again appeared was the shipping business. The need to transport millions of troops to Europe, along with their gear and weapons, while simultaneously supplying the Allies, packed every ship to the fullest extent possible. Still, the United States was seriously lacking ocean transportation, and prices rose accordingly. Woodrow Wilson angrily singled out the shippers for a tongue-lashing. He reproved the "almost insuperable obstacles they have been putting in the way of the successful prosecution of this war. They are doing everything that high freight charges can do to make this war a failure. . . . I do not say that they realize this or intend it. . . . I am not questioning motives. I am merely stating a fact."[81]

The War Department tried to fill the discrepancy by acquiring cargo carriers and passenger liners to convert to troopships. Both needed extensive remodeling: cargo vessels needed sanitary lines cut into their decks and passenger liners needed their elegantly appointed cabins converted into hospitals, offices, dining areas, and so forth. The government lacked the facilities to refurbish these vessels, so the conversions had to be carried out in private shipyards. The yards took advantage of this development, and prices rose substantially.[82]

The problem was not only high profits earned by shipbuilders during the war—Bethlehem Shipbuilding Company, for example, netted $70 million before taxes during the war, including a return of 57 percent of its invested capital in 1918—but also that the profits continued after the Armistice. Each ship that had been converted to military use had to be reconverted and returned to its owners after the war, again at great expense. Bethlehem Shipbuilding accordingly was able to earn another 57 percent on its invested capital in 1919. The shipowners, moreover, expected that very extensive, and therefore very expensive, repairs would be made. But how extensive must be the repairs in order to return a vessel in a seaworthy, prewar condition? There were many disagreements. By 1919 the army had reconditioned seventy-six ships, at a total cost of $4.6 million. This figure was far below the owners' request, which was $10.1 million.[83]

The government also had to compensate the shipowners for the lost profits that would have been earned by the ships if they had remained in civilian hire. This toll was known as a "demurrage" and was calculated on a daily basis—in effect, a kind of imputed rent. But how much would these ships have earned? Opinions differed considerably. Shipowners thought that they should be compensated at the comparatively high rates that pertained during the war, but the government felt that peacetime tariffs were appropriate.

The history of the SS *Manchuria*, a vessel owned by the Atlantic Transportation Company, exemplified both problems. When the *Manchuria* was returned to its owners after wartime service, they requested $1.06 million in repairs. An independent marine surveyor declared this figure to be "grossly excessive," and the government cut it to $450,000. During its two years of military service, the army paid the owners $2.35 million in demurrage, a figure that the army later admitted was considerably less than it would have earned in civilian traffic.[84]

In determining the "reasonable cost" of reconditioning a ship, or the "reasonable demurrage," government administrators were necessarily exercising their individual judgment, as they had when they decided that 10 percent was a "reasonable profit" on a cost-plus shipyard contract. Unfortunately, there was no national criterion of "reasonable profit," and this deepened the profiteering disagreement. In May 1917, the General Munitions Board made an attempt to establish a national standard. The Board appointed a committee intended to represent "various geographic sections of the country" and charged it vaguely with ascertaining the "proper percentages" of profit in each locality and the "usual percentages in various kinds of industries." As a consequence of this work, a governmentwide estimate of 10 percent as a measure of "reasonable profit" emerged. This idea, unfortunately, was more easily conceived than it was implemented.

There were several problems with the 10 percent paradigm. In a cost-plus-fixed-fee contract, the fixed fee was based on 10 percent of the expected cost of production. The true cost, as shown by the Liberty motor contracts, could turn out to be significantly below the estimated cost, producing a much higher yield. In a cost-plus-percentage-of-cost contract, the contracts were not always written at the 10 percent level. The Dayton-Wright Aircraft Company, for example, built Wilbur Wright Field near Dayton, Ohio. This company received a fee of 15 percent of all costs, including leasing land for the base, for constructing the buildings, and also for grading the soil. On the other hand, the Engle Aircraft Company built airplanes in Cleveland, and its contract stipulated a 10 percent fee. There was no clear or cogent reason why there should be a disparity between building an airfield and building an airplane, or between working in Dayton or working in Cleveland.[85]

A second question that had to be decided was the taxability of the 10 percent markup. A group of shipbuilders argued that these cost-plus contracts ought to be tax-free, since they had little choice but to accept the government offer. Secretary Daniels would have none of this, and the profits remained taxable. There was also the sticky question of subcontractors. The Marlin-Rockwell Company of New Haven, Connecticut, was an old-line arms manufacturer. It agreed to build a new plant to load bombs, organizing for this purpose a subsidiary called the Marlin-Rockwell Loading Company. The latter received a contract of the cost-plus 10 percent variety. The Loading Company nevertheless relet the contract to the Fred T. Ley Company for cost plus 3.25 percent. The difference of 6.75 percent remained with Marlin-Rockwell simply for supervising the subcontractor. Some doubted that this was reasonable, particularly when the officers of the Ley firm doubled their salaries for the duration of the contract.[86]

Besides deciding on what constituted a "reasonable profit," a further challenge in administering the cost-plus contracts was determining "reasonable cost." Each contract required a number of supervisors whose job was to monitor performance by including "reasonable costs" and excluding "excessive costs." Exactly which costs were reasonable and which were excessive was often a matter of considerable dispute. In every cost-plus contract, one or more government auditors had the responsibility of reviewing expenditures.

Cost-plus contracts became a bureaucratic aggravation. Certain costs were almost always allowable as "reasonable": these would include direct expenditures for labor and materials. The perplexing questions were raised about the contractor's overhead. Plant depreciation, repairs, taxes, insurance, power, fuel, and royalties were generally not disputed. Salaries of management and various "extra supplies" or "extraordinary costs" were not so easily acceptable.

Some corporations sought to gain advantage from the cost-plus contract by inflating salaries. The New York Shipbuilding Company, for example, increased its president's salary by 100 percent immediately upon receiving a cost-plus contract. At the Standard Aircraft Corporation, H. B. Mingle, its president, was paid $62,000 per year. He was a lawyer with no prewar experience in the aeronautical industry. The president of Bethlehem Steel, E. G. Grace, raised his compensation dramatically by paying huge bonuses. According to standing company policy, Bethlehem's president received a salary that was well below the norm for a comparable firm, but instead benefited from an unusually generous bonus system. Grace's prewar salary of $12,000 was far below the going wage for his level of management, which was about $75,000. When an avalanche of war profits poured in, on the other hand, Grace was in a very advantageous position; his bonus in 1917 was $1,501,532, and in 1918 it was $1,386,193. Grace's situation was unusual but not entirely atypical. At Atlas Powder the president went from $16,000 in 1914 to $56,179.94 in 1918, and at Goodyear Tire and Rubber Company the president went from $12,000 in 1914 to $75,000 in 1917. The army established an "interdepartmental cost conference"—a kind of salary police—to rule on the size of salaries in cost-plus contracts. In the case of the Marlin-Rockwell bomb loading contract, which was cost-plus, the government rejected as salary any payment above $15,000 for the general manager. In so doing, the government anticipated the salary control plan adopted during World War II by executive order of Franklin D. Roosevelt. It did not attempt to thwart the Bethlehem bonus program, however.[87]

Government watchdogs also attempted to control the padding of overhead charges. The Grand Rapids Airplane Company, for example, was cited for charging expensive dinners and high directors' fees against overhead and for seeking deductions for depreciation of office furniture. Auditors discovered that executives of the Duesenberg Motor Company sought to charge the government for $11,000 in unnecessary travel expenses that even included $171 for cigars. At the Dayton-Wright Aircraft Company, Signal Corps auditors disallowed charges for baseball uniforms for employees, for diplomas for expert shotgun practice for plant guards, and for free tires and gasoline for commuting workers. When the company objected, one beleaguered officer laconically observed, "We get along beautifully with the Dayton-Wright people as long as our officers approve all their purchase requests."[88]

Some of these expenditures were clearly lavish. The Standard Steel Car Company had a contract to make railroad cars. It constructed a $350,000 hotel featuring ornate private dining rooms for executives and employees. Accommodations were entirely rent-free, with operation of the entire building charged to the government. An immense powder plant built at Nitro, West Virginia, by the Thompson-Starrett Company was a complete city. Workers got double-time for overtime, plus free housing, medical care, pharmaceuticals, utilities, and laundry service. Tenants also received free cooking utensils, with the government replacing any broken dishes. Employees could also charge the government for free boots and coats, which could be replaced when worn out.[89]

Controlling spending under cost-plus contracts was a difficult and unpleasant task. Contractors chafed at the constant supervision. William F. Carey, a railroad builder, denounced cost-plus as "the most unsatisfactory kind of contract that ever was let" because of the constant interference. These conflicts sometimes reached foolish extremes. At its giant acetone complex at Muscle Shoals, Alabama, the Air Nitrates Corporation found the works invaded by twenty secret agents of the Bureau of Investigations. In retaliation for the raid, the company accused the Justice Department men of drinking the whiskey they confiscated on the premises and hired forty investigators to spy on the G-men. This expenditure was then billed to the Ordnance Corps, which approved the charge because it believed the Justice Department detectives were impeding production.[90] This was a classic example of the perennial clash between the conflicting goals of limiting defense contractors' rewards, as desired by civilians, and of rewarding production, as desired by the army. As usual, the military won.

The difficulties with cost-plus contracts furnished strong reasons to develop better arrangements. The government had several arrows in its quiver, but none was attractive. Instead of issuing a cost-plus contract, there was legal authority to issue a compulsory production order. This required a contractor to accept a contract at a price fixed by the controlling federal authority. The War Department issued about 1,000 such orders, and the Navy Department was even fonder of them, issuing 3,342 compulsory orders. Each of these directives carried with it a written pledge to the recipient: "You are assured of a *reasonable profit* under this order."[91] Since government regulators doubted their ability to compel performance, other approaches were the instruments of choice.

The government's favorite means of forcing business cooperation was the use of priorities. Since a firm had to be certified by the War Industries Board as engaged in vital war work in order to receive raw materials, power, or transportation, the WIB could strangle it by revoking its priority status. Chairman Bernard Baruch found no shortage of weapons to use against a recalcitrant firm. "You can choke it to death, deprive it of transportation, fuel, and power, divert its business, strengthen its rivals," he explained. Baruch later recalled how this had worked in practice. One reluctant firm, which had dared to refuse a military contract on the government's terms, was stripped of all access to coal, steel, and rail transportation. "That was pretty persuasive," he remembered smugly. "Then they acceded to our wishes."[92]

The weapon of last resort was commandeering. Under wartime rules the government could simply send troops to confiscate materials or even entire factories. In order to thwart hoarding, the navy seized the merchandise of 238 warehouses, 49 banks, 553 forwarding agents, and 223 exporters. The army commandeered six different defense plants: those of the Liberty Ordnance Company, Bigelow-Hartford Carpet Company, Hoboken Land and Improvement Company, Smith and Wesson, Inc., Mosler Safe Company, and the Federal Enamel and Stamping Company. Although the government used this weapon reluctantly, it was readily available. When the War Industries Board authorized the army to commandeer an acetone distillery in Terre Haute, Indiana, it simply stated as its reason, "due to the attitude of management."[93]

The War Industries Board was caught in a predicament. It was forced to award contracts either on a cost-plus basis or on a fixed-price basis. In the former instance, costs could quickly escalate; in the latter, costs would immediately escalate, since the government had no choice but to purchase and thus possessed no bargaining power. Compulsory contracts and commandeering were heavy-handed devices that were certain to inspire resentment and thus at least partially to defeat the purpose of obtaining efficient production. The War Industries Board sought to avoid this quandary by assuming the power to fix prices. Robert S. Brookings was appointed to chair the Price Fixing Board, and he remembered how his group had snatched power. "We had no authority from Congress to do other than commandeer for our war needs. . . . We arbitrarily exercised an authority which we did not have by law. We threatened to commandeer concerns unless they abided by our decisions as to prices for the civil population, as well as to prices for our war needs."[94]

The trouble with the formal price fixing program was that it established a single price for each commodity. As Brookings explained, "We had to fix a price that kept alive the least efficient plants, the least efficient producer, and, of necessity, that price was a more profitable price for the more efficient producer." The War Industries Board briefly considered a more complex, multi-price system, but rejected the concept. The reason was that the government would have to buy all the production of the low-cost producers, leaving the civilian market to the high-cost producers. The result would be a dangerous incitement to price inflation, a dismal choice for any leadership.[95]

As a consequence of these considerations, the Wilson administration endorsed a new and broader excess profits tax. The special 12.5 percent tax on income from munitions sales of September 1916 that fell heavily on the Du Pont powder company was expanded in March 1917 to include all American firms whose profits exceeded 8 percent of invested capital. This definition of what was an acceptable war profit was also only temporary. Only seven months later, Congress decided that the excess profits tax should be set at a rate of 60 percent of any net income that exceeded 33 percent of invested capital. Herbert Hoover, who then chaired the U.S. Food Administration, advocated this approach, arguing that it would serve to prevent low-cost producers from obtaining windfall profits. Others praised the excess profits approach for its ability to keep both high-cost and low-cost producers healthy and in production.[96]

Unfortunately, the high rates of the excess profits tax served as a stimulus to evasion. Corporations utilized a series of maneuvers to avoid returning the high profits to the Treasury. These included raising officers' salaries conspicuously, purchasing superfluous advertising, and scrapping and replacing machinery before it was worn out. Mining firms took this occasion to invest heavily in exploratory activity. Other expenditures of questionable necessity, but which were nonetheless tax-deductible, included heavy new investments in welfare capitalism. The excess profits tax, like the cost-plus contract, contributed to high salaries and thus did little to arrest resentment.[97]

After compulsory orders, price fixing, commandeering, personal and corporate income taxes, and the excess profits tax, the last line of defense against profiteering was to limit the citizens' ability to make use of allegedly ill-gotten gains. In 1917, in obedience to custom, Congress imposed a wartime luxury tax. This tax was in part aimed at the pleasures of the upper class. It affected club dues, railway tickets, and Pullman berths, on the assumption that the affluent were more likely to join expensive private clubs and to travel in comfort. Telephone and telegraph messages were similarly treated, since telephones were not then broadly available. Perhaps to restrict the frivolities of youth, since these items could hardly have had much attraction to the war wealthy, chewing gum and phonograph records were also deemed to be taxable luxury goods.[98]

After all these attempts to deter the growth of wealth were made, the unresolved question was whether or not they were successful. The actual increase in wealth holding by individuals in upper-income brackets during the First World War produced contention that persisted long after the war ended. The most fractious aspect of this lingering quarrel was the oft-raised charge that the war had produced a crop of 23,000 new millionaires.

This allegation became a staple of progressive spokespersons who issued ringing calls for redress and offered dire predictions of the collapse of social order if the outrage went uncorrected. Congressman John M. Nelson, Republican of Wisconsin, without checking his arithmetic declared, "Government statistics show that this war multiplied the millionaire class by five, where there were 8,000 there are now 30,000 millionaires and seventy-one million of our people do not even own their own houses. . . . What happens when injustice, lawlessness, and oppression work their way to their final end? Behold Russia. Do we want our fertile fields trampled down either by the hired hosts of the Wall St. Tsars or land taken by the maddened multitude of half-starved men and women under the leadership of the desperate despots of revengeful Bolshevism?"[99]

The claim that the war produced 23,000 new millionaires can be traced to a socialist organization, the research division of the People's Legislative Service. It discovered, correctly, that the number of Americans who paid taxes on an annual income of more than $50,000 increased from 7,500 in 1914 to 19,000 in 1917, an addition of 11,500 names. The People's Legislative Service assumed that all of these persons were millionaires, since $50,000 represented a return of 5 percent on an investment of $1 million. The Service then doubled the 11,500 figure to arrive at its final estimate of 23,000 new millionaires. Its justification for this doubling was its presumption that there were as many millionaires in the $30,000 to $50,000 bracket as in the higher level. It offered no evidence to substantiate this adjustment, nor did it consider the possibility that someone with an income of $50,000 might not be a millionaire. Nevertheless, these figures were published in a number of labor journals, in particular *The Call,* and the claim gained considerable currency.[100]

In 1924 the House Committee on Military Affairs investigated this accusation and, after discovering the weak assumptions on which it rested, determined that it had "no well-defined basis in fact." Although the 23,000 millionaire assertion was clearly an overstatement, an accurate estimate is more difficult to determine. The total number of taxpayers in the highest income bracket (above $50,000 per year) for the wartime period is as follows:[101]

Year	Number of Taxpayers
1914	7,500
1915	10,600
1916	17,000
1917	19,000
1918	14,000
1919	18,800

These figures are based on Internal Revenue Service records and do not include the effects of taxes or inflation. They may be interpreted in various ways. The greatest increase in wealthy individuals came in the years 1915 and 1916, the period before American entry. Since the United States was then a neutral nation selling to belligerent nations, it confuses the meaning of "war profiteering" to label these persons as such. In 1917, the first year of American participation, the number of wealthy taxpayers increased modestly from 17,000 to 19,000, but in 1918 there was a decrease to 14,000. Thus in 1918, the only year in which the United States was a belligerent for almost the entire twelve months, there were actually 3,000 fewer wealthy taxpayers than in the last year of peace. Of course, to make an accurate estimate of the number of war millionaires would also require analysis of the composition of the wealthy cohort; some persons might have fallen out of the select group to be replaced by others whose wealth derived from the war. Such an analysis is not possible, but there is little reason to believe that such a transposition took place. It can only be concluded, therefore, that although a substantial number of persons—at most 11,500—became very wealthy during World War I, they largely gained their wealth before the United States declared war.

The question of profiteering during the Great War should correctly be posed in a different way: To what extent was there a shift in the distribution of wealth that favored the wealthy? Analyzed in this fashion, it becomes highly doubtful that any general new unfairness among the civilian population occurred. The poorest stratum of wealth holders benefited greatly from the wage increase associated with war jobs. Their total wealth increased from nothing or less than nothing to a positive figure. It is an economic maxim that during a period of high employment, wealth is more broadly distributed than during a period of low employment, and this was clearly in effect in 1917-18. Although wealth was not evenly distributed in the war period, there is no evidence that it was more balanced then than before the war; in fact, as far as the civilian population was concerned, wealth was slightly more evenly distributed during the war than it was before or after it.[102]

The record of mobilization during World War I includes a number of examples of extreme accumulation by American businessmen. The $1.5 million bonus obtained by E. G. Grace in 1917 may well be the largest direct increase in wealth as a result of war in American history. Yet despite certain instances of unusual accretion, there were also examples of selfless sacrifice that were the equal of any in previous conflicts. The most frequently remembered instance of sacrifice is that of the famous "dollar-a-year" men, who allegedly entered government service for only a nominal wage. Unfortunately, however, the dollar-a-year men were not the best illustration of selflessness; some of them received allowances of $2,000 to $3,000 annually to compensate them for the cost of relocating to Washington, D.C.[103]

The real sacrifices came from firms that held the line on prices despite alluring opportunities to hike them. The cotton goods manufacturers, for example, disciplined themselves to avoid price gouging. "We took it on ourselves in the early days," recalled one executive, "to see to it that the industry did not profiteer with the Government. . . . If unreasonable offerings were made, we used personal pressure and various other kinds of pressure to have prices made on a reasonable basis." Copper, zinc, aluminum, and steel producers cut their prices upon American entry, as described, and certainly this was not a result of compulsion. Some firms offered prices so senselessly low that Howard Coffin of the Munitions Board feared that they could never be maintained in the long run. Coffin described these prices as "unsound from an economic point of view."[104]

Although Coffin's view would seem apologetic, there were instances of excessively generous contribution. John Baskerville, a Chicago property owner, allowed the Ordnance Corps to use 12,000 square feet of office space without charge for the duration of the war. This was a six-story building, and its free use continued for nine months after the Armistice.[105] Although Baskerville's wealth was certainly not harmed by this donation, such sacrifice, if taken on a grand scale, would have been ruinous.

Other forms of patriotic sacrifice included voluntary cuts in prices on war contracts. The Allied Silk Trading Company obtained a contract to supply 6.9 million yards of silk cartridge cloth at cost plus a 7.5 percent commission. Allied's president, M. C. Migel, observed that army procurement officers had been cooperative and that payments had been prompt, so he cut his fee to 3 percent. By this decision, Migel surrendered $500,000 in profits. "The spirit we are endeavoring to display we hope will be emulated by others," Migel explained. It was an offering that would have gladdened the heart of George Washington.[106]

A similar decision was reached by the Yale and Towne Manufacturing Company of Stamford, Connecticut. This firm was a lockmaker before the war, but upon American entry it converted its production to artillery fuses. Yale and Towne was the low bidder at $68.95 each on a contract to make 4,000 artillery fuses. This contract was not cost-plus, but was based upon information obtained from the Frankford arsenal. After Yale and Towne commenced production, it found that it could make the fuses much more cheaply than it had estimated. It sent the Ordnance Corps an unsolicited offer to renegotiate the contract, and after discussing the matter with army procurement officers Yale and Towne cut their price in half.[107]

The Ford Motor Company was another firm with a responsible record. World War I was the first conflict in which steel helmets were employed, and General Pershing requested their immediate manufacture. The lowest estimate obtained by the army was 31¢ each, but Ford agreed to make helmets at cost plus 10 percent, while guaranteeing that the price would never exceed 31¢. Ford made 955,516 helmets at an average price of 10.33¢ each, and after receiving payment returned to the army a check for $197,000. Ford engineers had discovered that a slight alteration in the specifications would actually strengthen the helmet while cutting production costs dramatically. Ford also held a cost-plus contract to produce artillery caissons that was expected to bring in about $13 million. At the time of the Armistice, about 75 percent of the caissons were finished and Ford could have collected handsomely—about $500,000—when the contract was terminated, but the firm waived this fee. This was not unusual, either. On Armistice Day, the War Department had about 30,000 contracts in effect, and most of them did not contain termination clauses. Most contractors simply released the government from any obligation to complete the contract. This may not have been a very significant concession, however, since some lawyers believed that the contracts were not enforceable against the government in any case.[108]

The wartime record of no other firm provoked more angry quarreling than that of the Du Pont Company. In 1916, the year when Du Pont benefited most from Britain's dependency, its net earnings shot up to more than ten times prewar levels. The primary source of this bonanza was powder sold to the Allies for one dollar per pound, or approximately twice the prewar rate. When the United States became a belligerent, the price of powder dropped markedly. Benefiting from the heavy investment in new production facilities, Du Pont was able to cut its prices to 51.3¢ per pound in May 1917 and to 43-46¢ in 1918. Du Pont prices were about the same as those charged by government arsenals and well below those of its competitors. This was no small achievement, considering that the price of cotton rose by two-and-a-half times during this period. As a consequence, the net taxable income of the Du Pont Company dropped precipitously. Taxable income, which was swollen by prewar sales, had been $127 million in 1916, but it fell to $78 million in 1917 and $13 million in 1918. Even so, this represented a handsome yield in relation to the untaxed prewar rate of about $5.26 million per year. From the point of view of the Du Pont family, by halving their prices they had demonstrated their patriotism amply. Irénée du Pont would later claim, inaccurately, that "we did not make any profit during the war and paid more in taxes than we made in profits."[109] From the point of view of Du Pont's critics, nevertheless, these returns were irresponsibly lavish.

To further complicate the issue, the Du Pont firm became engaged in a bitter dispute with Secretary of War Newton Baker over the cost of plant expansion. In 1917 the total U.S. capacity for the manufacture of powder was 480 million pounds, which was far below the expected need for 1919 of more than 600 million pounds. Plainly, new factories were necessary, and Du Pont proposed to build a huge new powder works with a capacity of one million pounds per day. This facility would cost $90 million, and it would become government property. Baker and Du Pont were unable to agree on an appropriate fee for building the factory, and their disagreement caused a three-month delay. Ultimately, the government decided to build two powder plants, one constructed by Du Pont and one by the Thompson-Starrett Company at Nitro, West Virginia. The latter firm was inexperienced in powder production, forcing it to rely on Du Pont designs for the new works. Du Pont graciously provided assistance to its competitor, but even so the Du Pont factory was completed before its rival. By the time the Nitro works were finished, Germany had surrendered, and there was no need for the plant.[110]

The controversy that swirled around the Du Pont firm summarized the defining themes of war profiteering during World War I. Du Pont was an eastern firm and therefore subject to western prejudices. It had earlier been convicted of violating federal antitrust laws, so its reputation was tarnished even before the war began. Its leading product, explosives, seemed at casual inspection to be destructive material rather than useful merchandise. The peculiar combination of Allied need and the German submarine blockade allowed Du Pont to reap enormous profits, and these gains had little if anything to do with the foresight of the company's managers, who nevertheless bathed in unseemly affluence. To many people, this newfound wealth was neither earned nor deserved.

Like other American manufacturers, Du Pont had charged foreign governments more for munitions than it charged American buyers, at least until the United States became a co-belligerent. Most of the Du Pont profits, therefore, were earned prior to American entry into the conflict. (For constructing five munitions factories for the U.S. government during the war, the Du Pont Company's net profit was modest—only .34 percent of the cost.)[111] Nevertheless, Du Pont had gained great wealth from the war, and it seemed reasonable to persons who did not distinguish between prewar profit and wartime profit that the firm must have yearned for war in order to fatten its earnings. When the firm became embroiled in controversy over how to build new production facilities, its reputation worsened. No longer merely the grasping "Powder Trust," Du Pont's name was now reviled as a bringer of war and a bloated scavenger that had gorged while patriots suffered. In the bitter afterglow of what was sarcastically termed the "war to end wars," Du Pont's achievements and its generosities would seldom be remembered. Its ample gains would seldom be forgotten.

8

Grave Objections

> I can vision the ideal Republic . . . where we may call to universal service every
> plant, agency, or facility, all in the sublime sacrifice for country, and not one penny
> of war profit shall inure to the benefit of private individual, corporation, or com-
> bination.
>
> WARREN G. HARDING, INAUGURAL ADDRESS (1921)

The great Armistice Day celebration of 1918 granted only a brief respite from
the war profiteering controversy. The mobilization of the American economy
for the Great War was a monumental task, and when Germany unexpectedly
surrendered, production was racing at full speed. The nation had paid too little
attention to preparedness until it was nearly too late, and it gave no attention
to the problems of reconversion until the war ended. Just as the United States
had never before simultaneously undertaken building projects as extensive as
the war demanded, it had also never attempted a smooth reconversion. The
process of returning to civilian production proved to be difficult and demanding. Old antagonisms that had been largely set aside for the duration of the
war reemerged among the unsettled economic conditions that followed the
Armistice.

The perennial war profiteering dispute appeared in several new forms in
the postwar decade. Besides the immediate questions of who should gain and
who should lose as a result of the sectoral shifts associated with demobilization, the legacy of the wartime profits dispute forced its way into the
Republican politics of the New Era, the decadelong quest for an effective arms
control policy, and even into the literary imagination of Jazz Age culture. The
postwar generation hoped to obliterate the scourge of war by legal action (the
Kellogg-Briand Pact), by parliamentary debate (the League of Nations), by
arms reduction (the Washington Naval Treaty), and not least of all by eliminating war profit. As Herbert Hoover, who as much as anyone personified New
Era culture, said in 1924, "The more evident it is that the whole nation will be

put in the storm and made to bear its share of sacrifice, the less likely we will be to go to war."[1] The crusade against profiteering became a prime example of the persistence of the Progressive voice in the 1920s and the 1930s.

The first profiteering issue of the 1920s was far less lofty and far more tangible than the larger question of eliminating war. It was the disposal of surplus military supplies. In late 1918 the cancellation of war contracts produced a national inventory imbalance. The United States government had become the possessor of vast supplies of military commodities. Disposal of these supplies would test the nation's ability to maintain the wartime cooperation between the government and business. The idea of building a cooperative state in the 1920s that would replace the friction which marked business-government relations during the Progressive era has become known as associationalism.[2] Exponents of associationalism would find the problem of controlling war profits as thorny as had the Progressives.

Many of the surplus military articles had considerable value on the civilian market. Others were of lesser utility or nearly worthless, and there was ample room for disagreement as to which were of use and which were not. The government had little notion as to how to dispose of its vast holdings efficiently. Should it simply flood the market with surplus commodities, or should it withhold some of its inventory in hopes of obtaining a better price? Should it sell its supplies on the wholesale market, or should it attempt to establish retail outlets? Whichever choice it made, some would gain and others would lose. As always, opinions differed on who deserved to benefit from the sale and who did not. One resentful doughboy saw fraud afoot: "The men whose chief object in the war remained, the traitor, the crook, the grafter, the profiteer, all those who had a covetous eye upon the teeming warehouses that Uncle Sam was to empty."[3]

As in the wartime period, copper was again a centerpiece of the debate. On Armistice Day in 1918, the U.S. Army Ordnance Corps had on hand an inventory of 100 million pounds of copper. This was the equivalent of one month's supply for the entire domestic market at prewar production levels. If the government sold this supply quickly, the price of copper would plummet. Since wage rates in western mining towns were linked to the price of copper, wage cuts would follow price cuts, and strikes would be highly probable. At the request of the copper producers, the government agreed to parcel out its copper holdings at a rate of five million pounds per month. The metal would not be sold on the open market but would be sold back to the copper producers. This would allegedly protect the government's interest by ensuring a good price, and at the same time it would protect the copper miners from wage reductions and the copper producers from depressed profits.

The problem with this plan was that copper was a highly concentrated industry. The transaction was handled by the Guggenheim-dominated United Metals Selling Company, and by using that firm the government was strengthening an anticompetitive practice.[4] The copper firms had benefited greatly by the unusually heavy demands of the preparedness period, and it was unclear to many why the government should help to preserve these monopolistic gains by protecting the copper firms from the vagaries of the peacetime period. Since the government also owned 50 percent of the total American supply of platinum, which it sold promptly at the market price, and since it did not dally in disposing of its inventories of lead, brass, antimony, tin, or steel,[5] the policy of delayed disposal appeared to Progressives to represent preferential treatment for both copper miners and copper monopolists.

Delayed disposal was also the pattern followed in the leather industry, to the great dismay of farmers who anticipated an opportunity to purchase harnesses cheaply. The quarrel over leather disposal originated during the early part of the war. The Council of National Defense organized two committees to coordinate the army's purchasing of leather goods. One of these had control over shoes, and the other supervised other leather equipment. Both committees were made up of executives of the leather industry, who fixed prices and assigned contracts. There was no competitive bidding, although huge amounts of leather products were purchased. The army took delivery of 32 million pairs of shoes, or nearly ten pairs for each soldier. The service also received 500,326 sets of double harness and 110,828 sets of single harness, although it owned only 580,182 work horses and of these only 67,948 were sent overseas and another 96,000 died. Procurement officers also purchased 945,000 saddles, 2.85 million halters, and 585,615 saddlebags, although the army owned only 86,418 cavalry horses. Army horses ate well, too; the army bought 2,033,204 nose bags during the war, which was roughly four bags for each horse! Indeed, the only limitation on the amount of leather products purchased was a shortage of hides; there were enough outstanding orders at the close of the war to have consumed 300,000 more hides than were available.[6]

When the war ended, this huge inventory of new harness owned by the army threatened to ravage the leather industry if it came on the market. Because leather executives were strategically placed to influence the process of demobilization, they moved decisively to prevent a glutting of the market. The key individuals were Colonel George B. Goetz and Major Joseph C. Byron. Both Goetz and Byron had been prominent leather executives before the war but had enlisted in the army in order to coordinate its leather purchasing. (In addition to his salary as an army colonel, Goetz drew a retainer of $100 per month from his former employer throughout his military service; he was a "dollar-a-year man" in reverse.)[7]

Goetz and Byron took active steps to thwart the sale of military harness. Their first step was to hinder its availability for sale by the Surplus Property Division. This was accomplished by simply refusing to certify that most of the harness was unneeded. To convince surplus property officers that the rest had no market value, Goetz and Byron kept a mud-splattered set of harness in their office to demonstrate its worthlessness as surplus. A third approach was an attempt to dump the government leather on the world market; all U.S. military attachés were ordered to try to find buyers, but none could be found. The real purpose behind these steps was revealed when a transcript of an address by Colonel Goetz to a convention of tanners in Chicago was published. "I have been very much disturbed," Goetz lamented, ". . .that some speculator should get hold of the material . . . and go and dump it in some one man's territory, maybe ruin his business for five years, so that what we are aiming to do is to pass this buck back to the consumer . . . and to spread it thin over the country, not dump it all in one man's territory."[8]

In order to "spread it thin," Goetz and Byron resigned their army commissions and organized a new firm, the U.S. Harness Company. This company was capitalized at $170,000, based on contributions of $10,000 from seventeen leather manufacturers. The U.S. Harness Company was to purchase the surplus harness, recondition it for civilian use (charging the customers a commission of 40 percent of the price), and resell it gradually to civilians so that the stability of the market was preserved.[9] Not only would no individual manufacturer suffer from a glutted market, but also the informal division of the market would be preserved against the threat of a rash of competition. Commercial self-protection through delayed disposal could also prove to be a profitable sideline.[10]

The actions of the leather executives in seeking to hold the harness off the civilian market aroused the ire of the nation's farmers, who were its principal purchasers. Farmers, not coincidentally, were among the chief critics of American mobilization for the war, and the farm economy had been heavily damaged in the postwar period. Farm belt congressmen complained, and when President Warren G. Harding learned of the matter, he canceled the contract between the army and the United States Harness Company by executive order. Harding's order was roundly applauded by the leaders of the American Legion, the newly formed and highly vocal organ of the millions of veterans of the American Expeditionary Force.[11]

The farmers and veterans were not fully appeased, and the recriminations continued. In 1923 the Justice Department indicted Goetz and seven others for conspiracy to defraud the government. At their trial in 1924 the leather executives presented a letter from former secretary of war Newton Baker showing that he had approved their plans. The presiding federal judge then directed a verdict of not guilty. Although the defendants avoided conviction, the damage to their business was permanent. During the immediate postwar years, the leather business became the least profitable industry in the United States, showing an average annual loss of 5.4 percent per firm. Delayed disposal, like other associationalist policies, failed when it ran afoul of such powerful interest groups as farmers and veterans.[12]

Samuel F. Colt (1814-1862), founder of the Colt's Patent Fire Arms Manufacturing Company, was the first American armsmaker to become a millionaire. Colt made his reputation during the Mexican War as the manufacturer of the highly regarded "Walker Colt" revolver. Although raised in New England, Colt hated abolitionists, and during the 1850s he sold guns indiscriminately to both Northerners and Southerners. Three days after the Confederate attack on Fort Sumter in 1861, Colt shipped his last order of five hundred guns to Richmond in boxes marked "hardware." Shrewdly perceiving that projections of a short war were wildly mistaken, he tooled up for a five-year conflict. Earnings of the Colt firm quadrupled from $237,000 a year to more than a million (courtesy Connecticut Historical Society).

(*Above*) Samuel F. Colt built his mansion in Hartford, Connecticut, and named it *Armsmear*. Upon his death, his holdings in the company that bore his name were worth about $2 million, and his entire estate was appraised at $3.2 million, an enormous sum for 1862. After the Civil War, earnings of the firm declined, but World War I brought new prosperity. By then a manufacturer of machine guns as well as pistols, the net income of the Colt firm soared from $322,930 in 1911 to $5,797,794 in 1917, a nearly eighteen-fold increase (courtesy Connecticut Historical Society). (*Below*) Every military conflict brings a shift in production, and for a nimble investor this reorientation offers tempting opportunities. At the beginning of World War I frenzied buying drove defense stocks skyward. The price of Colt's Fire Arms stock, for example, rose from a low of $160 a share in 1914 to a high of $940 in 1915. As this late 1915 scene reveals, some Wall Street clerks were so busy filling orders that they had to sleep in their offices in order to meet the demand. Great gains were indeed possible: five days before the U.S. war declaration in 1917, the Colt firm declared a dividend of $75 per share, then split its stock eight for one (*Current Opinion*, December 1915).

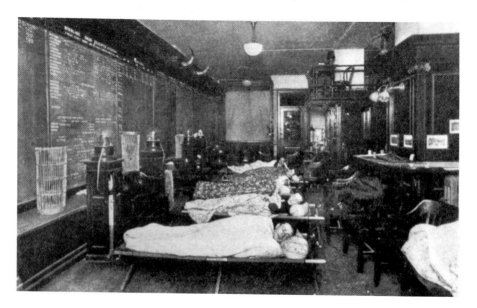

(Right) Marcellus Hartley Dodge (1881-1963) was a grandson of Marcellus Hartley (1828-1902), the Union's arms agent during the Civil War. Hartley, the founder of the Union Metallic Cartridge Company, held interests in fourteen other firms, and upon his death left to his grandson, then twenty-six, an estate valued at $60 million. In 1915 Marcellus Hartley Dodge controlled the Remington Arms Company, the leading American manufacturer of small arms, and this position enabled him to obtain a contract to produce the Lee-Enfield rifle for the British Army. With only this contract as an asset, Dodge organized the Midvale Steel and Ordnance Company, whose stock offering attracted investors in great numbers. Within a few days Dodge sold out his Midvale position, realizing a capital gain of $24 million (courtesy Remington Arms Company).

(Below) Eugene G. Grace (1877-1960), president of Bethlehem Steel Corporation, the nation's leading shipbuilder. Grace became a figure of controversy because of his gains during World War I. Although underpaid for his position, his prewar base salary was only twelve thousand dollars. Nevertheless, he received bonuses of $1,501,532 in 1917 and $1,386,193 in 1918—probably the largest single wartime wage increase ever recorded in American history. Evidently undaunted by criticism of his accumulations during the first war, when World War II stimulated steel sales Grace nearly doubled his salary from $272,224 in 1939 to $537,724 in 1941. This was the fourth largest salary in the United States. Grace's personal share represented 35 percent of the total increase in salaries paid to the sixteen top officers of Bethlehem Steel in 1941 (courtesy Bethlehem Steel Corporation).

Harold Gray, an angry former doughboy turned cartoonist, drew the immensely successful comic strip "Little Orphan Annie." A principal character was a guilt-stricken war millionaire, Oliver "Daddy" Warbucks, who converted a small machine shop into a very profitable munitions plant. Annie and "Daddy" appeared in 345 newspapers having a daily circulation of sixteen million copies.

ARE YOU THIS KIND OF PATRIOT?
—Morris inthe Sacramento *Bee*.

TWO KINDS OF AMERICANS.
—Kirby in the New York *World*.

Copyrighted by the New York Tribune Association.

OUT OF THE WEST.
—Darling in the New York *Tribune*.

OUR MODERN PAUL REVERE.
—Fitzpatrick in the St. Louis *Post-Dispatch*.

Profiteering has appeared almost unfailingly as a political issue during wartime. In "Are You This Kind of Patriot?" and "Two Kinds of Americans" (both 1917), a sharp contrast is drawn between the sacrifice of the wounded at the front and home-front gains based on fraud and greed. In "Out of the West" (1917) and "Our Modern Paul Revere" (1929), political cartoonists employ a porcine metaphor.

In the 1920s Edwin A. Link (1904-1981) *(above)* and his brother George T. Link (1897-1979) *(below)* of Binghamton, New York, invented the Link trainer or "Blue Box," a full-scale, working model cockpit used to train student pilots to fly by instruments. The Link trainer drew limited interest from the Army Air Corps until World War II, when the company's employment rolls expanded to fifty times its prewar figure. The Link brothers, whose salaries were but $12,000 and $15,000 in 1939, multiplied their salaries sixfold in 1941 to $72,000 and $90,000. Royalties paid on their jointly held patent rose from $87,000 in 1939 to $1,179,000 in 1941. In 1941 Edwin A. Link's net income of $849,976 was the seventy-fourth largest in the United States. In 1942 the firm paid salary bonuses to its employees in the amount of $352,000 (courtesy Edwin A. Link Archive, Roberson Museum and Science Center, Binghamton, New York).

(Above) Speaking for millions of aggrieved doughboys, during the 1920s the American Legion championed a scheme of "taking the profit out of war." The Legion claimed that if its plan had been adopted beforehand, the Great War would have been shortened by six months and its costs reduced by $10 billion. In this artistic rendition of the principle, Mars calls to service and demands equal sacrifice from every able-bodied man: soldier, sailor, businessman, farmer, and laborer. Significantly, the God of War fixes his gaze upon a desk-bound executive (*American Legion Weekly,* September 29, 1922).

(Right) Robert E. Gross (1897-1961), president, Lockheed Aircraft Corporation. As defense spending rose in anticipation of American entry into World War II, Lockheed sales skyrocketed, largely because of the success of its high-performance fighter, the P-38 Lightning. Gross's salary escalated from $27,400 in 1939 to $125,000 in 1941, an increase of 456 percent. Lockheed thrived during the early years of the cold war, and when Gross died the firm he had purchased for $40,000 in 1932 had assets of $532 million (courtesy Lockheed Martin Corporation).

(*Right*) Glenn L. Martin (1886-1955), president, Glenn L. Martin Corporation, Santa Ana, California. Martin was a pioneering designer and manufacturer of light bombers, including the B-26 Marauder of World War II. His salary rose from $33,970 in 1939 to $60,260 in 1941, an increase of 77 percent. Martin's net income in 1941, $779,476, was the eighty-ninth largest in the United States. Upon his death his estate was appraised at $14.3 million (courtesy Lockheed Martin Corporation).

(*Right*) Reuben H. Fleet (1887-1975), founder and president, Consolidated Aircraft Corporation. On the eve of World War II, his firm produced one of only two American four-engine heavy bombers, the B-24 Liberator. Consolidated also made the only long-range flying boats then in production, the twin-engine PBY Catalina and the four-engine PB2Y Coronado. From 1939 to 1941 the Consolidated payroll increased from 1,500 to more than 30,000 workers, and Fleet's salary more than tripled, rising from $20,000 to $62,500. Army and Navy leaders distrusted Fleet's autocratic managerial style, and they encouraged him to sell his interest in the firm. Anticipating rising tax rates and anxious to spend time with his young new wife, he did so in 1942, receiving nearly $11 million for his share, including a dividend of $880,000. Fleet's income for 1941 was the fourth largest in the United States. With the blessing of the Roosevelt admin-istration, Fleet's replacement was Tom Girdler, chief executive officer of the Republic Steel Corporation, whose superb managerial skills were rivaled only by his bitter enmity toward the New Deal (courtesy San Diego Historical Society, Photograph Collection).

The government's policy of delayed disposal did not entirely work against agricultural interests, however. The Quartermaster Corps possessed an inventory of $10 million worth of canned vegetables, and the National Canners' Association prevailed on the quartermaster general, R. E. Wood, to withhold this stock from public sale in order to support vegetable prices. Since food prices had risen significantly during the war, consumers understandably disputed this decision. Similarly, representatives of the lumber industry were afraid that the army would dump surplus lumber on the civilian market. They warned that this would force the closure of lumber mills. Reminding the nation of the radicalism of the lumberjacks, the mill owners warned that such a shutdown would lead to mass unemployment "and probably bolshevism." Although local lumberyards clamored for the early release of the surplus, in the worried days of the Red Scare of 1919-20, the government did not wish to risk labor trouble. In 1922, when the threat of Bolshevik violence seemed less imminent, eleven lumber executives were charged with conspiring to defraud the government by failing to use due diligence in disposal of surplus lumber.[13]

The army's disposal methods attracted scrutiny for several other reasons. The Holt Manufacturing Company, the parent of the Caterpillar Tractor Corporation, had successfully inserted in its contract to furnish 2,000 military tractors a clause that would have prevented the sale of these tractors as war surplus. Despite the illegality of this clause, the Ordnance Corps officer in charge accepted its inclusion. There were several other industries in which the army sold its inventory back to the manufacturers. These included mosquito bars, canvas duck, and cigarettes. In the latter case, the army discovered that it owned 12 million packages of Camel brand cigarettes. It sold these back to the R. J. Reynolds Company at 6.33¢ per package, only to find that it had underestimated its requirements by 2.5 million packages. The army then had to repurchase these cigarettes from the firm at a profit of 2¢ per package.[14]

The process of disposing of surplus airplanes unveiled new issues that produced misunderstanding. The army sold as surplus 4,608 aircraft engines and 2,716 training aircraft. The sale of these items brought a return of $2.72 million, which was only 13 percent of the manufacturing cost of $20.8 million. Most of the planes were sold back to the manufacturer, the Curtiss Aeroplane and Motor Company, at vastly reduced prices. One type of plane, the Curtiss J-1, had cost the government $4,250 each but brought only $200 as returned merchandise. The other, the famous Curtiss JN-4 Jenny, had cost $4,000 but again brought only $200 on the postwar market. Many prospective airplane buyers hoped to obtain these craft directly from the air service at firesale prices. They objected to the government returning the planes to the manufacturer, which would boost the price before the planes were made available to the public. The army rightfully refused to sell directly to civilians because it was afraid that the planes might be in a dangerous state of disrepair. By selling the planes wholesale, the army could ensure that the planes were reconditioned before being placed on the civilian market. The disappointed civilian aviators suspected a conspiracy to prop up prices.[15]

After the war, the United States government found itself the owner of factories of potentially immense commercial value. The army held title to a huge new smokeless powder plant at Old Hickory, Tennessee, which had been built by the Du Pont Company at a cost of $83 million. Unfortunately, there was little or no commercial market for this product, and the army had to sell the plant for a fraction of its price. The director of sales of the War Department, Ernest C. Morse, agreed to sell the Old Hickory facility to the Nashville Industrial Corporation for $3.5 million. Although this bid was nominally the highest, the sales contract contained certain provisions that reduced its value. Accordingly, Morse was indicted by a grand jury for fraud. Augmenting wealth by the acquisition of factories at bargain-basement prices was a new and highly controversial form of profiteering.[16]

The demobilization period also brought forth a new round of charges of profiteering by the provision of goods of substandard quality. The most consequential of these charges concerned the sale of rotten fish to the army. The summer of 1918 was unusually hot in Alaska, leading to the spoilage of about five million cans of salmon or 22 percent of the pack. The War Department subsequently forced the return of the tainted fish, recovering about $8,600,000 from the canners. Although a congressional committee investigated the incident, it was never able to determine conclusively whether the canning companies knew of the spoilage when the product was delivered. The Republican majority, sensing an opportunity to damage a Democratic administration, condemned the War Department for mishandling the purchase. Wilson's supporters, on the other hand, praised the military for discovering the spoilage and recovering the funds. The Democrats pointed out that the bad salmon never reached the doughboys, and they reminded the public of the "embalmed beef" scandal of the Spanish-American War, in which tainted beef had actually been served to the troops during a war managed by Republicans.[17]

Charges of bad products added fuel to the heated complaints of farmers about the management of the army's horses. The Henry Moss Company had supplied many of the army's branding irons, and it was discovered that these irons were adulterated with aluminum and zinc. This fact disclosed that the supplier had used scrap copper instead of pure copper in making the implements, and the presence of these impurities meant that the irons had no resale value whatsoever, even as scrap. Although this incident was minor in itself, some saw it as another facet of a broad conspiracy to defraud the American public through equipment of degraded quality.[18]

The complexity of the modern industrial economy was seldom so well illustrated as in the process of settling outstanding war contracts that were interrupted by the Armistice. Claims presented by contractors due to suspended defense contracts were adjudicated by the War Department's Claims Board, which was the province of the assistant secretary of war, Benedict Crowell. A debate developed within the government as to how carefully these claims should be examined. The army preferred to settle outstanding claims promptly, even if that meant generously, because to delay settlement until a careful review was possible would be administratively difficult. Key witnesses would disappear, and often the officers who were knowledgeable about the details of contracts, particularly verbal contracts, would leave the service. Accordingly, the army spent little time scrutinizing contractors' claims. By September 1919, less than a year after the Armistice, Major General George W. Burr, the director of purchase, storage, and traffic, testified that his office had already processed 24,000 claims for settlement of informal contracts. The army had about 30,000 "accountants and others" in the field to investigate the claims, and about 92 percent of the claims were approved.[19]

The army's haste at settling these claims kindled political opposition. One charge was that Secretary Crowell had approved contractors' claims at the rate of $35 million per day. Reacting to congressional and public criticism, the army commenced a supplementary review of the process. This review stretched into years, and its continuation provided an ongoing supply of fuel for the critics. By 1922, Brigadier General Kenzie W. Walker, the chief of finance of the army, reported that 17,000 contracts out of a total of 150,000 had been audited. General Walker estimated that $46 million was due to the government, but that only $15,138,000 had been collected. The War Department was endeavoring to collect the remaining $26 million; of this amount, $4.4 million in claims had been turned over to the Justice Department in hopes of recovering the funds by litigation.[20]

These audits produced some substantial, although disputed, charges of profiteering. One government auditor testified to a congressional committee that he disallowed about fifteen to twenty claims a month, saving an average of $500,000 during each period. The largest such disputes were in the aircraft industry, where wartime results had been highly disappointing. The chief of the air service, Major General Charles T. Menoher, told President Harding's secretary of war, John W. Weeks, that after auditing only six contracts he had found $16 million in overcharges and asked for authority to expand the audit. Eighteen months later, General Menoher reported that after auditing thirteen contracts he recommended the recovery of $28.9 million. The largest claims for repayment were against the Dayton-Wright Airplane Company, for $2,555,000, the Wright-Martin Aircraft Corporation, for $4,706,000, and the Lincoln Motor Company, for $9,188,561. The sheer size of these figures convinced some of the accuracy of the charges of profiteering.[21]

In seeking to recover profits from war contracts, Justice Department lawyers advanced an imaginative new concept of profiteering control. The government admitted that it had signed binding contracts for the production of war matériel, but it now claimed that these contracts could be voided retroactively. The grounds for such a suit were that contracts which produced a profit defined as "extortionate" were not enforceable against the government.

The test case for this concept was the Liberty motor contracts held by the Packard Motor Car Company. The various Packard contracts had cost the government $40.8 million, of which $28.9 million represented manufacturing costs and $11.4 million were profit to the contractor, a margin of 39.4 percent. After dispatching a corps of forty auditors to the Packard plant, the government sought to recover $6.58 million of these "extortionate profits." This would reduce the profit margin to 11.76 percent. Packard objected to the description of these profits as lavish, arguing that the firm's profits for the three years of the Liberty engine contract were below those for the last prewar year and the first postwar year. Packard also noted that it had been required to accept a flat-price contract at $4,000 per motor, although it had requested a cost-plus arrangement. The cost-plus contract that Packard preferred would actually have yielded a profit that would have been $2,311,511 less than the army's fixed fee system.[22] The courts ruled in favor of Packard. In most such instances, compromise solutions were found, with the companies usually returning part of the payments.[23] These compromises offered alluring opportunities for the critics of military procurement to practice their trade, and they seldom overlooked them. The performance of the Harding administration presented many more such openings.

The profiteering issue became intricately entwined in postwar politics. In 1919 the Republican-controlled Congress launched several wide-ranging investigations that scored the Wilson administration for allegedly excessive spending and other mismanagement of weapons procurement.[24] In their successful presidential election campaign of 1920, the Republicans charged further that the Democrats had failed to enforce laws that could have limited profiteering. President Warren G. Harding, in a successful attempt to win the support of the millions of former doughboys, promised to pursue the profiteers.

In his inaugural address in 1921, Harding deplored how the war had "uncovered our portion of hateful selfishness at home." He added nobly but unrealistically that "if . . . war is again forced upon us . . . I can vision the ideal republic . . . where we may call to universal service every plant, agency, or facility, all in the sublime sacrifice for country, and not one penny of war profit shall inure to the benefit of private individual, corporation, or combination. . . . There is something inherently wrong, something out of accord with the ideals of representative democracy, when one portion of our citizenship turns its activities to private gain amid defensive war while another is fighting, sacrificing, or dying for national preservation." Harding called for "universal service" in the next war so that there would be "no swollen fortunes to flout the sacrifices of our soldiers."[25] The new president's words perfectly described the hopes of the antiprofiteering forces for the next two decades. The performance of his administration illustrated well the difficulty of achieving them.

Despite his brief tenure in office, Warren G. Harding denounced profiteering on many occasions. "In all the wars of all time," Harding grandiloquently proclaimed at Arlington National Cemetery on Memorial Day in 1923, "the conscienceless profiteer has put the black blot of greed upon righteous sacrifice and highly proposed conflict." However, the president promised a Montana audience, "If war must come again—and God grant that it shall not—then we must draft all of the Nation.... It will be righteous and just ... if we draft all of capital, all of industry, all of agriculture, all of commerce.... When we do that, there will be less of war. When we do that, the contest will be aglow with unsullied patriotism, untouched by profiteering in any service."[26]

Although Warren Harding yielded to few others in his eagerness to denounce profiteering, the performance of his administration was considerably less stirring than his rhetoric. The pursuit of profiteers became the bailiwick of his attorney general, Harry M. Daugherty, who might have been a good choice as he had denounced the Wilson administration for failing to investigate profiteering. But Daugherty proved to have little stomach for the task, although the reasons for his timidity remain obscure. In part, Daugherty believed that the squabbles of the war years should be laid to rest, and in part he believed that to prosecute defense contractors aggressively would hamper recovery from the business depression that struck the nation in 1921. He also had secret reasons for delaying prosecution. "There are many things which I cannot talk about, and may never be able to talk about, which interfered with the work we have on hand," Daugherty wrote cryptically to a dissatisfied congressman.[27]

Nevertheless, Daugherty's foot-dragging angered congressional supporters of the returned veterans. Two were known for their unswerving loyalty to the political agenda of the American Legion, Royall D. Johnson of South Dakota and Roy O. Woodruff of Michigan. Johnson once declared, "I came back from France from a combat unit about two-thirds bolshevist myself." In what became identified as the "Woodruff-Harding controversy," Johnson and Woodruff called for Daugherty's impeachment and removal from office. Samuel Gompers and the American Federation of Labor echoed the veterans' demands. Reacting to the criticism, the attorney general agreed to expedite prosecution. In 1922 Daugherty obtained an appropriation of $500,000 to investigate 276 cases of profiteering, having a maximum potential recovery value of $192 million. Johnson and Woodruff considered the half-million dollar sum to be hopelessly inadequate, an election year ploy intended to head off a truly effective investigation. Nevertheless, the Justice Department established a War Transactions section that operated through 1926, pursuing profiteers. By 1923 the total of excess profits recovered was only about $4.4 million, and only seven profiteers had been indicted (one pleading guilty). These sums were far too meager to satisfy the indignant doughboys.[28]

The investigation continued until 1926. The collection of claims was afflicted by weak evidence due to the passage of time and by the financial instability of companies suffering from the postwar economic depression. Perhaps more serious yet, it suffered from a serious conceptual problem, the perennial difficulty of defining profiteering. "The line of demarcation between a bad bargain and an unconscionable sale is most difficult to draw," mused one investigator in 1924. "It is hard to better the general rule laid down in the leading case, which defines unconscionable consideration as 'that which shocks the conscience of the chancellor.' But such definition is most elusive when one seeks to apply it to a particular case."[29]

The results of the two and a half years of the War Transactions Section's work were meager. Of 25,000 total settlements entered into by the Army Ordnance Corps, the Section selected 3,600 for careful scrutiny. Of this group it chose 407 cases to consider for civil action. In eighty-five of these, the prosecutors decided the government had "no case." Forty-nine claims were compromised, twenty-five were deemed uncollectible, and in forty-four the contractors paid in full. Six cases were selected for criminal prosecution, resulting in four verdicts of "not guilty" and two convictions.[30]

The war frauds inquiry had even less success when it looked into the construction of cantonments, which had been the source of great wartime controversy. The Justice Department had serious doubts about reopening the cantonment dispute, but did so as the result of congressional pressure. Justice agents considered prosecution of the contractors who had built Camps Grant, Sherman, and Pike, but in each instance they decided against further action.[31] In 1926 the legal attack on profiteers concluded when its director happily announced that he could find only a "few instances of deliberate attempt to cheat and defraud the government."[32]

This could not be said of the Harding administration, which was renowned for its corruption, and the infamous Harding scandals involved the war profits issue in at least two incidents. Charles Hayden, a partner in the brokerage firm of Hayden, Stone and Company, wrote to Harding's secretary of war, John Weeks, to ask that something be done to "inspire confidence in a legitimate way" in order to bolster the stock market. Translated, Hayden hoped that Weeks would expedite an unpaid claim by the Wright-Martin Aircraft Company, of which Hayden was a director, for payment of $5.3 million for making aircraft engines. Weeks told Hayden to ask the air service for the money and hinted that if the army declined he would overrule it. Weeks may have acted improperly, but he was not indicted when the Harding scandals were prosecuted.[33]

A more serious matter concerned the attorney general himself. The Japanese-owned firm of Mitsui and Company controlled two American firms, the Standard Aircraft Company and the Standard Aero Company, through

Mitsui's American agent, Henry Mingel. These companies received payments during the war of $16.4 million, of which they could not account for $9.9 million. The War Department asked Attorney General Daugherty to attempt to recover $2.3 million of this. The Mitsui firm evidently attempted to evade payment by bribery. Gaston Means, a shadowy figure with a criminal record, claimed that he acted as the intermediary between Mitsui and Daugherty. According to the dubious Means account, he waited in a room in the Bellevue Hotel in Washington until a Japanese representative handed him $100,000 in thousand dollar bills. Means claimed that he then gave the money to Daugherty's agent, the scandal-plagued Jess Smith. The Justice Department dropped the case against Mitsui, possibly as a result of bribery. Although the details of this incident have never become fully known (Mingel was later found dead and Smith committed suicide), the odor of corruption was strong enough to leave the process of defense procurement permanently fouled. Oswald Garrison Villard, who had professed the Merchants of Death theory a decade earlier, now declared that "the postwar disposal of surplus army supplies has oozed corruption and dripped mire."[34]

A far more sweeping decision by President Harding himself greatly intensified the profiteering controversy. Beginning in 1920, the veterans of the American Expeditionary Force began demanding additional compensation for their war service. Officially known as the Adjusted Compensation Bill, this plan was commonly known as the veterans' bonus or just "the bonus." The American Legion was the most vocal supporter of this scheme, which in theory would even out the economic sacrifice of the war. The Legionnaires argued correctly that army privates had been drafted to serve for a nominal wage of a dollar per day, while civilian workers had been allowed to earn market wages, which were considerably higher. The Legion demanded that each veteran receive an additional payment of a dollar per day of military service, rising to $1.25 per day for overseas service. In 1922 the Congress approved the bonus and sent it to President Harding for his signature.[35]

Harding was advised by his secretary of the treasury, Andrew Mellon, one of the richest men in America, that he should veto the veterans' bonus. Mellon argued that the $80 million cost of the plan was more than the United States could afford. Mellon was supported in his opposition by the National Association of Manufacturers, the *Wall Street Journal,* and the U.S. Chamber of Commerce, which orchestrated the antibonus campaign. Further opposition came from an organization called the "Ex-Service Men's Anti-Bonus League." This group was bankrolled by prominent New York business executives who feared that passage of the bonus would delay repeal of the heavy war taxes. The American people were thus presented with the spectacle of ex-servicemen being denied a bonus by business executives who, while exempt from military service themselves, had spent the war years comfortably, safely, and in some cases profitably. Torn between his antiprofiteering rhetoric and his friendship for business interests, Harding chose the latter. He vetoed the bonus, and the Senate upheld his veto by a narrow four-vote margin.[36]

Infuriated by its loss, the American Legion intensified its campaign against war profiteering, and the bonus matter became an issue in the congressional elections of 1922. Four leading opponents of the bonus were defeated in senatorial campaigns. Of twenty-one new senators elected in 1922, eighteen were pro-bonus and three were opposed. When the bonus was reintroduced in the next Congress, it passed easily. Continuing Harding's opposition, President Calvin Coolidge vetoed the bill, but this time Congress overrode the veto.

By 1924 a general reappraisal of the Great War was in progress. The cultural ambience of the 1920s has been described as disillusioned, cynical, and embittered, and this mood permeated its flourishing literature. At first, the theme of wartime profiteering seldom appeared in the literature, as the early writers of the Lost Generation preferred shocking portraits of the horrors of trench warfare.[37] But by 1924 the war profiteering issue was creeping into the literary imagination.

Two major literary works published in 1924 drew attention to the profiteering issue. The first was Laurence Stallings's novel, *Plumes*. As a marine lieutenant, Stallings had been machine gunned at Belleau Wood. He had spent two years in hospitals in agonizing pain before his right leg was finally amputated. "The Plumes have been in this country two hundred and fifty years," Stallings wrote, "and not one of them was ever worth as much as $25,000. 'Not one of them,' said Richard, sometime after the Harding inaugural, 'had anything to go to war about.'" Stallings caustically called the home front the "blood-money hearthside." He angrily described how civilians had inquired about his wounds. "'Did it hurt?' the blood-money hearthside asked the returning patriots. And the patriots . . . shouted 'Naw!'" Stallings summed up a generation's jaundiced view of the war when he said that "war is economic—political . . . the suckers who die are the least of considerations . . . and there will always be plenty of suckers."[38]

Stallings teamed up with the playwright Maxwell Anderson to write an even more rancorous play, the classic *What Price Glory?* There is little resentment of the home front directly expressed in the script, but economic themes indirectly appear. The play's symbolic hero, Captain Flagg, declares that he has no good reason for fighting Germans. He is paid but $8 per day. The issue of poorly paid warriors and ill-gotten gains is again raised when a stretcher bearer steals $800 and a gold watch from a wounded soldier.[39] Although much of the civilian public was shocked by the play's bloody frankness, ex-doughboys who saw the performance approved. The *American Legion Weekly* called this candor "great stuff." Ironically, *What Price Glory?* made its authors wealthy men. Anderson's income reached $1,000 per week in 1924, and the first of three screen versions brought him $28,000.[40]

In the election year 1924 the profiteering became a prominent political issue. In part to soothe the wrath and win the votes of the returned veterans, the House of Representatives held hearings to consider how better to share the next wartime burden. Conducted by the Committee on Military Affairs, these hearings were entitled *Universal Mobilization for War Purposes.* It was the first of several inquiries having a similar goal, including the famous Nye-Vandenberg hearings a decade later. The 1924 investigation was the result of persistent efforts by Representative John J. McSwain of South Carolina. McSwain, who considered himself a champion of rural America against the evils of big business, had been forty-six years old when the United States declared war and was therefore exempt from the draft. His younger brother, at thirty-six, had served in France and had been gassed twice, deafened in one ear, blinded in one eye, and returned with shattered nerves. The elder brother thus had firm evidence of the inequities that war engenders, and he had sought for three years to bring attention to the issue before he was finally successful in 1924.[41]

Most of the participants in the hearings on universal mobilization demanded a more pervasive wartime sacrifice. If the suffering became truly general, they reasoned, there would be a strong disincentive to go to war. This idea, that war itself could be exterminated by eliminating war profit, was a prime example of the continuation of the Progressive spirit into the 1920s. No witness better expressed the mood of the times than did Secretary of Commerce Herbert Hoover. As the wartime food administrator, Hoover was an experienced bureaucrat who understood the inequities of price inflation. In case of another war, Hoover appealed, the president must be given authority to fix prices, to commandeer private inventories, and even to "suspend habeas corpus, and generally [have] complete and absolute control over the whole population." In another dangerous declaration, Hoover said he was willing to accept "instantaneous court-martial" of a war profiteer. "War is an unhappy business," said Hoover, "and the great bulk of our ordinary safeguards of life must be forgotten, and the more evident it is that the whole nation will be put into the storm and made to bear its share of the sacrifice, the less likely we are to go to war."[42]

It was undoubtedly good politics for Republican leaders to promise a broader sharing of war burdens. Constant mention of the matter reminded the voters that the Democrats had been in command when the alleged profiteering occurred. Promises to repair the damage were sound ways of appealing to the huge new veteran vote, as the 1922 elections had shown. In 1924, both major parties endorsed "Universal Mobilization" in their platforms (meaning to draft capital, labor, and manpower). The Democrats hailed universal mobilization because it would "tend to discourage war by depriving it of its profits." In 1924, Calvin Coolidge, running for the presidency in his own right, charged that "totally inconceivable amounts of money were raised and expended with a lavishness . . . which now seems like some wild nightmare. . . . The least scrupulous became the greatest gainers."[43] His assistant secretary of war, Dwight F. Davis, assured the Legionnaires that their views had been incorporated into War Department planning. "Financial burdens must be equitably distributed and every resource utilized to the utmost to win the war," Davis declared. "No man should be allowed to make a profit out of war while his brothers are risking their lives in their country's service."[44] Like Harding, however, Coolidge's commitment did not extend much beyond mere rhetoric, as his veto of the veterans' bonus demonstrated.

By 1924 a distinguished list of Americans from diverse fields supported the concept of universal mobilization besides the American Legion. This group included Bernard Baruch and Grosvenor Clarkson from the wartime mobilization team, John L. Lewis, president of the United Mine Workers, the liberal Rabbi Stephen S. Wise, the feminist Grace Wilbur Trout, the presidents of Northwestern and Boston universities, and a variety of newspaper editorialists headed by the *Christian Science Monitor*. The liberal Federal Council of Churches also endorsed the idea.[45]

Despite the nearly unanimous desire of the witnesses who testified at the hearings, the 1924 attempt to equalize the burdens of war produced no successful legislation. Secretary of the Navy Curtis D. Wilbur, a lawyer, argued that the restriction of profit during peacetime was probably unconstitutional, as were the severe restrictions on profit proposed for wartime.[46] The prospect of amending the Constitution during peacetime in order to restrict profits during wartime was too enormous an obstacle for its many sympathetic advocates. The conservative, uncritical public attitude toward business profits during the New Era, moreover, was anything but conducive to such a sweeping reform. Since the labor draft was unpopular with the American Federation of Labor, the Democrats were divided in their support for it. (The Democrats dropped the universal conscription plank in 1928, although the Republicans retained it.) Perhaps even more important than any of these obstacles was the general lack of interest in defense planning when there were no war clouds in sight. In the 1920s, Americans had at best a tepid interest in utopian schemes to equalize sacrifice. The mood of the times emphasized the acquisition and enjoyment of consumer goods, the compulsive pursuit of leisure-time diversions, and leadership by safe nonentities rather than serious consideration of the idealistic plans of Wilsonian visionaries.[47]

In 1925 the cultural reappraisal of the Great War broadened. That year marked the appearance of one of the most influential motion pictures ever produced, Metro-Goldwyn-Mayer's *The Big Parade*. Based on Stallings's novel *Plumes* of the previous year, the film offered a doughboy's view that struck a responsive chord. Earlier cinematic attempts at portraying war themes had proven modestly popular, but *The Big Parade* was wildly successful and immensely profitable. It inspired dozens of examples of the new "war movie" formula, including such major films as *What Price Glory?* (1926), *Wings* (1927), and *The Dawn Patrol* (1930).[48]

Most of these films were heroic action pictures that paid little if any direct attention to the issues of the conflict. This studious disregard for the issues nevertheless represented a subtle break with earlier depictions. Early films simplistically portrayed the war as merely a campaign against German degeneracy, a crusade to exterminate the venomous Hun. But the new motion pictures expressed a willingness to recognize the basic humanity of the enemy, thus contributing to a reappraisal of the meaning of the war.[49] Perhaps the best of the critical films was *All Quiet on the Western Front* (1930). Filmed from the German point of view, it was sympathetic to the enemy in ways that would have been impossible a decade earlier. As a sound picture, it was enormously more compelling than its silent predecessors. In one memorable line, a German infantryman, pausing to ponder the meaning of the war, muses that "it must be doing *somebody* some good."[50] Millions of people attended these screenings, which satisfied a desire to reconsider the experience of the war. When viewers saw a supposedly monstrous enemy wonder who was gaining from the war, they were prompted to ask themselves a similar question: Who, indeed, did gain from the war?

Popular films were only one form of the reconsideration of the war, albeit a potent one. Best-selling war memoirs, plays, and novels were another fountain of criticism. The late 1920s and 1930s witnessed an outpouring of literary aversion for war. Since war in general is profoundly uncivilized activity, and the Great War with its submarine attacks, machine guns, and poisonous gas was particularly heinous, postwar writers felt compelled to try to preserve civilized life by indicting the conflict.[51]

There were many antiwar authors whose work displayed a loathing of profiteering as the implied origin of the war. Upton Sinclair asserted the socialist view in *Jimmie Higgins* (1919) that "all governments were run by capitalists, and all went to war for foreign markets and such plunder."[52] In William March's *Company K* (1933), a working-class soldier describes the views of his better-educated comrades: "They'd been to college, and they could talk on any subject that came up. But mostly they talked about war and how it was brought about by moneyed interests for its own selfish ends. They laugh at the idea that idealism or love of country had anything to do with war. . . . Fools who fight are pawns shoved about to serve the interest of others."[53] In *Johnny Got His Gun* (1939), Dalton Trumbo warns that "when armies begin to move and flags wave and slogans pop up, watch out little guy because it's somebody else's chestnuts in the fire not yours."[54] In the 1920s and 1930s Americans were asking questions about whose chestnuts were really in the fire in 1917.

The pervasiveness of the skeptical attitude toward the war's purposes was demonstrated by its appearance in at least one enormously popular newspaper cartoon. In 1924 the ex-doughboy Harold Gray began drawing the comic strip

"Little Orphan Annie." In October 1924, he introduced a fabulously rich—and aptly named—principal character called Daddy Warbucks. Warbucks explains to Annie that when war came, "I was too old to fight but I wanted to do my bit so I made munitions—Well, I made a fortune, too, and everybody hates me for it—Maybe I did wrong, Annie, but I did the best I knew." To this confession, Annie replies: "Don't you worry—I love you, 'Daddy.'" Ultimately Little Orphan Annie and her guilt-stricken benefactor appeared in 345 newspapers having a daily circulation of sixteen million copies.[55] "Daddy's" effect on public views, while not subject to measurement, could have been nothing less than substantial.[56]

Although the 1924 attempt to establish the principle of universal sacrifice failed in Congress, the festering memory of unseemly war profits affected foreign policy. The Covenant of the League of Nations included a statement that obliquely called for public ownership of defense industries. This provision vaguely declared that "the manufacture by private enterprise of munitions and implements of war is open to grave objections." The basis supporting these "grave objections" was a widely shared assumption that private arms firms would seek to stir up business by starting wars and indeed had already done so. In 1924 Congressman Cordell Hull of Tennessee, who became famous later as Franklin D. Roosevelt's first secretary of state, declared at the McSwain committee hearings, "I think a great many people believe that the last war had its fundamental cause in the profiteering going on in the world."[57]

In France, Paul Fauré, the general secretary of the French Socialist Party, lent weight to the arms conspiracy theory by charging in the House of Deputies that the Schneider arms factories at Creuzot opportunistically but unethically sold arms both to the French government and to its potential enemies. According to Fauré, a nefarious ring of bankers and munitions makers led by Eugene Schneider controlled arms production in France, Hungary, and Czechoslovakia. This ring supposedly supplied arms not only to these nations but also to Greece, Yugoslavia, Bulgaria, Rumania, Turkey, Russia, Spain, Italy, Argentina, and Mexico. When news of the French accusations reached America, the charges served to confirm the plausibility of the existence of an international arms conspiracy.[58]

In 1925 an international conference on the "Arms Traffic" met under League auspices at Geneva, Switzerland. This conference sought to curtail the allegedly provocative actions of the "arms traffickers" by calling for the prohibition of arms sales except under license by national governments. Most of the American arms industry was not adamantly opposed to licensing (which they preferred to nationalization),[59] but the arms makers wanted the licenses to be issued by the pro-business Department of Commerce rather by than the more restrictive Department of State.[60] The arms makers further desired a blanket exemption of pistols and revolvers (which were big sellers in Latin America) from export licensing as war matériel. The United States, though not a member of the League of Nations, participated in the framing of the Arms Control Convention by assigning Joseph C. Grew, its minister to Switzerland, to attend the talks. Nevertheless, despite the urging of Presidents Hoover and Roosevelt and (by then) Secretary of State Cordell Hull, the United States refused to ratify the convention until 1935, because it would limit the American ability to sell arms to Latin America.[61]

In the 1920s some Americans became concerned about the existence of a cooperative relationship between American industry and the American military. In later years, this relationship would be labeled (with considerable hyperbole) the "military-industrial power complex." Some cooperation dated from the period of naval rearmament in the 1890s, but it was so limited as hardly to deserve the roguish reputation it later received.[62] In the 1920s the dependency of military strength on industrial might was tested by the growing military importance of the airplane.

Although the lesson was imperfectly absorbed, the failure of the American aircraft industry during the war demonstrated that modern warplanes could not be designed and manufactured on short notice. In 1919 the air service began campaigning for a national policy of government subsidies for the civilian aircraft industry to ensure that in case of war a healthy, technologically advanced aircraft industry would be available for conversion to military purposes. As one Air Corps colonel argued, "Our aviation must be put on a commercial basis so that it will never be jeopardized from now until eternity. . . . Then when you go to war, you will have your manufacturing facilities, etc." The wisdom of this argument was clear to the farsighted congressman James A. Frear, who warned: "In case of future war of the country, why it seems to me it would be a tremendously important element, the supremacy of the air . . . and the way I view it, instead of this Government being hindmost in this question we should be foremost."[63] In the 1920s, nevertheless, the U.S. government only modestly subsidized the nascent aviation industry, in part to ensure its availability in the event of war, but congressional qualms about profiteering kept the aviation business from receiving as much assistance as it needed. This lack of profitability stifled technological development during the interwar period. In World War II the policy of subsidizing aircraft amply proved its wisdom, but in the 1920s even limited assistance was unpopular in isolationist circles.[64]

The history of the arms industry in the 1920s offers other examples of the growing cooperation between government and private industry. In 1919 the Du Pont Company, citing excess capacity in the industry, sought to abandon the production of military ammunition, but army officers successfully pleaded against this step. In 1920 Congress took further action to nurture the relationship between industry and the military. The National Defense Act of 1920 gave the War Department authority to plan for the mobilization of the economy in case of war. As a result, military officers visited thousands of American factories, assessing their productive capabilities and assigning them hypothetical tasks in time of war. In 1926 the army asked Congress to assign many of these firms "educational orders" so that they would know how to produce a military component if the need arose. Educational orders would in theory also provide a benchmark for determining reasonable prices. Between 1926 and 1928 a War Department Business Council operated, but it met only three times and more than a third of its members were absent on each occasion. There were other instances of military officers exhibiting their growing dependency on civilian industry. In 1931 the president of the Electric Boat Company, the nation's leading producer of submarines, declared his intention to retire. This brought a flurry of objections from the secretary of the navy, admirals, congressmen, and even the White House.[65] In the absence of an immediate threat from abroad, many Americans in the 1920s doubted the need for such a snug relationship. Even Congressman Frear, who a decade earlier had endorsed the plan of government subsidies of the aviation industry, in 1928 scoffed at "Jingoes" in the military-industrial complex who concocted the fantasy that Japan . . . "is prepared to come over and capture our country."[66]

The collaboration between the American military and American arms makers led to several questionable practices. Since it was in the interest of the army and navy to have a strong defense industry in wartime, military officers had a parallel interest in encouraging its growth in peacetime. The needs of the small peacetime forces maintained by the United States in the 1920s and early 1930s were inadequate to support a large arms industry, so military officers encouraged foreign sales to maintain its strength. The Curtiss-Wright Export Company successfully solicited endorsements of its "Falcon" warplane from Captain Ernest J. King (later a fleet admiral in World War II), and this helped its sales in the Dominican Republic. The Driggs Ordnance and Engineering Company arranged for the USS *Raleigh* to visit Turkey to demonstrate its cannon to Turkish naval officers, and the New York Shipbuilding Company arranged for the navy to send a cruiser built by the firm to Rio de Janeiro to be inspected by the Brazilian navy in hopes of obtaining sales. These efforts met with some success. By 1925 the United States, with 20 percent of the market, was the number two exporter of arms in the world (behind Britain). To many critics, the spectacle of American officers serving as unofficial sales representatives of arms contractors was repugnant, since in this way they could indirectly encourage an outbreak of war—the very thing they were paid to prevent.[67]

The modest doubts about the allegedly dangerous and improper influence of the munitions industry were given a massive injection of vitality by a startling revelation. In the 1920s a close working relationship developed among the nation's three largest builders of warships: the Bethlehem Shipbuilding Company, the Newport News Shipbuilding and Dry Dock Company, and the New York Shipbuilding Company. These firms enjoyed an oligopolistic, semi-competitive control of the production of capital ships. They organized a jointly owned, nonprofit subsidiary named the Marine Engineering Corporation, which designed a common set of blueprints for a six-ship class of cruisers. Each firm then undertook to bid on the ship contracts in a way that basically rotated the work among the companies as opposed to competing head-to-head for the contracts.[69] This would not have been greatly surprising or of stunning significance but for the behavior of their jointly paid lobbyist, William Shearer.

Shearer was an unstable, thoroughly unreliable person who possibly suffered from mental disease. He was also flamboyant and outspoken in his hatred of Britain and in support of an imperialist America. His personal life was shady at best: he had been arrested for forgery in England, had received a less-than-honorable discharge from the U.S. Navy, had been implicated in a French jewel theft, had impersonated a naval officer, and had admitted under oath that he was a plagiarist.[69] Nevertheless, from 1926 to 1929 the Big 3 naval shipbuilders paid him to lobby Congress for an accelerated schedule of merchant and warship construction. Shearer received $7,500 for his lobbying efforts, which, while somewhat questionable, were neither illegal nor highly unusual.[70]

The irregular aspect of the Shearer affair concerned his appointment by the shipbuilders to attend the Geneva naval disarmament convention of 1927. In 1929 Shearer filed suit against the Bethlehem Shipbuilding Company, seeking $258,000 in unpaid back wages.[71] He presented a sensational claim that he had been paid $25,000 to wreck the disarmament conference. The motive of the alleged plot was to prevent a cutback in warship construction. In reply, C. L. Bardo, a vice president of Bethlehem Shipbuilding acting on behalf of the Big 3, stated that he had assigned Shearer to attend the conference as an "observer only."[72]

Whether Shearer's original instructions were to act as an agent provocateur or merely as an observer has never been firmly established. At the very least, the shipbuilders wanted to measure the strength of the disarmament advocates. Bardo and his associates were also evidently fearful that the Geneva convention might endorse the nationalization of warship production. In any event, Shearer's indiscreet and disruptive behavior drew the ire of Secretary of State Frank B. Kellogg, who successfully demanded his dismissal. The suit was ultimately settled out of court for $37,500, with the stipulation that the amount be kept secret. The failure to contest the case in court left the truth of Shearer's charges unresolved.[73]

Shearer's shocking claim that he was paid to obstruct the cause of peace rightfully drew an outraged reaction. President Hoover demanded an explanation from the shipbuilders, prompting a flurry of editorial comment. Shearer delivered a lecture to four hundred New Yorkers in Carnegie Hall, was the subject of a congressional investigation, received extensive press coverage, and even appeared as a character in an antiprofiteering novel, *The Budapest Murders*.[74] Although Shearer's claim that he destroyed the conference has not and cannot be established,[75] his audacious lawsuit and its secretive settlement lent credibility to the string of conspiratorial accusations that emerged after the war. For a decade after Shearer unveiled his claim, the affair would be cited regularly as compelling evidence of the infamous machinations of munitions makers.[76]

The 1920s were a decade that gloried in scandal and sensationalism, which was an ideal setting for the pursuit of profiteers. The profiteer hunt thus thrived as a denoting feature of the politics and culture of the Jazz Age. In the angry and frightened mood of the Great Depression, the antiprofiteering campaign would go far beyond the comparatively mild effort of the early postwar period. During the 1930s, the attempt to expunge war profits would take on a life-or-death urgency that marked it as far more resolute than its predecessor.

9

Profits or Peace?

> If we face the choice of profits or peace, the Nation will answer—must answer—we choose peace.
>
> FRANKLIN D. ROOSEVELT, CAMPAIGN ADDRESS (1936)

The probe of war profits that surfaced periodically in the 1920s became a more pressing matter during the troubled 1930s. Commencing in the first year of the Hoover administration, a contentious debate on how to limit profits in a future war proceeded throughout the Depression. The renewed scrutiny of war profits was partially an aftershock of the Great War, but in the 1930s worldwide economic collapse and the rising danger of American involvement in a new and even more frightening European war gave the dispute fresh urgency.

In the 1930s there were two major focal points in the review of the extent and effects of war profiteering. The two inquiries were the Hoover administration's War Policies Commission of 1930 and the more contentious Senate Munitions Investigation of 1934-35 (the Nye-Vandenberg committee). Both took place against the setting of the more general question of how the nation should mobilize for war. Both investigations revealed the lingering effects of the Great War, and both demonstrated the difficulty of reaching agreement on national defense policy in its aftermath. The principles of that policy would be shaped by the costs of each option, as well as by who would profit and who would forfeit in each instance.

The problem of restricting the profits of munitions firms often reflected other public concerns of the period in which it was considered. During the Progressive era, the campaign against munitions profits had been an important part of the struggle against the evils of monopolistic capitalism. In the postwar period, Herbert Hoover's associationalist attempt to improve business-government relations also addressed the issue of war profits in its management of reconversion. In turn, the problem of regulating defense profits would be affected by the economic policies of the New Deal.

President Franklin D. Roosevelt's limited Keynesianism included deficit spending on public works as a countercyclical device. One aspect of Roosevelt's recovery plan included an accelerated program of naval construction. This inescapably raised the issue of how much the shipbuilders and other defense contractors should properly gain from the New Deal's anti-Depression medicine. The New Deal also featured a policy of redistributing wealth, and the fact that some Americans had gained immensely as a result of the Great War became a major justification for Roosevelt's redistributional policy.

The embittered veterans, committed pacifists, disaffected intellectuals, and agrarian radicals who had long fought against war profiteering renewed their efforts during the 1930s. They were joined in the Roosevelt years by a new clamor of angry voices: dejected workers, determined isolationists, younger feminists, and aggressive publicists all worked toward the idealistic goal of constructing a defense policy that was completely unaffected by selfish interest. They remembered the carnage in France, harbored grudges from countless quarrels between capital and labor, and spoke and wrote with a biting style fashioned in the acrimony of the Progressive era. The perennial struggle to control war profits reached an angry crescendo in the 1930s. The campaign against war profits was a striking example of the perseverance of the Progressive voice in the New Deal setting.

The historian C. Hartley Grattan furnished clear evidence of the continuation of the Progressive persuasion. Canadian-born and descended from a Swiss-Scottish family with an antimilitary tradition, Grattan had studied at Clark University under Harry Elmer Barnes, a leading revisionist scholar. Grattan began writing as a literary critic with socialist affinities, and he deeply admired the acid pen of H. L. Mencken. On 11 November 1929—exactly eleven years after the Great War had ended unromantically on the eleventh hour of the eleventh day of the eleventh month—Grattan published an important book, *Why We Fought*.[1] Commercially, *Why We Fought* was a dismal failure, selling only about a thousand copies. But intellectually the book became a seminal part of the reappraisal of the Great War. Its central argument that the Great War had been fought for economic reasons became incorporated into far more successful texts and eventually into the thinking of common citizens.[2]

By 1930 there were many signs that the nation was ready for such a reappraisal. In January a Conference on Causes and Cure of War convened in Washington, D.C. As the title implied, the organizers thought of war as a kind of disease. As with epidemics of bacterial origin, progress toward a treatment might be made by the exchange of views of scientific experts. Senator Gerald P. Nye told the conferees that wars are "caused by fear and jealousy coupled with the purpose of men and interests who expect to profit by them."[3] Nye would lead a decadelong search for proof of this theory.

The renewed inspiration for the antiprofiteering campaign came from several other developments. The most important of these was the world economic collapse that produced the subsequent strength of fascism in Germany, Italy, Japan, and Spain. The Depression heaped anger on American business and its supporters, and defense contracting attracted an extra share of the wrath. In 1930 war between Japan and China began to produce sales of American warplanes to China, prompting fears of American engagement. In 1931 the administration of President Herbert Hoover recommended spending $30 million to repair three battleships while it opposed spending $25 million on direct relief for the unemployed.[4] Displaced workers resented Hoover's spending priorities.

In June 1933, following Hoover's landslide defeat by Franklin D. Roosevelt, Congress approved an administration proposal to stimulate the economy by spending $238 million in Public Works Administration funds on the construction of twenty-four new warships. Roosevelt reasoned that 80 percent of naval spending went to labor, and he also argued that a naval buildup was necessary in the face of a growing Japanese threat. To the expanding pacifist element in the United States, naval spending as an antidote to the Depression was extremely odious. It seemed likely to initiate a dangerous new naval arms race with Japan and Britain, and it seemed wasteful and unworkable. The warships constructed were hardly needed to defend Hawaii and the Panama Canal, and they were far too few to defend the Philippines or to protect American interests in distant Asia.[5]

A great but often unspoken fear of the Depression years was that defense contractors would inflame a new war in hopes of restoring their firms' profitability. This concern lurked behind both federal investigations of war profiteering of the 1930s. Senator Homer Bone of Washington, a close friend and collaborator of Senator Nye, was a frank advocate of this position. Bone believed that because of the technical advance of weaponry a new world war would spell the end of civilization. Bone was one war early, but his participation in the crusade to end war profiteering was inspired by nothing less than an earnest belief that the future of the world was at stake. "There may be a social explosion," warned the senator, "and just blow stable government all to bits."[6]

Even if there had been no Depression and no Hitler, there would still have been a searching retrospective on the Great War. Ten years had passed, which is approximately the minimum period before an event can be assessed with at least some degree of historical detachment. An event as devastating as the Great War begged for historical analysis; a thorough examination by historians was inescapable, and their audience would surely welcome it with earnest attention. The surviving veterans of the American Expeditionary Force had begun to reach the stage of life in which they wished to rethink the actions of their youth. Aided by willing writers eager to guide their reappraisal, many of the men of 1917 came to the conclusion that their perilous adventure in France had been a reckless act. The carefree doughboys of the AEF were now fathers raising children amid the hardships of the Depression. Many became determined that their own sons would not be sent into mortal danger.

The Depression was particularly harsh on American farmers, bringing about a renewal of agrarian radicalism. This appeared in several different forms, but one of these was a resurgence of isolationism in foreign policy.[7] Agrarian radicals broadly believed that the United States government was unresponsive to the interests of working people; that foreign trade principally benefited big business; that eastern financial interests benefited improperly at the expense of farmers and other working people; that the United States had no need for colonies or military bases abroad; that the United States should not intervene militarily in the Western Hemisphere, particularly in Mexico, Haiti, or Nicaragua; and that the eastern press was too influential and too much under the control of financial interests. They also specifically discounted any threat from Japan, and they saw little moral difference between English and German societies. A particular annoyance was the decision of the United States to reduce or reschedule debts owed by European governments; farmers reasoned that they had gone into debt to aid the nation in the World War and therefore their debts should be reduced as well.[8]

These factors combined to force a review of American defense policy in the early 1930s. In those years, conventional military thinking held that the navy was the first line of defense. Unless the United States became involved again in a European war—which the various veteran, pacifist, intellectual, and radical elements bitterly opposed—there was little need for more than a coastal fleet and the nucleus of a modern army. On the other hand, if the United States intended to protect its shipping lanes to Europe and Asia, then a powerful navy would be necessary. This meant in practice a fleet as large as that of Britain, supported by an extensive system of supply bases. In sharper detail, and with a new sense of urgency, the old debate between "big navy" men and "little navy" men reopened. Senator John S. Williams of Mississippi said that the alternatives were either to employ a military force big enough "to whip anybody and everybody" or to maintain minimal forces "and, when war comes, submit ourselves to the immense strain necessary, with the extravagance of blood and capital both necessary."[9] In the 1930s, the nation chose the latter course, as the predominant feeling was that large forces were not only needlessly expensive but also dangerously provocative. This would prove to be a costly decision.

The essential practical question on which this debate turned was whether the United States should build up its fleet. The goal of navalists was to maintain parity with the Royal Navy and to stay ahead of the Imperial Japanese Navy, as provided by the Washington Naval Treaty of 1923. Naval critics believed that a smaller and much cheaper force was sufficient, and they were half right. In 1933, following the rise to power of Adolf Hitler in Germany, Britain ceased all planning for a possible war against the United States. The War Office informed the American military attaché that the United States was no longer considered a threat to the British Empire. In 1933 the U.S. Navy was at 65 percent of its permissible strength under the Washington Treaty, which would have been sufficient to defend the Atlantic frontier now that the Royal Navy was not credible as a potential enemy.[10] Retaining the American fleet at its existing level seemed reasonable, and maintaining a "little navy" would certainly be much cheaper than constructing a larger force. In the view of naval critics, it would also be much less provocative.

In military planning, unfortunately, being only half right can have fatal consequences. The deciding factor shaping positions on the question of naval strength in the 1920s and 1930s was the extent of the area that the U.S. Navy ought properly to defend. "Little navy" proponents saw the American defense perimeter as continental in scope, meaning only the Western Hemisphere. These continentalists believed that American defenses in the Pacific Ocean should follow a line that extended from Alaska, through Pearl Harbor in the Hawaiian Islands, to the Panama Canal. This was termed the "Nome-Honolulu" line.[11] The Atlantic line proceeded from Panama northward through the Caribbean Sea via the Virgin Islands, finally terminating on the coast of Maine. Prominent among the "little navy" counselors were the prominent historian Charles A. Beard and the Socialist Party chieftain Norman Thomas. President Herbert Hoover was less outspoken but still inclined toward the "little navy" side.[12]

Their opponents, who wanted a larger or "treaty" navy, argued that the United States must be prepared to defend distant sea frontiers in the western Pacific. During the 1920s, American naval spending rose only modestly from $322 million in 1923 to $374 million in 1930. Meanwhile, the Imperial Japanese Navy swelled to 95 percent of its quota under the treaty.[13] Defending against Japanese sea power would require a mighty fleet and supporting facilities sufficient to defend the Philippines against Japanese attack. Navalists demanded large vessels with long cruising ranges on the grounds that the United States had few suitable naval bases.[14] Strong advocates of a "treaty navy" included Presidents Calvin Coolidge and Franklin D. Roosevelt. The latter, whose views proved crucial, had been a navalist since his days as assistant secretary of the navy during the Wilson years. The young FDR had fretted that his chief, Navy Secretary Josephus Daniels, was unwilling to support the grand fleet that Roosevelt felt the nation required. During the presidential campaign of 1932, Roosevelt temporarily set aside his views to placate the powerful isolationist publisher William Randolph Hearst, but soon after his election he returned to his earlier position.[15]

This question of naval policy was closely related to the problem of controlling profiteering. "Little navy" proponents often questioned explicitly or implicitly the motivation of the "treaty navy" advocates. Opponents of naval spending argued that their hawkish adversaries were motivated by greed for profit, not by patriotic purpose. Charles Beard, who could not imagine that

Japan could successfully attack the Nome-Honolulu line, believed that there was a conspiracy between the Roosevelt White House and Hollywood film producers to propagandize for naval expansion.[16] Unless these base motives were meticulously exposed and brought under control, they might involve the United States in another ghastly military blunder.

In 1929 Patrick J. Hurley, Hoover's assistant secretary of war, ordered a review of how to mobilize for war. This study, under army auspices, was led by Major General Van Horn Moseley, Hurley's executive assistant. Moseley had been General Pershing's supply officer for the American Expeditionary Force in 1917-18, and by 1930 he was deputy chief of staff of the U.S. Army. He was an able if intolerant soldier, but he selected as his principal assistant one of the army's most promising young officers, Major Dwight D. Eisenhower. Moseley would write, accurately, of Eisenhower, "He has a remarkable mind and an equally talented pen, enabling him to present a subject simply and clearly." Always a careful observer, three decades later Eisenhower would term the relationships he studied in 1930 the "military-industrial complex."[17]

The army mobilization plan was far from satisfactory to everyone. Besides Moseley, it was developed by General Hugh S. Johnson, who was an aide to Bernard Baruch when Baruch chaired the War Industries Board. Moseley disliked Democrats in general and Baruch in particular. General Johnson, like Baruch, wished to bridle wartime demand by fixing prices, whereas Moseley preferred to use the tax power. But the army plan also gave its officers sweeping control over civilian industry. At least 14,000 civilian managers had been commissioned as reserve officers and 15,000 factories had been surveyed to determine their suitability for war production. Baruch saw this as a return to the errors of the past; military officers were not businessmen, and to give them control of the economy would hamstring production again. Mobilization was now too important and too complex to be left to soldiers.[18]

Furthermore, the army plan paid scant attention to the problem of controlling profiteering. General Douglas MacArthur, the chief of staff, argued the timeworn military position again: "In our attempts to equalize the burdens and remove the profits from war, we must guard against a tendency to overemphasize administrative efficiency and underemphasize national effectiveness. . . . The objective of any warring nation is victory, immediate and complete." MacArthur was correct that the immediate national objective was victory, but like generations of officers before him he failed to understand that a secondary national aim was to win at an acceptable price. A victory that led to a generation of backbiting, resentment, bitterness, and alienation was a hollow triumph. The army plan rather simplistically assumed that all wars are popular, that a unified and committed civilian population would compel restraint by subjecting profiteers to the glare of publicity, and that an efficient procurement system would allow few opportunities for profiteering.[19] By 1930, by contrast, there was a growing belief among some civilians that a mobilization plan that promised large contracts with little restraint on profits would serve to invite the war that the nation was seeking to avoid.[20]

In 1930 Congress created the War Policies Commission, to be chaired by Secretary Hurley, for the general purpose of placing planning for wartime mobilization under civilian control. Its more specific purpose was to consider mechanisms for restricting profiteering: commandeering of war matériel, excess profits taxation, price fixing, and other means of restricting civilian income. The creation of the commission was the realization of a principal goal of the American Legion, which had sought such investigations since 1922 and was unappeased by the hearings of 1924. By 1930, WWI veterans were an ever more powerful political force, and it was probably not coincidental that President Hoover approved the legislation establishing the commission on the eve of the congressional election of 1930. Some veterans of the AEF were already in Congress, and six of the eighteen members of the War Policies Commission were Legionnaires.[21]

The War Policies Commission was composed of political heavyweights. Although chaired and dominated by Patrick Hurley, now promoted to secretary of war, the commission included five Cabinet members, four senators, four representatives, and five men from outside the government. Despite the prestige of its membership, the commission labored under serious handicaps that practically doomed its success in advance. In part, it was a commendable (if unusual) attempt by the federal government to peer far into the distant future, to anticipate a crisis, and to take responsible action in advance—in other words, to plan for war while still at peace. In part, it was an attempt by the civilian bureaucracy to prevent the army from controlling mobilization policy. In part, it was an attempt to placate powerful interest groups (the American Legion, the Disabled American Veterans, the liberal churches, and peace activists), who wanted to redistribute wartime sacrifice, to nationalize the defense industry, and/or to eliminate war profits. (The executive director of the commission, Robert H. Montgomery, another protégé of Bernard Baruch, sought to restrain the more radical demands of the Legionnaires while also restricting the army's ability to dominate mobilization.)[22]

The problem with these goals was that few Americans in 1930 were enthusiastic about planning for the next war, and some were dead set against doing so. Except for a very few concerned individuals, such as Bernard Baruch, most were content to let the army do whatever minimal planning was necessary. Military planners, on the other hand, were initially suspicious of civilian interference in what they considered their area of expertise. The army concluded, eventually, that the net effect of the investigation would be helpful to its institutional interest and so decided to be cooperative. Army leaders shrewdly observed that the composition of the commission was largely conservative, and therefore it was unlikely to threaten military prerogatives. For the commission to appease the American Legion fully, Congress would have to vote to modify the Fifth Amendment to the Constitution, and the army judged correctly that such a drastic change was unlikely. The army also hoped, incorrectly, that the commission would attract national attention to the problem of mobilization, and this would encourage civilian cooperation with military plans.[23] Nevertheless, the army never abandoned its traditional reluctance to address the question of war profits seriously.

The War Policies Commission was decisively limited by its charge from Congress. Primarily the creation of the American Legion, the commission was supposed to consider means for implementing the universal draft, a prime goal of the Legionnaires since 1922. As Congressman J. Mayhew Wainwright, a strong supporter of the commission, explained, there was a "burning sense of injustice felt by those who served." The universal draft would mean not only conscription of young men for the armed forces but also conscription of both capital and labor for the next war effort. Fairness demanded, Wainwright said, "no slackers, no inequalities, no special or financial advantage to anyone . . . equality of compensation for the man with the rifle and the man in commerce, business, or industry."[24] The fundamental problem with the concept of the universal draft was that conscription of capital and civilian labor was thought to be constitutionally questionable. The War Policies Commission was therefore expected to consider whether a constitutional amendment would be necessary to achieve equity. To block this possibility, the enabling act was amended, at the request of the American Federation of Labor, so as to prohibit the commission from discussing conscription of labor in any way.[25] This precluded discussion both of the constitutional amendment and of a principal method of achieving equity. If one device was not subject to consideration, other related devices would be difficult to discuss if overall equity was to be the final result.[26]

In 1933, following Hitler's seizure of power in Germany in the spring, the collapse of the London economic conference in the summer, and the failure of disarmament efforts in the fall, a war scare swept Europe. Prices of defense-related stocks soared in France. Despite the depressed world economy, sales and profits of English airplane manufacturers actually increased. At the same time, prices of almost every other stock and commodity traded on world markets were falling precipitously. "Europe is again thinking and talking about conflict as it did between 1905 and 1914," lamented Frank Simonds, an American observer.[27]

The obvious peril of a new and terribly destructive war was profoundly disturbing. The looming threat of a return to slaughter staggered again the old nineteenth-century liberal belief that wars were only a temporary and aberrant deviation from the ongoing advance of human civilization.[28] The awful new evidence indicated that war was rather customary human behavior, striking periodically like a ghastly scourge. Its origins must then be found in mundane matters—such as the endless pursuit of profit, the institutionalization of greed.

In the autumn of 1930 the International Red Cross had circulated a warning that in the next war the civilian population would be the target of poison gas attacks delivered from the air. Following the international publicity surrounding the transatlantic air crossing by Charles A. Lindbergh, the frightening prospect of what one writer termed an "aero-chemical war" seemed entirely plausible.[29] American women had gained in political influence during the 1920s, and in the 1930s their ability to have an impact on American foreign policy was strengthened. As likely victims of an intercontinental aero-chemical war, women had a vital new stake in the maintenance of the peace.

The development of modern weaponry meant that in the event of war American women could reasonably fear a direct attack upon their homes. The frightening possibility was that enemy planes would drop canisters of poison gas on the civilian population, placing everyone at grave risk. "The civilian population is going to be attacked," warned Dorothy Detzer, leader of the International League of Peace and Freedom. "The defenseless women, the old, the sick, the helpless are at the mercy of the new war method."[30] It is a peculiarity of modern warfare that poison gas inspires extraordinary alarm among both soldiers and civilians. This apprehension considerably exceeds the reality of the danger, but potential targets nonetheless tend to perceive poison gas as a fearful menace against which there is no defense. The mustard gas of the Great War had had appalling effects, but new compounds that were odorless, colorless, and horribly effective had been developed. Poison gas was therefore additionally hateful as it symbolized a perversion of scientific knowledge.[31]

The war scare of 1933 provoked a new outpouring of pacifism. In the early 1930s there were so many manifestations of disgust at the prospect of a new world cataclysm that no inventory of them would be complete. Any attempt to gauge the extensiveness of pacifism can never be precise, but the historian Robert Dallek estimates that the peace movement at its peak had about twelve million participants and a following of between forty-five and sixty million Americans.[32] In the spring of 1933 the National Student Federation cast 22,612 ballots at sixty-five colleges in twenty-seven states. Most of the students declared themselves to be complete pacifists or willing to fight only if the continental United States were invaded, and only 28 percent would support any war declared by Congress.[33]

Pacifists vied to find clever ways of expressing their opposition. A group called World Peaceways assembled a massive antiwar tome weighing more than a ton and measuring seven and one-half feet square entitled *War—The Super Racket.* College students organized units of the "Veterans of Future Wars" and demanded advance payment of their bonus for future war service. A companion organization of "Future Gold Star Mothers" asked for prepayment of pensions for prospective war widows and trips to Europe to select potential gravesites for their husbands and sons. The United Anti-War Association of the University of Chicago collected six hundred signatures on a

petition that read, "Whereas, we are the future cannon fodder for senile states-
men, capitalists, and generals—all of whom 'die in bed,' Be it resolved that
we . . . demand an immediate investigation." In November 1935, the *New York
Herald-Tribune* published a poll disclosing that 75 percent of the American
population favored a national referendum before the United States entered another war. Only 10 percent of respondents wanted to use military measures
in association with another country in order to check aggression. There was
great sympathy for the slogan chanted by protesters at the annual meeting of
the Army Ordnance Association in Philadelphia in 1935: "2 + 2 = 4, Gun +
Gun = War."[34]

A particularly influential event was the publication in March 1934 of a dis-
dainful article entitled "Arms and the Men" in the influential magazine
Fortune. This periodical, the brainchild of the highly successful publisher
Henry Luce, intended to carve its niche in American journalism by dramatiz-
ing the inner world of American business. "Arms and the Men" (which took its
title from George Bernard Shaw's famous play *Arms and the Man*) was un-
signed but written by a young *Fortune* staffer, Eric Hodgins, with the supervision and suggestions of Luce himself. "Without a shadow of a doubt," wrote
Hodgins darkly, "there is at the moment in Europe a huge and subversive force
that lies behind the arming and counter-arming of nations. . . . The control of
these myriad companies vests . . . in not more than a handful of men whose
power in some ways reaches above the power of the state itself." Most of the article dealt with the European arms trade, but this point was lost on most readers. When the powerful publisher of *Time* and *Fortune* placed his stamp of
approval on the call for arms control, the cause gained new respectability and
exposure. The article was reprinted extensively, appearing in such prominent
publications as *Time,* the *New York Times,* the *New York Herald-Tribune,* and
the *Congressional Record. Fortune* claimed that it was presenting objective journalism (even computing in businesslike fashion that the cost of killing each
soldier in the Great War was $25,000), but in fact it was presenting advocacy
journalism.[35]

The atmosphere of zealous pacifism of 1933-35 established the basis for the
most sweeping consideration of the war profits issue in American history. This
was the investigation of the munitions industry conducted by a special committee of the Senate. It was first called the Nye-Vandenberg committee after its
founders, Republican senators Gerald P. Nye of North Dakota and Arthur
Vandenberg of Michigan. After Nye was elected chair, it became known simply
as the Nye committee.

Although its fame has been dimmed by more recent inquiries, the Nye-
Vandenberg committee pursued issues of the gravest national consequence. It
was arguably the most earnest and influential political investigation of the first
half of the twentieth century. The hearings have been frequently reconsidered,
but the investigation's overall consequence remains unresolved. Writers have

generally disagreed with the committee's findings while remaining sympathetic to its purposes.[36]

The specific factors leading to the investigation of the munitions industry
included dissatisfaction with the work of the War Policies Commission, mis-
givings about American policy toward arms sales in South America and China,
disapproval of the Roosevelt administration's military expansion, disagreement about the threat of war with Germany and Japan, and pent-up rage lingering from the mobilization for and entry into the Great War. The probe was
supported by an unlikely coalition of city and farm, Right and Left, Republican and Democrat. Religious groups such as the Federal Council of Churches
and the World Alliance of Friendship through Churches were prominent supporters.[37]

The Nye-Vandenberg inquest resulted from a Senate compromise. Nye was
a former newspaper editor from North Dakota who initially supported
American intervention but came to oppose it. Nye shared the views of the
Non-Partisan League, which denounced the fixing of wheat prices at $2.20 per
bushel during wartime while industrial prices rose unchecked. He declared his
distaste for eastern bankers and war profits as early as 1920. In personal characteristics, Nye resembled William Jennings Bryan—both were doctrinaire,
simplistic, and tended to personalize issues. Neither Bryan nor Nye was a
farmer, but both came from rural backgrounds and modest circumstances,
and both had powerful oratorical abilities.[38] Born in Wisconsin and a longtime
admirer of Robert M. La Follette Sr., Nye was always a steadfast "little navy"
man. After reaching the Senate, he persistently inveighed against profiteers
whom he believed had brought on the Great War. His main interest in investigating the munitions makers was to establish their culpability once and for all.
This would lead to a mechanism that would foil their yearning for another war
for profit.[39]

When the United States declared war on Wilhelmine Germany, Arthur
Vandenberg was, like Nye, a young and ambitious midwestern newspaper
editor. From Grand Rapids, Michigan, he spurned enlistment and never re-
gretted his decision to avoid service, although he often supported positions advocated by the American Legion. As a member of the Senate and friend of the
Legion, Vandenberg served on the War Policies Commission in 1931, and thereafter continually supported equalization of the burdens of war.[40] Early in 1934,
when Nye proposed the establishment of a committee to investigate the munitions industry, Vandenberg introduced a resolution endorsed by the American
Legion to take the profits out of war.

The Senate considered Nye's request for two and a half months. In the interim, the Vinson-Trammell naval shipbuilding bill was approved, authorizing
the construction of 102 new ships. Nye and others sought to amend the naval
appropriation to limit profits to 8 percent of the cost of each warship constructed in a private shipyard and to force construction of half the ships in

government shipyards. To placate these and other foes of defense spending, the Senate agreed to create the special committee to investigate the munitions industry. In return, Nye withdrew his amendment to the shipbuilding bill.[41]

Credit for the plan to combine the Nye and Vandenberg resolutions and to create the investigating committee went to Dorothy Detzer of the Women's International League of Peace and Freedom, an organization founded by Jane Addams which in the 1930s could legitimately claim to lead American secular pacifism.[42] Detzer's pacifist commitment was deepened by personal experience with the carnage of war. Her twin brother was gassed in France and died in 1925, and two other brothers also served. A Hull House alumna, Detzer had visited postwar Europe during three years' service with a Quaker relief organization, the American Friends Service Committee.[43] She became a major figure in the campaign against war profits, and in helping to create the pacifist feeling of the 1930s, Detzer earned a place among the most important women influencing American foreign policy in the twentieth century.

Detzer hoped that Senator George Norris of Nebraska would head the probe, but Norris balked and recommended Gerald Nye. Norris chose Nye because there was no defense industry of significance in North Dakota and because Nye would not be up for reelection for four years. This would insulate the chair from political pressures. Norris's decision to recommend Nye for the leadership of the committee would have a fateful effect. Norris described Nye correctly as "rash," adding optimistically, "but it's the rashness of enthusiasm."[44]

Besides Nye and Vandenberg, the Munitions Committee included five senators, whose membership defined the orientation of the committee as mostly western and mostly isolationist.[45] Bennett Champ Clark of Missouri and Homer T. Bone of Washington teamed with the cochairs to form a solid isolationist majority.[46] W. Warren Barbour of New Jersey and Walter F. George of Georgia were moderates, and only James P. Pope of Idaho was devoted to collective security.

The actual driving force behind the investigation was its chief staff investigator, Stephen Raushenbush. The son of Walter Raushenbush, a famous Christian pacifist and political activist of the Progressive years, Stephen Raushenbush had served as an ambulance driver during World War I. He had come to believe that the Allied cause was no better than that of Germany and that the United States had been fooled into entering the conflict. Raushenbush's appointment came because he was an experienced, aggressive investigator with few ties to government. He assembled a staff of young, determined, liberal lawyers that included Alger Hiss, who gained notoriety a dozen years later because of his activities as a communist agent during the 1930s.[47] Another special assistant, Josephine Joan Burns, who believed that U.S. participation in the Great War was due to the House of Morgan, became Mrs. Stephen Raushenbush. With her new husband, she would write an extended defense of the committee's work.[48]

The Nye committee could have made little impact were it not for at least qualified support from Franklin D. Roosevelt. The president was a longtime "big navy" man, but he was well aware of the national and international strength of the pacifist-isolationist bloc. During the campaign of 1932, he had temporarily abandoned internationalism in order to placate foes of the League of Nations. Yet despite his passing bow to isolationism, the nation's shipbuilders recognized that Roosevelt's basic friendliness to the navy continued. One shipbuilder, soliciting funds for Roosevelt's 1932 campaign chest, commented that "if we are to have a treaty-strength navy, we must have someone other than a pacifist in the White House," meaning Herbert Hoover.[49]

Besides being a "big navy" advocate, Roosevelt was also a persistent foe of war profiteering, among the most vigorous in American history. As early as May 1917, while serving as assistant secretary of the navy, he had complained about yacht and fishing boat owners who refused to lease their vessels to the navy except at "outrageous prices." In 1918, while serving as acting secretary of the navy, FDR disallowed as excessive several monetary claims submitted by shipbuilding firms. When the contractors asked that their demands be referred to the U.S. Court of Claims, FDR refused. Nearly twenty years later, these events remained vivid in the president's memory.[50]

Although he remained determined to rebuild the fleet, Roosevelt was willing at least temporarily to support the munitions investigation, and he assured Nye and Vandenberg of his cooperation. In May 1934, FDR sent a message to the Senate denouncing the arms race and endorsing the proposed investigation.[51] By supporting the munitions inquiry, he hoped to win pacifist and isolationist support for his domestic program, to which he gave top priority.[52]

Roosevelt also wanted the United States to approve the still unratified 1925 League of Nations agreement to suppress the international arms trade. In June 1934, an American delegation in Geneva presented a draft of a new convention that would modify the original plan to make it acceptable to the United States. The American plan would create a Permanent Disarmament Commission, which would monitor, inspect, and license arms improvement and replacement in each country. This concept was a near duplication of the League's successful Drug Limitation Convention of 1931, suggesting that pacifists equated arms with narcotics. The Nye-Vandenberg investigation would bring publicity to the arms control issue, and this prospect served the president's purpose of winning support for an initiative of the League of Nations, which remained unpopular with the isolationist element. The Senate eventually ratified the treaty on 17 June 1935, thereby endorsing by implication the theory that unchecked private sales of arms for profit threatened the peace.[53]

There was ample reason for Roosevelt to become disenchanted with the work of the Nye committee, however, and his early support for it was short-lived. Politically, both Nye and Vandenberg were Republicans, both were believed to be ambitious, and both were being discussed as possible contenders for the Republican presidential nomination in 1936.[54] Besides furnishing publicity to his potential rivals, the line of inquiry pursued by the Nye committee became quite distressing to Roosevelt as the months wore on.

During the hearings, Raushenbush organized the munitions investigation cleverly by focusing on a number of practices identified at the close of the Great War by estranged critics of the international weapons procurement system as improper. In 1919 these distasteful methods had been generally labeled by the Covenant of the League of Nations as subject to "Grave Objections." (It became a matter of considerable dispute later whether the founders of the League believed these practices to be *facts* or simply *assertions.* Critics of the weapons industry argued that Wilson intended the League Covenant to prohibit the private manufacture of arms completely but that his wording had been diluted at the Versailles conference to the "Grave Objections" expression. Although there were no official minutes or other documentary records evidently Wilson's more stringent locution was changed at the suggestion of British military commanders. They warned that Britain could not depend on the United States government to sell them arms, but they sagely advised of a potential need to obtain weaponry from private American firms if another emergency developed.)

These "Grave Objections" were detailed by Philip Noel-Baker, formerly a professor of international relations at the University of London and Labour Party member of Parliament, and Manley O. Hudson, Bemis Professor of Law at Harvard University. Noel-Baker and Hudson charged that munitions makers kindled war scares, bribed government officials, spread false rumors, manipulated public opinion, formed international cartels to foster the armaments race, and conspired to fix prices.[55] These allegations became known as the "famous six,"[56] and bearing the imprimatur of the League of Nations, they formed a promising avenue of investigation for the Nye committee.

During the 1920s the accusations proliferated, and by the 1930s charges against the arms industry were plentiful. The copious body of antiwar literature reached a peak in the early 1930s, and numerous charges against the weapons makers wound their way back and forth throughout it. The investigators needed only to page through the many books and articles to find promising lines of attack. Even this effort was hardly necessary, as the Nye committee employed as an adviser a leading critic of the arms industry, the deeply committed and extreme isolationist John T. Flynn.[57]

The various charges floated against the arms makers were that they maintained congressional lobbies so as to make unneeded and unmerited sales; that they improperly influenced the press so as to frighten the public into purchasing arms; that they engaged in commercial bribery to generate sales; that they sold weapons to both sides in war, thus enabling and extending the bloodshed; that the U.S. government served as sales promotion agents for the armament firms; that arms makers were guilty of collusive, anticompetitive behavior that inflated prices; that profits on the arms trade were excessive; that the arms makers had, for their own profit, duped the United States into entering World War I; and finally and most important, that all of these malignant deeds fostered an arms race that menaced world peace.[58] Investigation of these many charges was a task that would stretch the limited financial and intellectual resources of the Nye committee.

One by one, the practices of the arms industry to which pacifists lodged "grave objection" were discovered to be either inaccurate or nearly harmless. An example of the weakness of the pacifist case was the charge that the arms lobby was controlled by defense contractors and that it was darkly influential. The Nye committee repeated the charge made twenty years earlier by Congressmen Henry Tavenner and Claude Kitchin and by the automaker Henry Ford that the Navy League was financed by warmongers and profiteers. Under close inspection, however, the committee investigators discovered that the Navy League received but $1,785 per year from shipbuilders. Most of its income came from members' dues and rental of real estate, and its membership was drawn from many different walks of life. The expenditures of the Navy League were nearly inconsequential when compared with the budgets of leading pacifist organizations. The annual budget of the National Council for the Prevention of War was $100,000, the National Peace Conference received $15,000 per year from the Carnegie Endowment, and the Women's International League of Peace and Freedom budgeted $50,000 for its 1931-32 campaign. Nye admitted that nothing adverse had been found against the Navy League.[59]

It was true that some defense contractors maintained lobbyists in Washington, but this was neither improper nor abnormal, if not entirely desirable. The principal builder of submarines, the Electric Boat Company, provided its lobbyist with a fifty-foot yacht, which he used to entertain naval officers and congressmen. While this was a conspicuous and certainly questionable privilege, it was insufficient to establish a case that entertainment provided by the defense industry was exceptionally lavish. Other organizations, such as the National Rifle Association and the industry-subsidized International Congress of Gun Makers, lobbied against nationalization of arms manufacture and restrictions on the overseas sale of sporting arms, but this was hardly illicit.[60]

Another line of inquiry proved even less productive. Raushenbush assigned an investigator to determine the extent to which the arms industry had exercised control over the nation's press to bring about intervention in Europe in 1917. Editorials in sixty-nine papers in twenty-five midwestern and southern states were evaluated for influence by the arms industry. The anonymous investigator reported confidentially, "I doubt seriously that any of these newspapers were bought." Most midwestern papers favored preparedness before

American entry. In Iowa (where military business was negligible), a Nye survey discovered that there were eight times as many daily papers favoring preparedness as opposing it. The fact was that there was almost no arms industry in the United States in 1914-15, American public opinion became pro-Ally even before arms sales surged, and most U.S. loans to Britain and France were to support purchases of nonmilitary goods. There was also no evidence that the fledgling munitions industry defended its stake in the Allies.[61]

Still another charge that proved to be weakly grounded was the claim that in the 1930s the international arms trade fomented war for profit. In the first place, the military trade was not a large part of most arms producers' business. In the early 1930s, no major U.S. firm depended exclusively on military contracts. The division of the Du Pont company that made powder for military use was last in sales of the firm's ten divisions. Its sales constituted only 1.08 percent of its parent firm's sales for the decade 1923-32. Bethlehem Steel's military sales were only .36 percent of total sales, although Midvale Steel sales were 12.74 percent military. In the second place, overseas sales were simply not a major portion of the arms industry; only 3 percent of world arms sales were in the export trade, only ten countries exported arms, and only three of them (Britain, France, and the United States) controlled 75 percent of the world market. Yet even Britain, the world's leading arms exporter, sold only 10 percent of its arms production in foreign markets. Of American firms, the Colt's Patent Fire Arm Company sold about 10 percent of its production abroad, and the Remington Arms Company sold only .5 percent overseas. Of course, the small present size of the world arms market did not preclude a greedy arms maker from seeking to enlarge it by setting off a war, but evidence that this was happening was missing.[62]

There was evidence, however, that some of the practices of arms salesmen were suspect. During the early 1930s, American arms firms were busily serving markets in places of international strife, most notably in Asia and in South America. These brushfire wars abroad, now mostly forgotten, raised troublesome issues that have endured. Questions arose as to when, under what circumstances, and in what degree the United States should become involved in disputes that originate abroad.

An early example was the conflict that developed between China and Japan. This commenced with Japan's seizure of Manchuria in 1931, followed by a decade of intermittent fighting. China moved to build up its military forces to meet the Japanese threat, and this reopened opportunities for arms sales—and profits—to American producers. (As early as 1915, a consortium of six American firms contributed $1,000 each to hire a former naval officer, Captain I. V. Gillis, as their sales representative in China.)[63]

The question was what should be the American policy toward Japanese aggression. Liberals, internationalists, and some pacifists demanded that trade sanctions be imposed against Japan, and their views were voiced by the

Scripps-Howard chain of newspapers. Farmers, importers, exporters, and financiers opposed trade cuts, because they were disturbed about the detrimental effects of reductions in foreign sales on the depressed economy. They were joined by religious pacifists, who worried that sanctions would invite military retaliation.[64] Their greatest fear was that the selfish interests of the arms exporters would decide this issue of potentially great national consequence. In December 1931, Representative Hamilton Fish Jr. of New York introduced a resolution to embargo all sales of arms to belligerents, a move that was intended to cut off sales to both Asian powers. In 1932 lobbying by the arms industry helped to kill the Fish resolution. In 1933 the U.S. Army Air Corps cooperated with the Curtiss-Wright Export Corporation to sell 122 warplanes to China, and Curtiss-Wright began to build an aircraft factory there. This led to the pacifist charge that American munitions makers were attempting "deliberately to exacerbate Japan's war psychosis."[65]

In the early 1930s fighting flared up in South America. The most notable of the conflicts pitted Bolivia against Paraguay over control of the Chaco Boreal, but in 1933 a second clash developed between Colombia and Peru. Since none of these nations maintained a significant arms industry, all needed to purchase arms abroad. These requirements tested American policy, as many Americans, supported by the League of Nations' Chaco Commission, believed that a U.S. embargo would throttle the belligerents and force an end to hostilities. It was troublesome and embarrassing to discover that while the war raged, no fewer than five American arms firms advertised in a single Bolivian newspaper in search of arms contracts.[66]

The Chaco war led Congress finally and belatedly to ratify the Geneva Arms Traffic Convention of 1925. This reversal also led to the establishment by the Neutrality Act of 1935 of a National Munitions Control Board, which soon began to issue licenses for arms sales abroad. Even with licensing going into effect, the Nye committee suspected that American munitions firms opposed a boycott of belligerents in order to enhance their profits. It began an all-out effort to document the sinister practices of the arms industry by exposing its behavior in South America. The Nye committee discovered a story that was much more complicated than it had imagined.

In 1920 the United States had commenced a program designed to bolster the defenses of South America by encouraging those nations to boost the size of their fleets. The theory was that if South American countries maintained powerful navies, the U.S. Navy could be relieved of part of the cost of defending the Western Hemisphere. Accordingly, on 5 June 1920, Congress formally permitted officers of the U.S. Navy to accept payments from South American countries. This would allow them to serve as advisers, while passing the cost to the South Americans. Pursuant to a policy approved by Secretary of State Charles Evans Hughes and by Secretary of the Navy Curtis D. Wilbur, a United States Naval Mission to Peru was appointed.

These American officers were expected to serve as agents who would encourage the Peruvians to purchase U.S.-made ships (which could operate in close collaboration with the U.S. Navy). The Peruvians were indecisive, so the American naval representatives tried to arouse public interest, encouraging the Peruvians to borrow to finance naval spending. American shipbuilders were nevertheless reluctant to enter the Peruvian market, fearing (for good reason) the strength of Peruvian credit. Thus the origins of American arms sales to South America had come at the request of the United States government, not the shipbuilders, who were in fact hesitant to enter the market.[67]

Even more disturbing than these and other conflicts of interest were some of the sales practices used by the arms merchants.[68] One such questionable practice was the alleged willingness of American dealers to sell to both sides in a struggle, which would needlessly amplify the carnage. Curtiss-Wright was supposedly willing to sell warplanes to both Colombia and Peru, and the Electric Boat Company was reputedly ready to sell submarines to both Argentina and Brazil. On careful inspection, however, this double-dealing was difficult to establish, since a belligerent nation was usually unwilling to patronize a firm that also supplied its enemy. I. J. Miranda, a partner in the American Armament Corporation, wrote that in the dispute between Colombia and Peru, "We can't sell to both." He advised choosing Colombia, "because they have money." (In 1930 Peru had purchased twenty-six military planes from the United Aircraft Company for $745,000. When the Depression struck, the Peruvian government defaulted, and United Aircraft was forced to accept payment in the form of 60,000 tons of guano delivered over five years. Miranda understandably did not want to be paid in guano.)[69]

Arms salesmen were also prepared to engage in commercial bribery and scare tactics, Nye investigators learned. Bribery was so rampant in dealings with South American governments that the only remarkable discovery was that the United Aircraft Export Company alone refused to employ this stratagem. Meanwhile, a Du Pont agent bribed the son of the president of Argentina, Federal Laboratories used bribery in Costa Rica, and Colt's Patent Fire Arms firm employed bribery throughout Latin America. The Nye committee was shocked when they learned that the American Armament Corporation representative casually referred to bribery money as "palm oil" and used the verb "to grease" when he meant "to bribe." American arms salesmen were also willing to use the tried-and-true tactic of warning against enemy purchases. For example, in order to sell airplanes to Peru, one firm told the Peruvians that the Colombians were "still purchasing heavily."[70]

Although evidence of commercial bribery in any form was repugnant to most Americans, the unpleasant reality was that it was the modus operandi in much of the world. For that reason, it was not entirely inexcusable, as some pointed out. To the moralists of the Nye committee, this was no justification. They argued that the real danger loomed not from the act of bribery itself but from the creation of a class of people who lived by bribing governments into augmenting their arsenals. This allegedly promoted arms races, which in turn invited the outbreak of war.[71]

The Nye committee was unable to produce proof of this link and in fact discovered evidence to the contrary. Nye summoned the president of Colt's Patent Fire Arms Company, Samuel M. Stone, to testify and insinuated that Colt's had known of the coming of the European war in 1914 and had encouraged it. When Colt's records were examined, however, they disclosed that Colt's general sales manager had indeed spent the early months of the war attempting to drum up business on a long sales trip, but the expedition took him not to Europe but to South America! In an August 1934 report to Arthur Vandenberg, Raushenbush admitted the failure of his investigators to find the proof they sought: "We have not yet found the man we are looking for, who is going around the South American countries selling them wars . . . but we have some good half-cousins of his." Unfortunately for Nye's reputation, half-truths were the best they could ever collect.[72]

Another area in which Nye's hounds thought they caught a scent was in price collusion among naval shipbuilders. The prime evidence of collusive bidding, which had been often charged, was that contracts advertised for the construction of destroyers and cruisers in private shipyards had yielded offers that were suspiciously close in price. Bids submitted in the early 1930s by four separate yards for the construction of cruisers varied less than 1 percent, which the investigators assumed was due to collusion. Collusion was easy to infer but difficult to prove. To reduce costs, the U.S. Navy frequently ordered that several ships be produced from the same design. Since the cost of materials was similar for each builder, and since the blueprints were the same, major differences in construction costs were unlikely. Moreover, the existence of a market with only one customer (the navy) and only three or four firms capable of building large warships could never be perfectly competitive. These facts did not greatly trouble the investigators, who searched industriously for bid-rigging. But a confidential memorandum circulated within the committee reported that investigators found "no evidence of collusion on the bids for destroyers." The investigator added that "frankly, after reviewing all the material, it doesn't seem to show PROOF of collusion, but it would seem that some general understanding was in existence." The basis for this allegation of informal collusion was that each shipbuilder figured the cost of construction and then added an identical profit margin—10 percent.[73] This was a very weak reed on which to hang a charge of conspiracy, but that did not bother the Nye committee, which never reported to the public its failure to establish collusion.

The Nye inquiry did uncover the complexity of the issues of warship construction when it determined that the average cost of constructing a cruiser in a private shipyard was $1.3 million less than in a government-owned facility. This was because wage rates in government yards were 20 percent higher than in private industry, the government granted twice as much vacation time, and the machinery in government yards was obsolescent. When the private ship-builders added on their profit margin, however, their prices exceeded the cost of construction in government yards. It was far from clear, nevertheless, that private shipbuilders had been overcharging the government, because in 1935 the government shipyards needed renovation in the amount of $23.6 million because of their failure to include depreciation as a cost of construction. Nor was lobbying for shipping contracts the sole province of private shipbuilders. Workers in public shipyards aggressively lobbied their elected representatives to obtain contracts. Complete nationalization of warship construction, as most of the Nye group preferred, would not necessarily have eliminated super-fluous building.[74]

The squall that finally capsized the Nye investigation developed early in 1935 when the munitions committee centered its inquiry on American entry into the Great War. As developing events in Europe and Asia became increas-ingly ominous, Nye directed the committee's attention toward preventing American participation.[75] The centerpiece of the plan was to establish that the United States had been fooled into entering the war by greedy businessmen.

The chief obstacle to making this case was that it required a grave indict-ment of the Wilson administration. Franklin D. Roosevelt, whose patronage the Nye committee needed, was only one of many Old Wilsonians who still participated actively in the nation's affairs. Woodrow Wilson had a legendary ability to inspire affection among his subordinates, and their devotion was as lasting as it was deep.[76] Wilson loyalists rankled when the upstart Republican from North Dakota criticized their captain.

Former secretary of war Newton D. Baker was among the first to point out conspicuous flaws in the variant of the Merchants of Death theory espoused by the Nye group. Baker noted that when war broke out in Europe there was no munitions industry in the United States that could dictate policy. In fact, one of Baker's main problems had been a major shortage in defense contrac-tors capable of producing the implements of war that the nation required.[77] Investigations by the Nye committee confirmed Baker's analysis, but this in-convenient discovery was kept confidential.[78]

Not troubled by the absence of a munitions industry to impeach, the mu-nitions committee turned to bankers, who were an easy target in the depths of the Depression. The argument here was that the bankers had loaned millions to the Allies, and the presence of these debts deterred Wilson from imposing a trade embargo.[79] It was an undeniable fact that the United States had accumu-lated a huge credit against the Allies and that the British representative in the United States, the House of Morgan (which Nye deeply distrusted),[80] had ac-tively encouraged the extension of these loans. The awkward aspect of this sce-nario was that *before* the grant of credit (as Secretary of State Robert Lansing noted at the time) public opinion had "crystallized in favor of [Great Britain] to such an extent that the purchase of bonds would in no way [compromise neutrality] or cause a possibly serious situation."[81] The Wilson administration had thus considered the potentiality that the loans could commit the United States involuntarily to the Allies, but it had rejected this danger because public opinion already supported them. The bankers, moreover, were pro-British before their profits began to rise due to war sales. Lansing argued in his memoirs that American businessmen who were profiting from sales created by the war in Europe *opposed* U.S. entry. This stood the Merchants of Death theory on its head.[82]

The Nye committee uncovered additional embarrassing evidence that did not fit its munitions-maker conspiracy thesis. Secretary of State Lansing had published contemporary letters that revealed his state of mind at the time he became an interventionist. "Everywhere German agents are plotting and intriguing," wrote Lansing. "From many sources evidence has been coming until it would be folly to close one's eyes to it. . . . German absolutism is the greatest menace to democracy." This confirmation that the Wilson administration was doing in private exactly as it publicly stated was part of a confidential memorandum by Stephen Raushenbush circulated within the Nye committee. Since it did not fit the munitions-maker hypothesis, it did not become part of the committee's findings.[83]

The Nye committee's most serious blunder was its ad hominem attacks on Old Wilsonians. The first person subjected to reproach was Bernard Baruch. Since a major goal of the Nye committee was to destroy the view that the War Industries Board had worked cooperatively with selfless dollar-a-year men, Baruch as an obvious target. Baruch, moreover, disagreed with the fundamental direction of the Nye investigation. He testified that in his opinion high taxes during wartime might interfere with war production, which was the view held by most departments of the Roosevelt administration, including the War Department. As a result, Nye investigators sought to examine Baruch's income tax returns for 1917 and 1918, evidently in order to establish a claim that the WIB chief was a war profiteer himself. The tax returns, which were almost two decades old, had been destroyed. The Nye committee made this public, implying that Baruch had concealed his ill-gotten gains. Baruch denounced this testimony appropriately as "cheap and unjust."[84]

Although Baruch was Nye's most prominent opponent in the defense contracting debate, he was a minor personage compared with the major quarry, Woodrow Wilson himself. Nye and Raushenbush hoped to demonstrate that Wilson had lied to the American people about the reason for American intervention in 1917. The committee released evidence purporting to show that Wilson and Lansing had been informed in April 1917 of the existence of the famous "secret treaties" linking victory to territorial gains for the British and French empires. Since Wilson and Lansing later testified that they had been unaware of the treaties' existence, this would establish that the Wilson administration had lied to the American people about the fundamental purpose of the war.[85] If the publicly stated reasons for American entry could be demonstrated to be untrue, Nye's revisionist explanation might be more acceptable.

Nye's charge of falsification was technically accurate (more recent leaders have been indicted in federal courts for less flagrant falsehoods),[86] but it was nevertheless politically imprudent. Wilson's many living admirers, including Secretary of State Cordell Hull, rallied to defend their fallen leader's reputation. The Old Wilsonians proved to be far more powerful than the brash young Republican from North Dakota. Early in 1936 the Democratic leadership in the Senate discontinued funding for the investigation, and Nye had no choice except to wind up the inquiry.[87]

The bitter controversy over American entry into World War I obscured the impact of the Nye committee on the central problem it was originally commissioned to investigate, namely, war profiteering. The Nye committee's recommendations on this question were, to use Nye's apt characterization, "drastic."[88] Based on the ideas of the isolationist author John T. Flynn (who later was a leading member of such isolationist organizations as the America First committee and the Keep America Out of War Congress), they called for the enactment of no less than three constitutional amendments. The first would have allowed the government to commandeer industry, the second to tax without uniformity (so as to soak up war profits directly), and the third to tax interest on tax-exempt securities in wartime. Flynn believed that the prospect of heavy taxation would make American businessmen oppose war, and they would exert their influence to keep the United States out of the coming European conflict. His hopes were echoed by a liberal journal, the *Nation:* "Business men would become our leading pacifists." The constitutional amendments, as well as two bills to regulate profits in warship construction, died in congressional committees.[89]

Despite its failure to gain congressional approval, the Flynn/Nye plan suggests the fundamental radicalism of 1935, which was the high-water mark of the antiprofiteering campaign. The main recommendation was that the highest personal income allowed during wartime would be $10,000. The federal income tax would commence at an annual income of $1,000, gradually increasing to 100 percent on income over $10,000. All business profits above 3 percent on invested capital were to be confiscated by a tax of 100 percent. This tax plan, of course, was hardly new; it was very close to what Nye had advocated well before the investigation commenced.[90]

The most immediate effects of the Nye investigation came in 1935. The Nye inquiry contributed to the national mood that produced the Neutrality Act of 1935, which established the National Munitions Control Board.[91] Chaired by the secretary of state but including representatives of the War, Navy, Treasury, and Commerce Departments, the Munitions Control Board was granted authority to license peacetime arms sales abroad. This innovation, which would eventually have passed without the Nye committee, was a variation of earlier proscriptions against trading with the enemy in that trade with a *potential* enemy was now impermissible. The Munitions Control Board held its first meeting on 24 September 1935, and in the next two months approved 132 licenses for arms export to thirty-four countries, excluding Italy and Ethiopia. In one instance, the Curtiss-Wright Export Company was denied a license to sell pursuit planes to Bolivia. The firm was permitted to fly the planes with bomb racks removed to Chile, from which they illegally flew on to Peru. Curtiss-Wright was indicted, and when the Supreme Court upheld the constitutionality of the Munitions Control Board, the firm had to pay a $282,000 fine. Curtiss-Wright officials were fined $2,000 each, and some were jailed for a year and a day.[92]

The long-term effects of the general inquiry into war profiteering during the 1930s and of the Nye committee have proved more difficult to gauge. The profiteering allegations touched off an impassioned national and even international debate on the causes of war. Millions of Americans joined Nye in embracing the Merchants of Death theory of the coming of World War I.[93] On 6 April 1935, the eighteenth anniversary of America's war declaration, 50,000 veterans paraded through Washington in a march for peace. Six days later, 175,000 college students staged a one-hour strike against war, and by 1935 the groups and individuals endorsing the Merchants of Death theory to at least a limited degree included the Veterans of Foreign Wars, the railroad unions, the American Federation of Labor, the National Grange, the National Education Association, the Women's International League of Peace and Freedom, the Democratic Party, President Roosevelt, and the Congregational Church. Other supporters included influential newspapers such as the *Baltimore Sun, Chicago Tribune,* and the Hearst and Scripps-Howard chains, influential journals of opinion such as the *Nation,* the *New Republic, Living Age,* the *World Tomorrow,* and the *Christian Century,* and such prominent scholars as Charles Beard, Charles Tansill, and Merle Curti. Of these, the most important was undoubtedly the Democratic Party, which pledged in its 1936 platform "to work for peace and to take the profits out of war; to guard against being drawn, by political commitments, international banking or foreign trading, into any war which may develop anywhere." Evidently seeking Nye's endorsement in his campaign for reelection, President Roosevelt issued a ringing manifesto. "If we face the choice of profits or peace," the president declared at Chautauqua, New York, "the Nation will answer—must answer—we choose peace." One week later, FDR met with Nye at Hyde Park, New York, but Nye, who had been passed over as a potential Republican nominee because of his isolationism, refused to endorse the Democratic presidential nominee.[94]

Against this powerful array of influence, a smaller group rejected the Merchants of Death analysis. This included the War Department, the National Rifle Association, Old Wilsonians such as Bernard Baruch and Cordell Hull, industry journals such as *Business Week* and *Army Ordnance,* business spokesmen such as Pierre du Pont and Thomas Lamont (of the Du Pont Company and the House of Morgan, respectively), and such prominent scholars as Norman Angell, Charles Tansill, and Jacob Viner.[95] Angell, a leading British pacifist for twenty years, was the most vigorous spokesman, writing or reissuing such volumes as *The Great Illusion, The Fruits of Victory, The Unseen Assassins,* and *Does Capitalism Cause War?* as well as delivering numerous addresses.

Pierre du Pont, however, presented the most concise and cogent rebuttal by arguing that wars are fundamentally bad for business. They are short, and therefore a business cannot depend upon war profits for survival. During peacetime, only about 2 percent of Du Pont's gunpowder production was for military purposes. During wartime, a firm must expand hastily and expensively during a tight market for both credit and labor. Civilian production lines will not produce military goods and must be scrapped. Heavy taxes absorb profits. Thousands of untrained workers must be hired, although the quality of their work is dubious and their accident rate is high. Extra security guards must be hired to protect against sabotage. Wars end abruptly, causing the military market to vanish almost overnight. The greatest gains go to raw materials producers—Du Pont was least accurate on this point—who reap windfall profits but are not blamed for stirring up the conflict, as are processors.[96]

Despite their weakness in numbers and prestige, Nye's opponents eventually carried the day, largely because the Merchants of Death theory implied support for isolationism, which became increasingly out of vogue as the danger of fascist aggression loomed closer.[97] By 1939 German, Japanese, and Italian belligerence offered ample evidence to persuade all but the most faithful believers in the simplistic theories of Nye and others.

Although the devil theory of war did not hold up, the war profiteering quarrel had other substantial effects, both in the United States and abroad. In Britain, Prime Minister Ramsay MacDonald ordered the establishment of a counterpart to the Nye committee called the Royal Commission on the Private Manufacture and Trading in Arms. After holding hearings during 1935-36, the Royal Commission rather cautiously reported that modern arms increased the fear of war, that arms makers were sometimes cynical, and that they employed too many former military officers. The commission also found that arms makers did not foment war scares nor did they control the press. Since Britain had desperately needed foreign arms in order to defeat Germany (and might do so again), the commission concluded that it was best to keep the arms business private.[98]

In France, the Chamber of Deputies also conducted a munitions investigation. Premier Pierre-Etienne Flandin argued, like Franklin D. Roosevelt, that taking the profits out of war "is the best means to stop war."[99] Although the antiprofiteering, antiwar climate led to its action (rather than the munitions in-

vestigation alone), France nationalized its arms industry. The unhappy effect
of that decision appeared in 1940: when Nazi armies invaded France, French
defenders were fatally impaired by underspeed planes and underweight tanks.
Investigations of defense contractors also followed in Canada, Argentina, and
Sweden, without decisive results.[100]

The war profiteering controversy of the 1930s and the Nye committee in
particular had some considerable effect upon the mobilization for World
War II. Unquestionably, the Nye inquiry and the general clamor for a reduction in war profits strengthened the determination—already great—of the
Roosevelt administration to thwart war profiteering as much as was practicable. In fact, Roosevelt became the most frequent critic of war profits of all
American presidents.

The Nye/Flynn plan for preventing war profits eventually gave way to the
fundamental concepts advocated by Bernard Baruch and the U.S. Army for the
new mobilization. Baruch, the leading critic of the Nye/Flynn plan, argued that
taxation, while it should be high, should not be confiscatory, as that would
have disincentive effects. It would also be impossible for firms to pay 100 percent excess profits taxes on profits deriving from appreciation of inventory and
raw materials, as that would force them to sell off their stock. Baruch likewise
pointed out that the Flynn plan would not allow high-cost producers to contribute, although their products would be greatly needed. Finally, the Flynn
plan would favor companies with large debts over those with substantial cash
on hand. Baruch argued that price control was the indispensable ingredient to
the control of profiteering.[101]

Still, the Nye committee affected efforts to control war profiteering during
World War II. Among the Nye committee's most controversial qualities was its
self-righteous, crusading style. Although Nye's desire to avoid another horrific
slaughter was probably well intentioned, his committee's methods approached, if they did not exceed, the limits of propriety. Besides overlooking
embarrassing facts whenever truth was inconvenient, Nye and his committee
engaged in demagoguery and character assassination that were eerily close to
the methods of Senator Joseph R. McCarthy, who investigated the U.S. Army
two decades later. Nye charged, for example, that "the munition-makers . . . are
welded in a shrewd and secret internationale of intrigue . . . without scruple or
conscience. . . . They can rule and ruin us." Nye claimed also that "in this investigation we are penetrating the heart of the competitive, profit-seeking business system of the world." Whereas McCarthy investigated the army, Nye said
that he was placing the War Department on trial.[102]

Few if any speakers have exceeded McCarthy's skill at colorful tongue-
lashing, but Nye came close. Nye charged that munitions makers engaged
in "racketeering . . . it means going out and fostering war spirit, building the
spirit of hate and fear and suspicion among nations." He denounced "public
enemies . . . who threaten and kill for profit": "Public Enemy No. 1 should be

the munitions maker who wants to sell his powder and poison gas. . . . Public Enemy No. 2. is the banker who raises money to pay for the munitions. . . . Public Enemy No. 3 is the industrialist who knows that the only way to get fascism established in America is to get the country into a war. . . . Public Enemy No. 4 is the American who goes into the war zones to make money."[103]

Just as McCarthy was not alone in abusing opponents, Nye's colleagues were partners in misconduct. Senator Homer T. Bone called upon his associates to "assume an attitude of sportsmanship." Bone nevertheless referred to Sir Basil Zaharoff as "the gentleman who has made mass murder the pastime of the world."[104] An angry exchange between Senator Bennett Clark and Irénée du Pont went as follows: "du Pont: Do you mean that President Wilson was dragged into the war by the heels at the insistence of the munitions makers? Clark: I do not. I mean you set in motion a series of events that led us into war." When du Pont protested this allegation, the senator interrupted and refused to allow him to defend himself against this most serious charge.[105]

Senator Bone remained the uncrowned master of hyperbole: "Everyone has come to recognize that the Great War was utter social insanity, and was a crazy war, and we had no business in it at all. Oh yes; we heard a great deal of talk then about 'freedom of the seas.' Whose seas? The seas upon which were shipped munitions of war which served only to enrich a comparatively small group of men, and whose enrichment cost this country a staggering price. . . . Freedom of the seas! Out with such nonsense! For the sake of this fantastic theory that could at best serve the few and not the many, thousands have died, and our hospitals are filled with insane boys who had a right under God's providence, to live their lives in peace. What a distortion! . . . National interests! In God's name, whose interests were they? They were the 'interests' of the business profiteers and bankers. They were not the 'interests' of those boys who are now in the insane asylums. . . . They were not the interests of the boys whose broken bodies lie in French soil. They were in the potential interest, if not the direct interest, of the 23,000 new millionaires whose fortunes were soon to be coined from widows' and orphans' tears." Not to be outdone in the demagogic defense of women and children, Nye added that "women and children who did lose their lives were, in effect, camouflage for covering up shipments of munitions of war."[106]

The arguments of the munitions committee and other Great War revisionists had a deep and lasting effect on public opinion. In January 1937, nearly a year after the conclusion of Nye's investigation, the Gallup poll asked Americans if they thought it had been a mistake to enter the Great War. A rousing 70 percent of the respondents declared that it had been a mistake to participate, and 82 percent favored prohibition of the sale of munitions by private parties. More than two years later, the persuasive power of the Merchants of Death theory was still very evident. In October 1939, after Europe's muskets had begun to flame again, Gallup asked the question a second time. Sixty-eight percent of Gallup's respondents still said that fighting the Great War had been a mistake. When Gallup asked why the United States had entered the war, only 18 percent of Americans said that it was in defense of a just cause. Thirty-four percent of Americans answered that the nation was a "victim of propaganda and selfish interests."[107]

Nye, his committee, and the Merchants of Death theory have not lacked defenders. Dorothy Detzer, anticipating the defense of Joseph McCarthy, claimed that "no Senate committee ever rendered to the American people a more intelligent or important service. It was the nation's loss that it did not comprehend it." Charles Beard chaired a committee supporting Nye's reelection campaign in 1938. More recent scholars such as John E. Wiltz, Charles DeBenedetti, Anne Trotter, Paul Koistinen, and Anthony Sampson have treated the inquiry sympathetically, if not uncritically. Robert James Leonard likened Nye to Thomas Jefferson and William Jennings Bryan, who were allegedly well-meaning supporters of "the cause of Agrarian Radicalism." In 1987 Joan Hoff-Wilson maintained that the Nye committee findings confirmed the economic analysis of the origin of war advanced by Jeannette Rankin, the congresswoman who voted against American participation in both world wars.[108]

It was an ex-doughboy, however, who became the most notable student of the Nye committee. During World War II, Senator Harry S Truman of Missouri, who had served in France as a captain of artillery, headed a committee bearing his name that investigated abuses in defense mobilization. Truman reviewed the Nye committee hearings, denouncing them with considerable accuracy as "pure demagoguery." He became determined that the Truman committee would not follow the disgraced path of Senator Nye. As a consequence, the Truman committee was comparatively cautious in its investigations of profiteering during World War II, and it has generally received high marks from historians.[109]

Finally, the findings of the profiteering inquiries of the early 1930s certainly helped to form public support for the Roosevelt administration's antiprofiteering effort during the Second World War with its high nominal income tax rates, but that was not quite the same thing as controlling profiteering effectively. The massive mobilization for World War II caused profit-control problems that were far more complicated than the misguided investigators of the 1930s ever imagined.

10

Penning the Warhog

I don't want to see a single war millionaire created in the United States as a result of this world disaster. I think everybody is entitled to a reasonable profit.

FRANKLIN D. ROOSEVELT (1940)

By the late 1930s widespread military aggression gave warning to the world that another deadly maelstrom was forming. In the United States, internationalists and isolationists passionately debated the appropriate American role in the coming misfortune, and the ongoing campaign against war profits came to assume life-or-death proportions. Opponents of American military engagement battled to eradicate what they believed was the most sinister danger facing the nation—trading in arms for profit. Internationalists campaigned to reinforce the nation's defenses against what they thought was the paramount threat—open attack by malicious enemies.

As Americans debated participation in what would become the most expensive war in their history, circumstances were uniquely favorable for a successful campaign against war profiteering. The social memory of profiteering during the Civil War and the Great War still gripped the popular imagination. The antiprofiteering constituency was large, determined, and well organized in Congress. The White House was occupied by an experienced, able, and popular leader who spoke eloquently and often against profiteering. The reservoir of support for antiprofiteering measures was therefore broad and deep.

Only twenty years had passed since new techniques of controlling profits were developed for the Great War, and the experience remained fresh in the memories of executives seasoned during that mobilization. During the interwar period, there had been ample opportunity to reconsider how to manage sacrifice equitably, and the chance had not been entirely overlooked. Economic knowledge was more advanced in 1941 than in 1917, and as the United States prepared for battle the procedures for controlling wartime profits were becoming better developed than ever before. The United States could and did benefit from the experience of Britain and Canada, which had a head start in learning to share the sacrifices imposed by mobilization.

The New Deal Congress and Profiteering

The campaign against profiteering during World War II may be dated from a late-night session of Congress in March 1934. The issue once again under debate was the appropriate size and cost of the nation's military forces. The result was a compromise known as the Vinson-Trammell Act, which appropriated $470 million for naval construction. This measure realized a long-standing aspiration of Franklin D. Roosevelt and other internationalists for a treaty navy. The U.S. Navy would finally become the full equal of the British navy as provided by the Washington and London naval treaties of 1922 and 1930.[1]

The naval buildup was intensely disliked by the isolationist element in Congress. Amid cries of excessive profits in naval and aircraft construction contracts during the long session, the "little navy" faction, in alliance with a leftist contingent known as the Mavericks, forced the inclusion of profit restrictions in the Vinson-Trammell bill. Designed to thwart profiteering, this change limited profits on shipbuilding contracts to 10 percent of the cost of the contract and profits on all aircraft construction contracts to 12 percent of the cost. The limits applied to all contracts valued in excess of $10,000, and the unallowable profits were seized by means of imposing an excess profits tax of 100 percent on the compensation that exceeded the limits. This levy was the first true peacetime excess profits tax in American history (disregarding the munitions tax of 1916). The restrictions on profits on defense contracts became known as the Tobey amendment after its isolationist author, Republican senator Charles W. Tobey of New Hampshire.[2]

The discrepancy between the acceptable profit levels on ship and aircraft contracts (10 versus 12 percent) resulted from a political compromise—military critics in the House of Representatives wanted 10 percent on both ships and aircraft whereas the Senate held out for the 12 percent figure for airplanes. A faction of western senators, known as the Sons of the Wild Jackass,[3] preferred spending defense funds on aircraft rather than on warships. They regarded airplanes as a cheap means of building a continental defense, and they perceived naval vessels as expensive subsidies to northeastern financial interests. Warships were also judged undesirable because ships seemed more likely to be employed in distant imperialist wars.

The Tobey amendment, hastily attached to the Vinson-Trammell Act almost as an afterthought, was never debated by the appropriate congressional committees. It proved to be exceptionally ill advised: one New Dealer later described it aptly as "a snare and a delusion." It applied to only two industries—shipbuilding and aircraft—leaving all other military contractors unaffected. Hence a contractor could manufacture a machine gun for a tank without being subjected to a profit restriction, but if a machine gun were installed on an airplane, the 12 percent limitation applied. The limitation on profit to a percentage of cost also encouraged wasteful charges. The higher the contract price, the greater the allowable profit. The profit limitation bore no relationship to the contractor's actual investment or risk, and it was an administrative nightmare. Every subcontractor for minor parts on ship and aircraft contracts had to establish an elaborate and expensive cost accounting system that could differentiate between that part of his business which was covered by the Vinson-Trammell Act and that which was not. He also had to admit to his factory a corps of federal auditors who would police his costs. In a perplexing borderline case, there was even a question whether an amphibious aircraft legally fell under the 10 percent restriction as a boat or under the 12 percent restriction as an aircraft. Finally, as an excess profits tax the Tobey amendment was a poor producer of revenue. In its first six years of operation, the Treasury Department recovered only $3,700,000 on a total of 2,577 contracts and subcontracts.[4]

Despite their clumsiness, the profit restrictions remained in effect until 1940. Although the restrictions partially satisfied the isolationist appetite, they severely damaged American defense readiness. They did not seriously impair the ship construction program, which proceeded on schedule, but the limitation on profits did serious damage to the aircraft industry, as Jacob Vander Meulen has shown and as the Roosevelt administration realized. Throughout the 1930s the aircraft industry was starved for funds and basically survived due to the dedication of its leaders and to export sales. This lack of support for military aviation by the Congress exacted a terrible price; when the United States entered the war, many brave pilots lost their lives in slow and obsolete aircraft.[5] Only the Curtiss P-40 even approached the ability of state-of-the-art enemy planes, and it arrived belatedly. For once in American history, profits in the defense business were much too low.[6]

The lesson of the price of technological obsolescence during the Great War was there for anyone to learn, but it was not the only lesson that was disregarded. In an extraordinary irony, the provision of the Vinson-Trammell Act that allowed contractors a greater profit on sales of warplanes to foreign governments proved to be counterproductive. Intended to rein in the "Merchants of Death," the profit restriction actually encouraged military contractors to go abroad seeking markets for warships and warplanes. It invited American firms to overlook the needs of the United States military in preference for more lucrative foreign markets, often in Latin America.

Besides the Vinson-Trammell Act, the antiprofiteering campaign of the late 1930s featured a continuation of the American Legion's campaign for a universal draft. First approved by the Legionnaires in 1922, the universal draft idea appeared in various forms for the next fifteen years. The versions differed slightly, but the plan intended to equalize wartime sacrifice through a combination of excess profit taxes, price fixing, and conscription of manpower, capital, and labor. The House Committee on Military Affairs held hearings on the plan in 1924, and it remained before the Congress continually until 1938.[7] Although the excess profits taxes, price fixing, and conscription of manpower and capital provisions were broadly acceptable, one aspect of the universal draft concept repeatedly led to its defeat. The American Federation of Labor was resentful of the wartime inflation of 1917-18, and it used its considerable political influence to thwart any form of the universal draft plan that included conscription of labor.[8]

In 1935, at the peak of pacifist enthusiasm, the House of Representatives voted to approve a variation of the universal mobilization concept that met the AFL objection. The Legion plan emerged almost intact from the Committee on Military Affairs, whose chairman, Congressman John J. McSwain of South Carolina, was its longtime champion. But to placate the AFL, the House, after a six-day debate, deleted the provision establishing a labor draft, enabling the measure to win overwhelming approval. Without conscription of labor the "universal draft" was hardly universal, so the plan died quietly in the Senate Finance Committee. Although the universal draft never again came close to passage (partly owing to McSwain's death), its continuing advocacy brought publicity and support for its major features—excess profits taxation and price fixing. These provisions formed the heart of the antiprofiteering campaign during World War II.

FDR's Campaign against War Profits

During the prewar period Franklin D. Roosevelt led the nation with great political courage and exceptional wisdom. He rightfully called upon his often reluctant countrymen to resist the menace of fascism. He led a successful campaign to begin to build the tools necessary to defeat the Axis—50,000 planes each year. But he also led in asking for the conversion of the economy to be managed in such a way that no one gained an unfair advantage. If the sacrifices could be borne equitably, as a great democracy should, there would be no bitter aftertaste when peace was restored. There would be no angry investigation by a second Gerald Nye.

By 1937 the emerging danger of war compelled increased attention to American defenses. In January 1938, following new Japanese attacks in China, Roosevelt asked Congress to accelerate the ship construction program by 20 percent. In October 1938, reacting to Adolf Hitler's announcement of increased spending on fortifications, the president asked Congress to appropriate an additional $300 million, to be spent mostly on aircraft. In January 1939, he raised the figure to $500 million.[9] These increases were only the beginning, as expenditures rose massively in every year until 1945. Appropriations of this magnitude forced a general reconsideration of military contracting procedures, which commenced before the United States formally entered the war.

Although Congress ultimately approved Roosevelt's requests, it did not do so without opposition. The isolationist, continentalist, antiprofiteering persuasion held significant strength in Congress and the nation. The president therefore mounted an exceptionally vigorous campaign to neutralize his adversaries by promising to stop profiteering. He was ideally qualified to do so. Of all American presidents, Franklin D. Roosevelt had the greatest amount of direct personal experience with the administration of military contracts, having served for eight years as Wilson's assistant secretary of the navy. Only Herbert Hoover, who was Wilson's food administrator, could rival Roosevelt in experience.

Roosevelt's chief responsibilities as assistant secretary were to supervise the navy's shore installations. He was much involved in purchasing commodities consumed by the navy: coal, textiles, shellac, tobacco, and so forth. Roosevelt became a foe of collusive bidding and sought in various ways to break it up.[10] Charges that defense contractors had profited improperly were personally damaging to Roosevelt and politically damaging to the Wilson administration, and Roosevelt attempted to deflect them. In testimony before a congressional committee, Roosevelt disagreed publicly with his immediate superior, Navy Secretary Josephus Daniels, who was extremely suspicious of contractors' profits. Roosevelt privately asserted that "nine-tenths of these allegations are entirely false."[11] Roosevelt took the position that profits were lower than Secretary Daniels claimed. By World War II, however, Roosevelt had moved closer to Daniels's position.[12]

To an astute politician like Franklin D. Roosevelt there were several lessons to be learned from the Merchants of Death controversy of the early thirties. He realized that the Senate munitions investigations were immensely popular and that they could be used to defeat his political goals and to damage his reputation. Roosevelt was determined not to let that happen. In a May 1934 message to Congress pledging his cooperation with the leaders of the inquiry, Senators Gerald Nye and Arthur Vandenberg, Roosevelt very nearly endorsed the Merchants of Death theory. He declared that the arms race was a "grave menace to the peace of the world . . . due in no small measure to the uncontrolled activities of the manufacturers and merchants of engines of destruction."[13]

Yet only six months later, Roosevelt preempted Nye and Vandenberg by announcing that he had formed his own committee to study war profits. He commented that "those of us who served in the World War know that we got into the war in a great hurry. . . . As a result, we muddled through the war and did a lot of things we should not have done. . . . We have decided that the time has come when legislation to take the profit out of war should be enacted. . . .

Everybody in the country knows what munitions profits and other profits meant during the World War. . . . A good many things happened that, perhaps, headed us for the unfortunate ten-year period [the Depression], . . . such as overproduction, enormous salaries, enormous profits, and . . . unequal mobilization."[14]

Roosevelt's distinct reasons for organizing his own study of war profits remain obscure. He often veiled his real purposes, once remarking, "I never let my right hand know what my left hand does." The White House munitions "investigation" was a prime example of such practiced ambiguity. Conceivably, Roosevelt was looking for jobs for Hugh S. Johnson and Bernard Baruch, whom he appointed as the cochairmen, but more likely he aspired to undercut and co-opt the Nye investigation. He almost certainly distrusted the direction that the isolationists Nye and Vandenberg might take. Nye's ambition was to nationalize weapons production, and the Senate investigation might build momentum for nationalization. Roosevelt opposed this, and therefore he resolved to keep firm personal control over mobilization policy.[15]

Whatever his precise purpose, Roosevelt nevertheless helped the Nye committee obtain its initial appropriations, and he maintained that both the Nye committee and his own committee were working toward a "common objective—to take the profits out of war."[16] Yet although Roosevelt had publicly proclaimed his opposition to war profiteering, as early as 1934 the depth of his commitment to controlling profiteering became suspect. The Baruch-Johnson war profits investigating committee never met, nor did it issue a written report.[17]

As the possibility of war increased in the late 1930s, Roosevelt developed a new reason for opposing war profiteering. The immense popularity of the campaign against war profits demonstrated to Roosevelt and everyone else who thought about it that if the American people were to be persuaded to support larger military appropriations they would have to be assured that there would be no repetition of the profiteering controversy of 1917-18. In vetoing a 1935 bill to satisfy the doughboys' demands for a postwar bonus, Roosevelt wrote, "I have much sympathy for the argument that some who remained at home in civilian employ enjoyed special privilege and unwarranted compensation. That is true—bitterly true—but a recurrence of that type of war profiteering can and must be prevented in any future war." But although FDR demurred, Congress was in a mood to rectify the error of the past war. When the bonus bill was introduced in the election year of 1936, both the House and Senate overrode the veto, and the doughboys finally got their cash bonus.[18]

FDR became convinced that the United States must assist England and France in their stand against fascism. He also urged the nation to rearm in hope of possibly deterring a future German attack. But he realized he could never hope to win his countrymen's support for rearmament if they believed that the sacrifices it would entail would be borne unfairly. Referring to the profits of the Great War, Roosevelt pleaded, "I invite the Congress and the veterans to join with me in progressive efforts to root a recurrence of such injustice out of American life."[19]

Throughout the prewar period, Roosevelt remained an articulate, forceful, and frequent opponent of war profiteering.[20] When he requested an increase in defense spending in 1938, he asked for companion legislation for the "prevention of profiteering and equalization of the burdens of possible war." In a statement following the Nazi invasion of the Netherlands in 1940, Roosevelt said, "I don't want to see a single war millionaire created in the United States as a result of the war disaster."[21] Shortly after the invasion of Norway, he declared in a radio address that "our present emergency and a common sense of decency make it imperative that no new group of war millionaires shall come into being in this nation as a result of the struggles abroad. The American people will not relish the idea of any American citizen growing rich and fat in an emergency of blood and slaughter and human suffering." Almost every month throughout the course of the war a similar sentiment emanated from the White House. The president publicly declared his determination to arrest war profiteering in press conferences, fireside chats, messages to Congress, executive orders, and public addresses—a total of at least thirty-four occasions. Privately, he confirmed his distaste for profiteering to his personal aide and close confidant, William D. Hassett.[22]

The president's antiprofiteering campaign had several effects. Every new act of fascist aggression weakened the theory that arms "traffickers" were manipulating the United States into war again, and certainly Roosevelt's constant assurances that war profits would not be allowed also contributed to the demise of the Merchants of Death hypothesis. As Germany's army swept across Europe and then launched a devastating air attack on Britain, a dramatic change in public opinion took place. A Gallup poll taken in November 1940 found that the American people, by a margin of 41 percent to 39 percent, had reversed their verdict of a year earlier: the Great War had not been a mistake.[23] The increasingly clear danger posed by Nazi Germany undoubtedly accounted for the switch, but Roosevelt's antiprofiteering policy helped to make his rearmament program more palatable to the numerous isolationists, pacifists, and other doubters who opposed it.

In keeping with Roosevelt's words, and in deference to the political strength of the antiprofiteering bloc in Congress, until 1940 the Vinson-Trammell Act remained the centerpiece of the effort to control war profits, despite its glaring deficiencies. In October 1939, Roosevelt successfully deflected an attempt by a contingent of liberal congressmen led by Jerry Voorhis of California to broaden Vinson-Trammell. The Voorhis plan would have made the Vinson-Trammell profit restriction apply to all munitions manufacturers, not just shipbuilding and aircraft. But in 1939 Roosevelt wanted neither to levy new taxes nor to do anything that would inhibit the prosperity deriving from steadily rising sales of arms to foreign customers. As he told his chief political aides early in 1940, "Let's be very frank. These foreign orders mean prosperity in this country and we can't elect the Democratic Party unless we get prosperity and these foreign orders are of the greatest importance."[24]

The Demise of Profit Limitation

It is one thing to set forth noble ideas, another to put them into effect. Just as the aircraft industry would strain mightily to fill the president's request for new planes, the nation's political system would be greatly strained to fulfill its democratic obligation of equal sacrifice. In 1940 Roosevelt's mobilization policy included a comprehensive program that would eliminate the cumbersome Vinson-Trammell Act, encourage business to enter defense production, and establish a new excess profits tax to distribute the economic sacrifice equitably. Such a program would be difficult to obtain at any time, but in 1940 Roosevelt's ability to achieve prompt action was hampered by two very important factors: a multitude of voices were competing with little direct guidance to shape the new antiprofiteering program, and it was an election year.

If the Vinson-Trammell Act had been an efficient encouragement to American firms to produce military goods for Britain and France, it would have been consistent with American policy, but it was ineffective even in this respect. Two important provisions of Vinson-Trammell worked to discourage military production for any market, domestic or foreign. The most important stipulation concerned the depreciation of tools and facilities used for the manufacture of ships and airplanes—they could not be written off against tax liabilities any more quickly than facilities used for the more stable civilian markets. The second (and less burdensome) provision specified that all contracts for ships and planes had to be let through competitive bidding.

During the National Defense period, American businessmen were hesitant to expand or convert their plants to defense production. They nourished a reasonable and well-founded resistance to the idea of tooling up to serve a temporary market that would abruptly evaporate when the emergency ended. This apprehension originated in unpleasant experiences after the Spanish-American War and again after the Armistice of 1918. More recently, painful memories of the thousands of business failures during the Great Depression magnified their nervousness. In 1932, at the low point of the Depression, steel production in the United States fell to only one-sixth of capacity. A steel executive who had experienced such a decline would be imprudent if he were not cautious about building new mills. Although American businessmen generally backed Roosevelt's internationalist foreign policy, they understandably did not relish exposing themselves to new charges of being fiendish merchants of death. Eugene Grace, who had been vilified by the Nye committee, remained in charge of Bethlehem Steel. Along with other influential steelmen and most of the nation's industrial leaders, Grace was militantly hostile to Franklin Roosevelt's social policies.[25]

The Vinson-Trammell Act presented several serious obstacles to conversion. Military contractors regularly complained that under Vinson-Trammell they could never bid a contract accurately because the Internal Revenue Service refused to specify in advance which production costs would be accepted as legitimate tax deductions. In particular, a provision of Vinson-Trammell specified that tools used for the production of weapons could not be depreciated more quickly than those used to make civilian products.[26] This restriction created an odd bottleneck. Products covered by Vinson-Trammell tax rules were frequently produced more slowly than those which were not encumbered by Vinson-Trammell depreciation rates. Since a weapons system usually required tools of both types, its components would often be finished at substantially different rates, delaying final assembly until all elements were available. In peacetime, this impediment was a costly and dangerous nuisance, but once war was actually impending it became intolerable.

The problem of depreciation rules on defense contracts uncovered a serious deficiency in the Roosevelt administration's antiprofiteering program. The ancient dispute between civilian leaders who wanted to manage wartime sacrifice equitably and military commanders who wanted to obtain war matériel quickly reappeared in a new guise in the early mobilization for World War II. This time it took the form of a bureaucratic clash between New Dealers in the Treasury Department and more business-friendly executives in the War and Navy Departments over control of depreciation policy.

The issue arose in November 1939 in respect to a request by the Consolidated Aircraft Corporation to write off in a single contract two-thirds of the $1,800,000 in new facilities needed to build patrol bombers for the navy. The navy cut Consolidated's request to 60 percent of the cost of facilities, and then asked the Treasury Department to approve the write-off. Treasury refused, on the reasonable grounds that the new buildings would not deteriorate into worthlessness after only one contract, and therefore to approve the accelerated rate of depreciation "would open the door to a negation of the Vinson-Trammell Act." In a companion decision, Treasury rejected a request from the Electric Boat Company to write off in one contract half the cost of a new building to be used for the construction of twenty-three submarines.[27]

Shortly after Treasury Secretary Henry Morgenthau Jr. rejected the Consolidated request, he and Secretary of the Navy Charles Edison became engaged in what Morgenthau described as "a very heated argument" in a cabinet discussion of the matter. Roosevelt sided with the Treasury position, asking Edison, "Charlie, what do you want to do, just give Consolidated a handout?" Edison replied in a succinct statement of the services' oft-expressed willingness to overlook law and equity when these considerations interfered with a pressing need to obtain weapons. "Well," the secretary protested, "I can't find any planes." Edison and the president continued to wrangle, but in such an unequal contest the naval leader could not prevail. The regular depreciation rules remained in effect until the following year.[28]

Soon after declaring war on Germany, Britain and France attempted to place large orders with American defense contractors. They found that American tax amortization requirements severely limited their ability to obtain contracts. Arthur Purvis, the chief of the Allied Purchasing Mission, sought to place a large order for gunpowder with the Hercules Powder Company. In order to fulfill this contract, Hercules executives would have to invest $1,600,000 in a new factory, and they were unwilling to do so unless they were able to depreciate their investment quickly. Purvis asked Secretary Morgenthau for help, protesting that "they [Hercules] say they will not take any risk whatsoever." Without congressional approval, the Treasury Department quite properly refused to suspend its depreciation rules, on the grounds that to do so "would cut the heart out of the Vinson-Trammell Act."[29]

When the German blitzkrieg of May 1940 ended the limited fighting known as the Phony War, the replacement of the Vinson-Trammell Act became only a matter of time. One of its chief weaknesses was that its restrictions on profitability applied only to sales of aircraft and ships to the American military services. The restrictions on profits did not apply to arms sales abroad. Initially, this discrimination in favor of foreign sales did not trouble the Roosevelt administration greatly, because after 1938 it was American policy to assist the British and French rearmament programs in order to deter German aggression and to create jobs at home. But when the Wehrmacht smashed the French army, the need to accelerate American production became more pressing. Meanwhile, Congress, in an election year mood, actually *reduced* the profit margins allowable under the Vinson-Trammell Act from 10 percent to 8 percent, thereby making the defense business even less attractive than before. By early 1940, Vinson-Trammell was seriously interfering with the vital production of ships and planes for the American military services.[30]

The President Compromises

The fall of France in the summer of 1940 followed by the impending invasion of Britain persuaded most of the nation that all impediments that hampered military production must be removed. In practice, this meant that the selective profit limitations and the restricted depreciation rates of the Vinson-Trammell Act had to go. Yet, as Roosevelt kept insisting, there was general agreement that no new crop of warhogs should be propagated. In place of Vinson-Trammell, Roosevelt demanded a comprehensive antiprofiteering program that did not impede military production. The defense program should not be burdened by unwieldy profit limitations or by restrictive amortization rules, but it also should not allow defense contractors to make unfair profits.

In 1940 a compromise that would embody these principles began to take shape. The Vinson-Trammell Act would be replaced by a new, more comprehensive excess profits tax that would feature substantially heavier corporate income tax levies. In return, the depreciation period on tools, land, and facilities used in defense production would be reduced to five years. This accelerated amortization rate would be an attractive incentive for business to convert to defense production, since the normal rate of amortization was ten years for tools and twenty for buildings. The five-year depreciation period was a minor concession by the defense contractors, who had requested a four-year write-off. They also surrendered on another point. They had hoped to simplify the process of obtaining approval of the faster depreciation schedule by having the War and Navy Departments issue the required certificates of necessity, but the government insisted that the certifying authority be shared between military officers and civilian administrators.[31] The fabrication of this compromise in 1940 established the basic policy that shaped both the achievements and the shortcomings of Roosevelt's overall antiprofiteering program.

In 1939-40 Roosevelt approached mobilization policy cautiously. He wanted to encourage defense production, but he hesitated to alarm the public about the great disaster in the offing. During the early 1930s the War Department had developed an "Industrial Mobilization Plan," which envisioned a rapid conversion of industry from peaceful to military production. The IMP recommended the establishment of a War Resources Administration that would be granted broad powers after "M-Day," the day when war was declared. But both the all-powerful superagency and the concept of M-Day were unpopular in 1939 and 1940. The M-Day plan had been drawn up in consultation with business, but labor representatives had been excluded, and they correctly feared and aggressively opposed its grant of power to the military. The M-Day plan and the superagency were too drastic for a nation that remained uncertain of its proper role in the world conflict, and Roosevelt shelved them.[32]

Roosevelt preferred a less threatening strategy. A few days after the presidential election of 1936, and fully five years before the United States actually declared war, Roosevelt began to consider how to organize economic mobilization. By 29 August 1939, three days before Germany invaded Poland, he had already worked out in detail his own plan for managing the wartime economy, which he presented to the cabinet.[33] Its distinguishing feature was that the president would retain for himself primary authority over mobilization policy. In this decision Roosevelt departed from the precedent set by his mentor, Woodrow Wilson, who delegated great authority to manage the wartime economy to Bernard Baruch and the War Industries Board. Until 1943, in fact, Roosevelt remained unwilling to assign to any individual or any agency broad powers to make mobilization policy. "I am not going to give it to one man the way the newspapers want it," Roosevelt said. "It is too big a job."[34] A cohort of new war agencies, organized mainly according to economic function, would report directly to the president.

Until the United States actually entered the war, Roosevelt balked at fully revealing his mobilization plan or attempting to implement it. In May 1940, he cautiously reconstituted Wilson's Council of National Defense, composed of six cabinet officers, and he appointed a seven-person board with the un-threatening title of "Advisory Commission" as its operating arm. This agency had the real authority for organizing mobilization and was soon known as the National Defense Advisory Commission. The main advantages of the NDAC were that it could be formed without consulting Congress (the enabling legis-lation remained in place), it included no controversial M-Day to threaten a dramatic expansion of federal power, and it had no potentially controversial chairman. There thus appeared to be no danger of an all-powerful, unelected czar of the American economy. The most important executives were safe ap-pointments: William S. Knudsen, president of General Motors, its coordinator of war production, and Leon Henderson, formerly of the Securities and Exchange Commission, its price controller. Since all seven executives reported to the president, Roosevelt remained in control, against the advice of Secretary of War Henry L. Stimson and others who wanted a single executive in charge.[35]

The NDAC was short-lived, but it played an important role in antiprofi-teering policy. The Council of National Defense was defined by statute to in-clude six cabinet officers, but it excluded the secretary of the treasury. With the secretaries of war and the navy as leading players on the Council, the NDAC pressed for a shorter period of amortization of tools and facilities for tax pur-poses. This was what Consolidated Aircraft, the Electric Boat Company, the British-French Purchasing Mission, and Navy Secretary Edison called for and what Treasury Secretary Morgenthau opposed. Morgenthau reacted to this stratagem by agreeing to a shortened amortization plan only if it were com-bined with the replacement of the Vinson-Trammell Act by a broad new excess profits tax. As Morgenthau explained the politics of war profits taxation, "I don't know any better answer to the criticism that we are the war party than to take the profits out of war. That is the best answer I know." On this point, Morgenthau obtained the support of the president, who despite doubts about repealing the Vinson-Trammell Act during the election campaign, thought that the excess profits tax was popular and would help to defeat his Republican opponent, Wendell Willkie. The congressional leadership also supported the replacement of Vinson-Trammell with an excess profits tax.[36]

The Excess Profits Tax

In the heat of the campaign of 1940, the Treasury, Congress, and other agencies began to construct the tax program that would become the centerpiece of the effort to restrict or eliminate wartime profits. In forming the new excess profits tax, there was no shortage of time, expertise, or imagination. In late 1939 the Treasury's tax experts had begun planning for a new war profits tax by studying the systems used by fourteen foreign countries.[37] Besides its own staff, the Treasury Department also had the finest tax economists in the United States available for consultation.[38] Even if these experts made mistakes in formulating the antiprofiteering strategy, there would be ample opportunity for correction—the excess profits tax of 1940 would be followed by a second excess profits tax in 1941.

The perplexing riddle facing the tax planners was how to distinguish among large profits that arose as a direct result of war contracts, those that arose incidentally as a result of the heated-up civilian economy, and those that were derived legitimately from superior managerial ability. The object was to discriminate against windfall or unearned profits without punishing superior performance, and this proved to be most difficult. In order to measure the amount of increased profits that were defense-related, the Treasury Department adopted a four-year peacetime base period. Earnings that exceeded the average of those obtained during the base period (1936-39) were considered "war profits." But this method, which followed the contemporary practice of Britain and Canada as well as that of the United States in 1917-18, introduced a new source of inequity. Profits in some industries were more cyclical than others; companies that had been in the trough of their cycle during the depressed preemergency period (such as machine tools) would be unfairly penalized later when the economy strengthened.[39]

To avoid this difficulty, the Treasury constructed a second means of distinguishing between legitimate prewar profits and incidental profits that derived from the unexpected emergency. This was the "invested capital" method, whereby the legitimacy of a company's profits was related to the amount of capital invested in the venture. Secretary Morgenthau's assumption was that 6 percent on investment was a "normal" profit, and anything above that was an "excess" profit. There was some precedent for his choice: 6 percent was the rate that regulated utilities were customarily allowed to earn. As one Treasury staffer added, "six percent has a sort of a traditional sound that seems reasonable." According to the concept of excess profits, cyclical companies would be allowed to increase their profitability until they reached the normal (6 percent) level. The invested capital method had also been used in 1917 and 1918, when the definition of a legitimate profit was 8 percent.[40]

Unfortunately for this theory, some companies were managed better than others. The better-managed companies produced a higher return on invested capital during peacetime than did the less efficient. (An example of a highly profitable, well-managed firm was the large retailer, J. C. Penney. Because it rented all its stores, Penney regularly earned 30-35 percent annually on invested capital.) If the government began to tax the well-managed firm at a higher rate than the less efficient, it placed itself in the imprudent position of rewarding inefficiency and penalizing superior management. Younger, growing companies tended to have a higher return on invested capital than did older, more mature firms. Consequently, the use of the invested capital method assisted larger, mature firms at the expense of their younger, more dynamic competitors. Roosevelt liked the invested capital method best; his preference was to attach the earnings of any firm that "has had very high earnings on the money actually invested," regardless of the reason.[41]

The quandary in which the government found itself is illustrated by table 1. Fourteen representative firms were selected by the Treasury Department for close scrutiny. They are listed in order of the percentage of their increase in profits in 1940 over the average of the four preceding years. American Car and Foundry, whose profits increased nearly twelve times in 1940 over the base period, surrendered only 2.42 percent of its total profits to the Treasury as "excess." The Chrysler Motor Company, which had been quite profitable in the prewar years, increased its profits by another 55 percent in 1940, but paid excess profits taxes of nearly 15 percent. In effect, the government was treating the better-managed firm, Chrysler, more punitively than the less efficient company, American Car and Foundry. The tax treated American Woolen even more generously. A relatively unprofitable firm in the late thirties, its profits increased in the first year of defense production by four and a half times. Although this was certainly a war-related gain, American Woolen nevertheless paid no excess profits tax whatsoever. But in the case of Curtiss-Wright, a huge manufacturer of military aircraft, the tax worked approximately as intended. Curtiss's pretax earnings multiplied sevenfold as military contracts proliferated, but the Treasury was able to reclaim 55 percent of the profits through the excess profits tax. Smaller aircraft firms, more recently organized than Curtiss, suffered a considerable competitive disadvantage under the system, however.[42]

The Treasury Department recognized this inequity and attempted unsuccessfully to rectify it in both the 1940 and 1941 excess profits acts. The Treasury plan was to limit the amount of its prewar earnings that a highly profitable firm could use to evade an excess profits liability. A firm that was very profitable in relation to its invested capital would face enormous outlays if all earnings above 6 percent were subject to a 50 percent excess profits liability. For such a firm, the 1936-39 base period was welcome news. Although a firm might be extremely profitable, if its profits did not rise dramatically above the prewar period the excess profits tax would not be unduly burdensome.

The company that was most frequently discussed in this respect, and that played an important role in the politics of excess profits taxation, was the Coca-Cola Company. In relation to its invested capital, Coca-Cola was among the most profitable in the nation. In 1940 it earned $44 million on an $89 million investment—a striking 49.3 percent return. Coca-Cola profits rose sharply in that year—47 percent above the prewar average—but not nearly so steeply as those of firms directly engaged in defense production (see table 1). If Coca-

Table 1. Effects of the 1941 Excess Profits Tax on Selected American Corporations

	Average Income, 1936-39*	Net Income 1940	Percentage of Increase (1940/1936-39)	Invested Capital, 1940	Net Income as Percentage of Investment, 1940	Excess Profits tax 1940	Excess Profits Tax as percentage of 1940 Net Income
American Car and Foundry	512	6,579	1184.96	88,953	7.42	159	2.42
Curtiss-Wright	5,615	45,070	702.67	46,184	97.59	24,889	55.22
International Paper	4,083	23,184	467.81	184,625	12.56	7,554	32.58
American Woolen	719	3,958	450.49	66,554	5.95		
U.S. Steel	48,227	155,830	223.12	1,589,396	9.80	31,063	19.93
Binks Manufacturing	56.2	147.6	162.63	727.1	20.30	31.40	21.27
Du Pont	45,985	112,529	144.71	706,825	15.92	36,924	32.81
Colt's Patent Fire Arms	1,299	3,167	143.80	16,039	19.74	991	31.29
Indiana Steel Products	63.0	110.7	75.71	582.6	19.00	13.3	12.01
Dexter	50.3	87.0	72.96	732.8	11.87	8.4	9.66
Chrysler	41,811	64,806	55.0	188,785	34.33	9,708	14.98
Continental Can	7,964	12,236	53.64	101,580	12.04	1,729	14.13
General Alloys	44.9	66.8	48.78	375.3	17.80	3.3	4.94
Coca-Cola	31,187	43,876	46.69	88,953	49.3	4,325	9.86

*All figures in thousands of dollars.

Source: Memorandum of Dave H. Morris Jr. to Secretary of the Treasury Henry Morgenthau Jr., 31 October, 1941, Morgenthau diary, vol. 456, pp. 202-7, in the Franklin D. Roosevelt Library, Hyde Park, New York.

Cola were required to expose as excess all its profits above a return of 6 percent on its investment, its tax burden would be staggering. Therefore, the concept of taxing only the rise in profits, rather than the percentage of investment, was extremely attractive to large, well-managed firms like Coca-Cola.

On the other hand, if the excess profits tax did not harvest an augmented portion of the considerable profits of a firm such as Coca-Cola, the concept of equality of sacrifice in wartime would be compromised. While soldiers and sailors were being required to give their lives by the hundreds of thousands, was it appropriate for any person or corporation to escape with an increase of only 10 percent in its taxes over its previous liability, as did Coca-Cola? The Treasury Department thought not, and it recommended that a limit be imposed on a firm's ability to reduce its wartime tax liability as a consequence of its prewar earnings record.

This introduced a novel concept of controlling war profiteering, one which was original to World War II. Until 1940, the basic ethical assumption supporting the limitation of war profits was that while others were sacrificing their lives, no citizen should rise economically as a consequence of the misfortune. Therefore, time-honored tax policy encouraged the recovery of gains that arose due to wartime circumstances. But the question that was new to the latest conflict was whether a person or company ought to receive an uncommonly large income during wartime *for any reason whatsoever*. Proponents of imposing excess profits taxes on wealth per se argued that it was immaterial whether the income was directly derived from the war or not related to it at all. No one should live luxuriously while others sacrificed for the good of the community. As one assistant told Secretary Morgenthau, "You are . . . now going on the assumption that anything over 10 percent of invested capital is abnormal profit, [which] hadn't been previously made."[43]

During the Great War, the Wilson administration had imposed an excess profits tax that applied to any profit that exceeded 7 to 9 percent of invested capital. The primary purpose of that tax was to produce revenue rather than to equalize sacrifice. It worked well in respect to its central purpose, bringing in $2 billion in 1918, but it did not directly attach the gains of those firms whose incomes soared due to the emergency (although the 1916 munitions tax did so). The 1917 and 1918 taxes, in keeping with the precepts of New Freedom progressivism, struck a blow against monopoly capital rather than being clearly redistributionalist, as was the 1916 tax.[44]

The approach of the New Deal antiprofiteering campaign was distinctly different. Within the New Deal entourage there were many who wanted to use the federal income tax as an instrument of wealth redistribution as well as of revenue acquisition. This was particularly true of the Treasury Department, which swarmed with liberals. Secretary Morgenthau and his chief aides deplored the inability of previous New Deal tax policy to do enough "to curtail bigness" in American industry. Morgenthau wanted the 1940 excess profits tax to succeed where the New Deal had heretofore failed—by establishing "a tax against big business."[45]

The Treasury tax economists therefore designed an imaginative method of recapturing excess profits obtained by very large, highly profitable firms such as J. C. Penney or Coca-Cola. Neither of these firms benefited directly from defense contracts, so their profits would be unlikely to increase at the precipitous rates of large aircraft firms. Their sales and profits would certainly rise during the war emergency, but probably not dramatically in relation to the base years, 1936-39. These firms were very profitable in relation to their invested capital, so they would prefer to calculate their tax liability under the base years method as opposed to the invested capital method adopted by the Wilson administration. The Treasury plan aimed to prevent these very large, highly profitable firms from reducing their tax exposure by filing under the base years method. To accomplish this, Treasury recommended a provision that no corporation be allowed to escape excess profits tax liability on any profits that exceeded 10 percent of invested capital.[46]

This ceiling on tax exemptions was the Treasury's method of establishing the doctrine that in the midst of sacrifice and sorrow no firm should receive a very large profit for any reason. This novel concept gained the approval of the president, who recommended its adoption, but in the middle of the 1940 election campaign it was too unconventional, too liberal, and too politically risky for Congress.

In September 1940, very few congressmen wished to honor Roosevelt's request and vote for a new excess profits tax of any kind. The need to eliminate the Vinson-Trammell mistake and to shorten the period of amortization of new defense facilities was clear, but voting for the excess profits tax just before election day was emphatically not popular. Some congressional leaders sought to decouple the excess profits tax from the other measures, promising action after the election, but Roosevelt refused, declaring that "he had seen many glittering assurances that failed to materialize."[47]

Roosevelt was very anxious to obtain passage of an excess profits bill before the Congress adjourned, and thus he was willing to compromise on the details of the plan. This willingness sounded the death knell for the 1940 session for the concept of an exemption ceiling of 10 percent profit on invested capital. Powerful members of Congress, including the chairmen of the House Ways and Means Committee, Representative Robert L. Doughton of North Carolina, and the Senate Finance Committee, Senator Pat Harrison of Mississippi, opposed the ceiling. Indeed, both tax-writing committees had been bastions of conservatism during the late thirties. Only liberal Representative Jere Cooper of Tennessee, who chaired the Subcommittee on Taxation of the House Ways and Means Committee, remained a firm advocate of the 10 percent ceiling. Cooper had suffered multiple wounds in heavy combat in France as a captain in the AEF. He was subsequently elected commander of the Tennessee

American Legion, and he was much more committed to arresting profiteering than his more senior colleagues. In fact, Secretary Morgenthau once remarked, "I see Jere Cooper foaming at the mouth about the fellow who made a couple of hundred thousand dollars while he was oversea[s] getting gassed four times."[48]

The congressional leaders who opposed this plan argued that large profits were not necessarily excess profits. In effect, they rejected the idea that large gains obtained during a national emergency were ipso facto tainted. They preferred a narrower definition of war profiteering, one that limited recoverable profits to those earned as a direct result of the emergency. But by adopting this argument, they opened a large loophole in the excess profits tax through which a large share of the profits of large, very profitable firms could escape.

Roosevelt made a second effort to obtain passage of the 10 percent ceiling in 1941, but to no avail. Always a realist, Roosevelt harbored no optimism that he could gain congressional approval in 1941, but decided to try anyway, observing, "I am a woodchuck and I keep digging when there is a chance to dig." Meeting with Treasury officials, the president first stated that he would like to tax all income above $100,000 at the rate of 99.5 percent, arguing "Why not? None of us is ever going to make $100,000 a year." Roosevelt was surprised when he was informed that there were 1,100 taxpayers who were already included in that category.[49]

Then, at the Treasury's urging, Roosevelt complained to Ways and Means chairman Doughton that certain corporations "may be making 20 or 30 or 50 percent on their equity capital. It is my definite opinion that they ought to contribute to the cost of our great defense program far more heavily this year than last year or the year before. But just because they happen to have made equally large profits in recent years, they are called on to contribute no more . . . than they did before. That seems to me clearly a discrimination in their favor." When Roosevelt declared that "just because they happen to have made equally large profits in recent years" they are not required to contribute more, he was endorsing the principle that great wealth obtained in war was wrong, no matter how legitimately it was gained in peace. Doughton presented Roosevelt's request to the Ways and Means Committee, where it was again decisively rejected.[50]

Although the Ways and Means Committee and the Finance Committee had long been known for their disinterest in the New Deal's soak-the-rich tax creed, in 1940 and 1941 there were more tangible and immediate reasons why they were lukewarm about the 10 percent profit ceiling proposed by Treasury for the excess profits tax. This provision would place in jeopardy the profits of certain large, very profitable firms located in the states represented by powerful committee members.

Chairman Doughton of Ways and Means represented western North Carolina, and he was a longtime ally of the tobacco economy that dominated his state. Large tobacco firms had been exceptionally profitable during the 1936-39 base period, and they were also very profitable in relationship to their invested capital. As a consequence, they were certain to benefit substantially from the opportunity to use the base period method of calculating their tax obligation, and any ceiling imposed upon their ability to do so would be very costly. Doughton understood that imposing the ceiling would work against his constituents' financial interests.[51]

The tax exemption ceiling faced even stronger opposition from southern conservatives in the Senate. The most unrelenting foe was Walter George of Georgia, whose state was home to the highly profitable Coca-Cola Company. After Harrison, Senator George was the ranking majority member of the Finance Committee, and he endeavored to use his influence to protect Coca-Cola's interests. Although a Democrat, George was also a longtime enemy of the New Deal. Roosevelt disliked George personally, and in 1938 he publicly and personally rebuked the unfaithful senator in a daring attempt to oust him from the Senate.[52]

Roosevelt's attempt to purge Walter George and other conservative Democrats in the congressional elections of 1938 had failed miserably. George remained in office, and he never forgave the presidential affront. George had some credentials as a foe of profiteering, having spoken against it during congressional hearings in 1924 and having served as a halfhearted member of the Nye committee. Nevertheless, George and Roosevelt differed profoundly on wartime tax policy. Even early in the war, George believed that the tax burden on the middle class was too high, but not high enough on the lower class. From this perspective, George could and did use his considerable influence on behalf of Coca-Cola.[53]

In 1940 Coca-Cola was very profitable, netting nearly $44 million before taxes, and the presence or absence of a ceiling on excess profits tax credits was extremely important to the firm. If there were a ceiling on credits, Coca-Cola's total tax bill would be about $33.2 million, but without the ceiling it would fall to about $25.6 million—a savings of at least $7.6 million.[54] Senator George understood the great value of this loophole to profitable corporations like Coca-Cola and successfully forced the exclusion of the ceiling on tax credits from the 1940 excess profits tax. Congressman Cooper attempted to enlist Roosevelt's assistance in reestablishing the ceiling with a last-minute telephone call, but he was not successful.[55] In 1941 George became the chairman of the Finance Committee, from which position he offered minimal assistance to the administration's antiprofiteering program.

Roosevelt's Concession to Business

By mid-1941, as direct participation in the war neared, the Roosevelt administration's early antiprofiteering program was essentially complete. Its main feature, the excess profits tax, imposed steeply graduated rates on corporate and personal incomes, more steeply graduated estate taxes, and excise duties on luxury items. In February 1941, Roosevelt added a comprehensive system of price controls. By imposing steeply graduated corporate and personal taxes designed to arrest war profiteering, the president and Congress testified to a desire to continue or extend into wartime the redistributional policies that were oft-stated tenets of the peacetime New Deal.[56]

Congressional approval of the excess profits tax exacted a price. In order to gain its steeply graduated levies, Roosevelt had to surrender a much-shortened amortization schedule as an incentive for business to convert to war production. Each corporation could determine its liability by choosing the lesser of its profits in relation to the 1936-39 base period or to its capital investment, and there was no upper limit to the amount of liability that could be exempted under this provision. Both the amortization schedule and the removal of an exemption ceiling were significant concessions; the former reasonable and necessary, the latter less so.

Although temporarily concealed, Roosevelt's willingness to compromise the social goals of the New Deal in order to defeat the Axis eventually became public knowledge. In a December 1943 press conference, Roosevelt candidly admitted that "Dr. New Deal" had been replaced by "Dr. Win-the-War." Several students of the wartime period, including Rexford G. Tugwell, Frances Perkins, John Morton Blum, and Warren F. Kimball, have addressed the questions of when Roosevelt decided to compromise with corporate America, how deliberately he did so, and whether he gave up too much.[57]

As he often said, one of Roosevelt's social goals was the elimination of war profiteering. He followed this policy consistently from his election through 1939, when his opposition to war profits first showed signs of weakening. In July 1939, Roosevelt began planning for new war taxes, and by September 4 he had expressed his intention to use tax incentives to spur defense production. On 9 November 1939, Roosevelt acted on his plan by directing Secretary Morgenthau to "work out a policy of depreciation allowance by which abnormal investment in plant expansion . . . will be absorbed over the life of the contracts or during the emergency period." By late September 1939, when asked if he agreed that personal income taxes should be made less progressive by increasing rates on incomes in the middle and lower brackets, Roosevelt replied, "Perfect tax program; simple; perfect!"[58] These statements, in combination with his decision in 1939 and 1940 to appoint Republican businessmen to the key civilian positions managing the war economy, confirm that Roosevelt's decision to compromise with corporate America was taken quite promptly after the European war opened and also quite deliberately.

Writing in 1952, Rexford Tugwell, the senior member of Roosevelt's brain trust and hence a New Deal evangelist, was among the first to identify and comment on the president's compromise on war mobilization. Tugwell was generally an admirer of Roosevelt, but he was deeply ambivalent about Roosevelt's frequently evidenced penchant for compromise. Tugwell considered Roosevelt's peacemaking initiative to American businessmen in 1939-40 to be among Roosevelt's "more serious" departures from liberal precepts, and he questioned whether the compromise was either fully necessary or justified by the results it brought.

In any event, the new excess profits taxes of 1940 and 1941 produced a bountiful revenue. But as Tugwell understood and as Roosevelt surely knew, when defense contractors received tax incentives through accelerated amortization, and when very profitable firms were allowed to escape the full impact of the tax, the lofty goal of abolishing war profit could not be fully realized. Roosevelt never abandoned that goal; indeed, even as he signed into law the excess profits tax of 1940 he requested a study of its deficiencies.[59] In 1941 he attempted unsuccessfully to remedy the failings he found, and he continued to pursue his objective of restricting profits throughout the war years. Roosevelt did not survive to evaluate his efforts, but Tugwell did, and he was disheartened. Even a New Deal enthusiast could hardly have failed to feel some discouragement by the number of warhogs who evaded Roosevelt's efforts to send them to market.

11

A Prescription for Profiteering

If you are going to try to go to war, or prepare for war, in a capitalist country, you have to let business make money out of the process or business won't work.

SECRETARY OF WAR HENRY L. STIMSON (1940)

As Admiral Yamamoto feared, the Japanese victory at Pearl Harbor awakened a sleeping giant and filled it with an awful rage. The last of the dissension that had disabled the United States melted away, and strong public support for the war, mixed with a perceived danger of invasion, allowed the government to expand its role dramatically. The limited Keynesianism of the later New Deal gave way to a modified command economy, as the federal government obtained a domination of the economy unprecedented in American history.[1] The sweeping grant of power to an experienced national leadership with an expressed commitment to limiting war profits augured well for the success of the antiprofiteering effort.

Mobilizing American industry for World War II was an enormous task. America's role in the second great conflict of the century was certain to exceed its earlier participation by a large margin, but it was obvious that the United States would rely heavily on its traditional strategy of overwhelming its enemies with its vastly superior productive capacity. The economic problems associated with war mobilization would differ in degree, but they would generally correspond to earlier experiences. Many of the techniques used to equalize sacrifice and to prevent swollen fortunes during the Great War could be used again. These included excess profits taxation, price controls, luxury taxes, and rationing.

The junior executives of 1917 had become the senior officials of 1941. Memories of the Great War and its bitter aftertaste remained vivid. In case recollections dimmed, a host of doughboys and former mentors were only too eager to remind their protégés of past mistakes.[2] When the United States entered into hostilities, moreover, its problems of mobilization resembled those that had faced its British ally for more than two years. Thus the generation of Americans who managed World War II had a unique opportunity in American wartime history to learn from the experience and mistakes of others.

This opportunity was only partially grasped, unfortunately, as mastery of war profiteering proved to be more elusive than even the most resolute war leaders realized. The United States economy during World War II was substantially a command economy, but it was a democratic government that was in command. The power to make profit limitation policy was shared by several governmental agencies that differed in personnel, priority, and degree of commitment. In an intricate, fully mature industrial economy, the regulation of war profits was a problem of exceptional complexity. Although certain ancient modes of wartime wealth enhancement had been virtually eliminated, such as ransom and trading with the enemy, other familiar forms, such as price extortion and quality degradation, were more unyielding.[3] By World War II they were joined by new techniques: large corporate salaries, tax evasion, and trading on the black market. Restriction of war profits proved to be a problem capable of thwarting the best efforts of the nation's most able and committed economic planners.

Dr. Win-the-War

Although Roosevelt never directly renounced the goal of abolishing war profit, his admission that "Dr. New Deal" had been replaced by "Dr. Win-the-War" acknowledged implicitly that war profit would not be entirely eliminated. Roosevelt's initial insistence upon retaining personal control of mobilization planning proved to be a serious impediment to attaining his goal, substantially because his administrative abilities did not match his mastery of political leadership. Critics of the Roosevelt presidency have often pointed to the administrative confusion that persisted throughout most of his tenure in office. Henry Stimson, who held cabinet posts under two other presidents as well as serving as secretary of war during World War II, described FDR's administrative arrangements as "inherently disorderly." Bernard Baruch observed that Roosevelt's mobilization plan of 1939 deliberately and conspicuously rejected the central concept that had been found successful during World War I, namely, a chief of the civilian economy as developed in the War Industries Board in 1918. Baruch noted disapprovingly that Roosevelt instead reconstituted an organization that had failed in 1917 (the National Defense Advisory Commission), only to replace it with an organizational chart whose main feature was overall supervision by the president himself. In May 1943, Roosevelt, belatedly bowing to congressional pressure, created a superagency, the Office of War Mobilization. James F. Byrnes was its chief, but although Byrnes was dubbed the "assistant president" for the home front, his position never bore that full responsibility. In effect, Roosevelt needlessly retraced Wilson's serpentine administrative path. Even Eleanor Roosevelt commented that after 1940 her husband had lost his "zest for administrative detail."[4]

Roosevelt allowed the responsibilities of his many agencies to overlap and then failed to referee their dissension. This was conspicuously true of his attempt to formulate and implement an antiprofiteering policy in 1940-41.[5] Although Roosevelt spoke early and often about controlling war profits, he never gave antiprofiteering policy a clearly designated home, and he never identified a clearly recognized leader of the effort. Similarly, though he sought to maintain personal control of mobilization, he was often reluctant to burden himself with its details.

At various times in the war, several different agencies claimed a share of the authority to make antiprofiteering policy, but none had limiting war profits as its primary responsibility. This omission would have telling effect as the emergency developed. Although Henry Morgenthau and the Treasury Department seized an early lead, there were a number of late entries in the race to control profiteering policy. One of these was Harold Smith, director of the Bureau of the Budget. Smith's institutional responsibility was to limit the size of the deficit, so his primary interest was in augmenting revenue collection. Smith sought to enlarge his influence from how funds were spent to how funds were raised, and he became a champion of a national sales tax, which would have made the tax system even more regressive than it already was. Leon Henderson, Roosevelt's anti-inflation chief, also backed the national sales tax because a tax increase would siphon away spending power. Jesse Jones, as head of the Reconstruction Finance Corporation, was heavily involved in the construction of new defense plants. Since this involved depreciation write-offs, he too managed to force his way into tax policy, pleading for accelerated amortization. Other agencies, such as the War Department, and other personalities, such as James F. Byrnes, the economic stabilization administrator, and World War I economic coordinator Bernard Baruch, were also involved. A visitor to Washington who asked the simple question, "Who is making antiprofiteering policy?" would have had to do some searching to find an answer. The confusion was not caused by a defect in the Constitution, however. In part, it was the fault, as Warren F. Kimball has aptly dubbed it, of Roosevelt's "debonair" administrative style. When in 1943 Henry Morgenthau complained to the president about infringement on the Treasury's preeminence in tax policy, FDR ignored the jurisdictional dispute. He simply replied, "Aw Hen: The weather is hot and I am goin' off fishing. I decline to be serious even when you see 'gremlins' which ain't there!"[6]

Franklin D. Roosevelt had several good reasons for opposing profiteering and a longheld interest in doing so, but he also encountered several impediments to achieving his objective. One was personal: Roosevelt had a weak understanding of advanced economics, and he was not deeply interested in its details. Since control of war profiteering was largely a matter of tax policy, this deficiency was especially severe, because FDR found the intricacies of this dismal aspect of the dismal science particularly tedious. His lack of interest in the technicalities of tax policy appeared early and often.[7]

Roosevelt's most important tax adviser was his old friend, Treasury Secretary Henry Morgenthau Jr., who consistently and continuously despaired over the president's dislike of tax matters. After a meeting in November 1938 to discuss a special tax for defense purposes, Morgenthau wrote, "I came away from Saturday's meeting feeling that the president was rushing the whole thing terribly and really wasn't giving it anything like the time or thought it deserved."[8] After a 1942 visit to Roosevelt, Morgenthau recorded in his diary, "I could tell by the look on his face that he hadn't seen the budget, and didn't know and didn't care. He gave me the usual thing, and I said, 'Now when do you want to take this up, Mr. President?' He said, 'Not before Christmas.' You know what that means."[9] On another occasion Morgenthau reminisced about the experience of a former Columbia University law professor who signed on as a tax adviser: "I will never forget. I think that Roswell Magill must have spent a minimum of three months getting a statement ready for the president. He had fifteen items, and he never got further than one. He came to my bedroom up at Hyde Park after the meeting, and he said, 'God, I might just as well go back to teaching. This is terrible.' I said, 'No, this is just Roosevelt. You did all right.' He says, 'I never got beyond number one.' I said, 'What did you think you were going to do?' Gee, he felt awful."[10]

When Roosevelt did consent to discuss tax policy, he betrayed a weak grasp of the intricacies of the subject. The president was fond of dramatic and climactic events, not of technicalities. In taxes, he relished simplicity. Mark Leff, the historian of the New Deal tax credo, argues that Roosevelt substituted "vague moral principles for a coherent tax philosophy."[11] Roosevelt's expressed preference was for a *gross* income tax rather than a *net* income tax (a gross income tax would have overlooked the cost an individual incurred in earning his income).[12] The technical aspects of tax policy that did interest Franklin Roosevelt were chiefly those with clear political overtones. When FDR became determined to enact an excess profits tax in 1940, he did so because he believed it would help him win reelection, and he paid scant attention to its details. This haste introduced significant loopholes that weakened the bill.[13]

Another limitation on Roosevelt's ability to restrain profiteering, according to some writers, was that throughout the war he was so preoccupied with military and foreign affairs that he had little or no time for war finance and other domestic matters. But Roosevelt's control over the military apparatus was never complete, and the attention he accorded to matters of war strategy and tactics was never total; in fact, until 1944, much of Roosevelt's foreign policy was conducted by his closest adviser, Harry Hopkins. No person was more devoted to Franklin Roosevelt than his Judge Samuel Rosenman. In August 1942, when Rosenman was seeking presidential approval for an executive order freezing wages and salaries, he grumbled that "the thing that worries me is the President is so complacent. The President doesn't devote more than two days a week to the war. . . . I have been up to Shangri-La three times and he sits there playing with his stamps. . . . [War Production Board Chairman Donald] Nelson never gets to see him." There was ample time to give attention to domestic policy, if that had been Roosevelt's priority.[14]

A final—and perhaps the most important—constraint on Roosevelt's ability to restrain war profiteering was the mixture of goals the government pursued. Although Roosevelt and others forcefully articulated the government's determination to prevent excessive profits, they even more vigorously directed attention to the competing goal of war production. Roosevelt and the nation's military leadership understood that the victory depended primarily on filling the oceans with warships, the land with tanks and trucks, and the skies with planes. The question they now addressed was how to achieve an avalanche of production if taxes drained away all incentive and rules encumbered each stage of the process.

Wartime Politics

Even if Roosevelt had been a superlative administrator and willing to give meticulous attention to economic management (which he was not), political impediments would have barred a fully harmonious antiprofiteering program. Authority for resolving issues of public finance resided in the Congress, which had no intention of surrendering its role for the duration of the war. Although historians have devoted copious attention to the actions of the executive branch during World War II, the role of Congress has received comparatively little attention. Although Congress loyally rallied behind the president in the dark days after Pearl Harbor, it became increasingly independent as the immediate danger passed.

An aggressive antiprofiteering policy would have required the unified backing of Congress, but Roosevelt's support for such measures was fleeting. The war and the prosperity it brought gradually diminished approval for the pacifist and prolabor positions that would necessarily underpin an antiprofiteering campaign. The strength of the liberal Democratic bloc had begun to wane by 1938, three years before the Japanese attacked, and the trend continued throughout the war.

In the congressional elections of 1938, American voters sent eighty-one more Republicans to the House and eight to the Senate, destroying forever the comfortable margins that had powered the legislative victories of the early New Deal. The staunchly liberal bloc in the House was reduced by 50 percent. Roosevelt easily gained reelection in 1940, but his margin of victory was smaller than in his two previous efforts and he fully understood that the fundamental shift in the political climate compelled him to change his methods.

His mobilization plan marked a substantial compromise with business leadership. For the duration of the emergency, he toned down but did not entirely eliminate the blistering attacks on business he made during the heyday of class antagonism in 1936.[15]

After the congressional elections of 1942, the Republican comeback was nearly complete. The GOP increased its strength from 162 to 209 seats in the House, leaving the Democrats with only a thirteen-vote majority. When conservative Democrats were added to the Republican membership, the conservative bloc held a majority of both houses. Conservative Democrats controlled the key committees that dealt with profiteering policy: the House Ways and Means and Rules Committees and the Senate Finance Committee. Roosevelt's relations with Congress became increasingly strained, and in 1944 he was close to the mark when in disgust he described the national legislature as a "Republican Congress."[16]

The revival of the conservative opposition to the social policies of the New Deal had tangible effects on the antiprofiteering program. In 1942 delays by the House Ways and Means Committee temporarily blocked a Treasury plan to make the income tax more effective through regular payroll deductions, the now familiar pay-as-you-go system. When the Revenue Act of 1942 finally passed, it made the tax structure more regressive by imposing a flat 5 percent gross income tax. In 1943 the Congress went even further. Against Roosevelt's wishes, it approved the "Ruml Plan," which forgave 75 percent of an entire year's personal income tax. And in 1944, when Roosevelt asked for $10.5 billion in new taxes, Congress responded with only a minimal $2 billion contribution. Congress preferred to defer payment of the costs of the war until peace returned, when the nation's tax structure would be less progressive. Although Americans paid lip service to their desire to share the war burden equitably, as John Morton Blum has observed, they refused to tax themselves at a rate as high as did the Allies, and they evaded price controls more frequently.[17]

The interplay between the social goals of the New Deal, presidential preoccupation and indifference, and entrenched congressional conservatism is illustrated by Roosevelt's failed attempt to establish a wartime limit on executive salaries. The New Deal had often declared in favor of wealth redistribution (although its seriousness has been questioned), and the concept of limiting executive salaries in wartime to $10,000 had been proposed in Congress as early as 1935.[18] When inflationary pressures mounted in early 1942, Roosevelt resolved that he must seek a freeze on both prices and wages. Labor was naturally quite reluctant to accept wage controls unless management compensation was also frozen, and farmers would be more willing to accept controls on commodity prices if business incomes were similarly restricted. With the backing of Walter Reuther of the United Auto Workers and William Green of the American Federation of Labor, Roosevelt in May 1942 asked Congress for legislation that would limit executive salaries to $25,000 after taxes. But Roosevelt's support for the idea was tepid, and the Joint Committee on Internal Revenue Taxation rejected the plan.[19]

Following his defeat in Congress, Roosevelt acted boldly. In October 1942, just before the congressional elections, Roosevelt issued an executive order limiting salaries to $25,000 after taxes. (When it became apparent that this would work a "hardship" on certain movie stars and corporate officers, the regulations were amended to afford them relief.)[20] But the powerful Senator Walter George disliked the plan, arguing that it was "inadvisable for a liberal government since it penalized earned income and did not touch unearned income." It was also opposed by Senator George Norris of Nebraska, who thought it would aid those who were accusing the New Deal of dictatorship, and by Senators John McCarran of Arkansas and Robert A. Taft of Ohio, who thought it threatened the existence of Congress. Early in 1943, subsequent to the Republican electoral victory, an aroused Congress emphatically repealed Roosevelt's executive order. Henceforth, executive compensation could only be limited by taxes imposed by the legislative branch.[21]

Excess Profits Taxes

In early 1941, and even more so in the alarmed atmosphere after Pearl Harbor, Congress was willing, even eager, to expand the tax system to provide the revenue the war effort required. Although there was general agreement that revenues must be greatly augmented beyond the excess profits tax of 1940, Congress was much less willing to alter the tax structure sufficiently to make it a fully effective antiprofiteering tool.

The disincentive effects of high taxes on war production remained a controversial matter throughout the war. The question often raised was how high taxes could be raised without interfering with production. Before the war declaration, army and navy procurement officers complained of their inability to place contracts. Although Henry Morgenthau scoffed at these claims, demanding proof that any defense contracts had been impeded, the counsel of the military services carried great weight.[22] Following the outbreak of hostilities, government planners actually tried to calculate the highest possible tax rates that could be imposed without destroying incentive. President Roosevelt suggested that a rate of 100 percent be established and that a different incentive be devised to replace the monetary one. He also suggested that defense contractors be presented with a flag as sufficient reward for their efforts in the nation's behalf.[23] There was evidence, however, that a rate of 100 percent was indeed too high. This was the rate Britain had imposed on the war profits earned by its defense contractors, and many Britons believed that it had interfered with productivity.[24] American procurement officers agreed that the highest rate they dared to impose without destroying incentive was 88 percent.[25]

Another aspect of the tax question was enforcement. A theorem of public finance is that the more tax rates rise, the greater becomes the incentive to cheat. Compelling compliance with extremely high tax rates is a costly process, and it conflicted with other wartime goals. During peacetime, defense contractors were subject to a long, time-consuming process designed to guarantee low prices to the government. A contract had to be advertised, suitable formal bids had to be prepared, bidders' qualifications had to be evaluated, performance bonds often had to be posted, and pilot models commonly had to be submitted, all before work could progress. In wartime there was a general and legitimate demand that these and similar rules be relaxed. Little inclination existed to support the assembly of a corps of auditors large enough to ensure that no cheating occurred. Even if such an inclination did exist, there was doubt about its feasibility. In fact, there may not have been a sufficient number of accountants in the entire country to audit all the nation's defense contractors thoroughly. (By the close of 1943, the War Department employed about 6,400 accountants for this purpose.) There was even doubt as to whether the country could spare enough typewriters (which were rationed) and filing cabinets to enable the auditors to perform their functions.[26]

Price Controls

Price controls were a major tool employed in the struggle against swollen profits. They were originally imposed in February 1941, even before the United States declared war, and they remained in effect until well after the Japanese surrender. Data assembled by the economist Hugh Rockoff indicates that price controls were initially very successful. They arrested inflation without producing an excessive degree of regimentation or costly bureaucracy. In 1943, although there were more than three million firms doing business in the United States, federal prosecutors commenced action for price control violations against only 5,010 companies. It seems evident that price violations were negligible early in the war, when nearly all Americans shared a sense of peril and an acute spirit of national unity existed.

By 1945, however, the number of prosecutions had risen to 33,036. By then the considerable slack that had remained in the economy from the depressed levels of the 1930s had totally disappeared. The threat of invasion and fascism, though not eradicated, had plainly receded. These circumstances served to increase inflationary pressures and to reduce compliance with price control edicts. The effectiveness of the price control system gradually weakened, and for the war as a whole, about one-eighth of all American firms were found to be in violation of the price ceilings. The Prohibition era is usually cited as the period when federal law has been most widely flouted, and by that standard price control violations in 1945 were about half as frequent.[27]

In addition to its role of directly overseeing price control, the Office of Price Administration also monitored standards of quality, since producers could obtain imputed price increases simply by cutting back quality. By 1945 the OPA employed 64,517 people, or about .12 percent of the workforce, but even so it could only partially cope with the age-old tendency to trim quality as a means of evading price controls. The American economy had become so complicated in its industrial maturity that eliminating quality degradation defied human ingenuity. Few companies knowingly shipped substandard goods to the government, but there was much dispute about quality standards under price controls.[28]

An illustrative example was the scrap iron industry, where war-induced shortages led normally speculative markets to become exceptionally volatile. The steel industry was of critical importance to war production, and since about half of all steel forged originated in scrap iron, any price increase in the form of quality degradation in this basic material would permeate the entire economy. But what was good quality in scrap iron, and who could identify it? In 1941 there were twenty different grades of scrap iron, twelve principal processing points, 3,000 foundry customers, and between 100,000 and 150,000 peddlers selling scrap iron to some 12,000 to 14,000 scrap dealers. Important to this complicated business were scrap brokers, who obtained widely varying commissions from storing and processing the product for its ultimate use. Scrap metal prices were highly sensitive to transportation costs, which proved to be a difficult regulatory problem as operating railroads could transport their own scrap without charge. Heavy war demand also led to the utilization of unconventional forms of transportation for transporting scrap such as switching lines, electric interurban passenger systems, and street railways. The market price for these unlikely carriers was untested in peacetime. By 1944 there were thirty-four grades of steel scrap, thirty-three grades of railroad scrap, and ten grades of cast-iron scrap. Only an experienced scrap grader, a trade which few practiced, could tell the difference, and even he would have difficulty. Speculators often resorted to "top-dressing," in which a carload of low-grade scrap would be covered with a layer of the higher grade indicated in the bill of lading. If the Office of Price Administration had ever solved the problem of regulating prices in scrap iron—and it had only modest success-it would have had to turn its attention to similarly unstable industries with equally problematical quality levels such as scrap paper, rubber, glass, rags, and even bones.[29]

The Black Market

A novel governmental control intended in part as a means of spreading the sacrifices of war was the rationing of consumer items.[30] Inevitably (considering the size of the program) there was some unevenness in the application of the rules. Black markets existed in a number of commodities, perhaps even most. Since illegal activity cannot be measured accurately, it is difficult to estimate the size of the black markets closely. Estimates have varied widely. In 1944 the head of the Office of Price Administration estimated that only 4 percent of the nation's food supply was sold outside the rationing system. But industry sources challenged his figures. According to the meat industry, as much as 40 percent of the nation's meat supply was sold on the black market. The OPA admitted that 5 percent of the gasoline supply was purchased with counterfeit coupons, and estimates of the amount of black market gasoline sold ranged from 1 to 2.5 million gallons per day. The truth is probably somewhere between these estimates.[31]

The persistence of black markets favored affluent consumers as their larger buying power granted them access to scarce and prized goods. As a historian of the Office of Price Administration admitted, "Certain commodities especially in the luxury and variety store categories were pretty largely ignored and conspicuous violators dealing in them frequently went unmolested."[32] This is also illustrated by the black market in rationed meat. The cuts of meat that were most likely to be traded outside the rationing system were the tender variety preferred by the affluent.[33]

Another example was gasoline. Fuel was in short supply throughout the war, but it was considerably more available to businessmen. The Office of Defense Transportation had the authority to issue a "certificate of war necessity" to a civilian who could establish a war-related need for extra gasoline. The difference between war-related business travel and travel conducted strictly for personal purposes proved very difficult to define in practice, and it was hard to evaluate the legitimacy of thousands of requests. The government granted exemptions so freely that according to its historian the entire system of gasoline rationing nearly collapsed. In any case, the evasion of gasoline rationing during World War II was an extensive form of profiteering in terms of the number of defendants prosecuted. By only the third summer of the war, 1,300 persons had been convicted, 4,000 filling stations had lost their licenses, and 32,500 drivers were forced to forfeit their ration books.[34]

Despite its reputation as an epoch of common sacrifice, World War II was a time of flagrant evasion of price, rent, and ration controls. During the five years of its operation, the Office of Price Administration instituted 280,724 sanctions against cheaters—an average of more than 55,000 cases per year. The OPA employed a force of 3,000 investigators and 800 attorneys to prosecute violations, and still it was hopelessly understaffed. During the last two years of the war, 10,813 persons faced criminal charges for violating the price control and rationing laws. For the entire period the OPA operated, 41,724 persons lost their rationing privileges, and 16,153 were fined, imprisoned, or put on probation for willful violations. This figure exceeded the total for all earlier

American wars, and it excludes those cases which were settled informally. Only 2–3 percent of the cases reached the courts.[35]

The New Deal Subsidizes Corporate America

An example of the conflicting objectives pursued by the war managers was the role played by the War Department in defense contracting. Robert Patterson, the undersecretary of war charged with acquisition of war matériel, echoed the president's words by avowing that there should be "no recurrence" of the spate of war millionaires produced twenty years earlier. Yet Patterson also asked for "some relaxation" of the excess profits tax rates "for the benefit of production."[36] Patterson's office was charged with protecting the government's interest through the close pricing of defense goods. But military procurement officers were also under heavy pressure to increase production. Assigned conflicting responsibilities, they usually leaned toward the production goal. In fact, Patterson even established an agency whose sole purpose was to eliminate government regulations that interfered in any way with defense production (the legal division of the Planning Branch, Office of the Assistant Secretary of War).[37] The War Department was profoundly more concerned with production than with cost. This perspective had been a fundamental military tenet long before the War Department was formally organized; indeed, a similar attitude that persisted in Queen Anne's War was still recognizable three centuries later.

When World War II opened, the United States leadership was shocked to find that the American capacity to produce war matériel was woefully inadequate. In 1940, the entire stockpile of gunpowder was too small to meet the needs of a single day of battle at 1943 levels. The rate of rifle production was so slow that had it not been speeded up it would have taken fifty years to equip the army.[38] Conversion of existing civilian productive capacity was vital, but it would never be enough to meet the nearly insatiable needs of the armed forces. New facilities had to be built, but only the federal government was large enough to raise the capital or strong enough to assume the risk this expansion required.

To encourage the construction of new defense facilities, the government devised innovative techniques. In 1940 the device of choice was generous tax incentives. New defense plants received accelerated depreciation schedules: they could be written off in just five years, which was far short of the probable useful life of the factory. This program was large and lavish. Plant expansion in the amount of about $6.5 billion received beneficial treatment, and approximately 86 percent of all applications were approved. Of the plant expansion carried out before Pearl Harbor, 57 percent was privately financed. The program succeeded in stimulating construction, according to its historian, because of its "frank appeal to the profit motive."[39] In accepting accelerated depreciation, the New Dealers in the administration reluctantly bowed to the demands of the production managers and their friends in the military. The liberals exacted as their price for this concession a counterdemand that the terms of all defense contracts let under this program be open to public scrutiny, and they successfully imposed excess profits taxes of unparalleled severity.[40]

Despite its size, tax amortization was not enough, and the government turned to direct reimbursement. Under the Emergency Plant Facilities contract, a defense contractor built a factory or other facility and the government paid him for it. When peace returned, the contractor, at his option, could either buy the building from the government or require that it be demolished. Quite evidently, in such an instance the government would have little or no bargaining ability, and for this and other reasons the plan was abandoned. There were forty-one contracts written under this program, including one for the production of the famous B-29 bomber, but the total sum appropriated was low by World War II standards—only $350 million. The system was cumbersome and starved of capital.

In June 1940, Congress created the Defense Plant Corporation, an adjunct to the Reconstruction Finance Corporation. Under this program the Defense Plant Corporation could build an entire factory and then lease it to a defense contractor. This approach solved a number of bureaucratic tangles and enlarged the amount of capital available. The War Department alone spent nearly $3 billion in financing plant expansions under this plan. A single Chrysler factory cost nearly as much as the combined value of all the facilities built under the Emergency Plant Facilities program, and in total the Defense Plant Corporation was about forty times larger than its predecessor.[41]

The Defense Plant Corporation was so enormous that by June 1945 the federal government owned about one-sixth of the nation's industrial capacity. The Defense Plant Corporation alone had financed about 30 percent of the new facilities built during the war. The federal government owned 90 percent of the synthetic rubber, aircraft, and magnesium industries and 55 percent of the aluminum business. The federal inventory of machine tools was so large that it amounted to about twenty-five years of peacetime production of the American machine tool industry. As victory neared, the future of this huge block of capital became a matter of intense public discussion.[42]

"Legal Profiteering"

In the afterglow of the Great War, disheartened veterans asked an angry question: "Who got the money?" The divisive question of the late World War II period was "Who will get the capital?" Harold Ickes, Roosevelt's secretary of the interior and one of the staunchest devotees of the New Deal, proposed that government-owned factories be given to the returning servicemen. He recommended that Congress establish a vast holding company with several subsidiary corporations managing each facility. Veterans would own stock in the holding company and would receive dividends from any profits distributed. In effect, a grant of capital to veterans would be the twentieth-century counterpart to lands granted to veterans of the wars of the nineteenth century. Max Lerner, a socialist critic of the New Deal, applauded the Ickes plan. Like the Tennessee Valley Authority and its several Progressive era ancestors, the great public holding company was appealing for its ability to serve as a yardstick by which to measure—and thus to limit—profits earned in America's large and monopolistic industries. American businessmen expressed no enthusiasm for the idea, which they considered more of the New Deal's creeping socialism.[43]

Congress preferred an alternative plan authored by Bernard Baruch and embodied in the Surplus Property Act of 1944. The principal objective of this scheme was to return to a peacetime economy based on private enterprise as quickly as possible. This would be accomplished by selling the federally owned productive facilities to the highest bidder. Having this as its purpose, the plan was endorsed by the National Association of Manufacturers and the United States Chamber of Commerce. Its key backer was the ubiquitous conservative, Senator Walter George of Georgia, who chaired a special Senate committee on postwar planning. Senator Harry S Truman's Special Committee Investigating the National Defense Program advocated a slower approach to ensure that the government received a "fair" price.[44]

The program has been called a huge garage sale, and in that sense it was the biggest such sale in history. Sales began even before the war ended and continued for several years. The most desirable property sold first; in 1946, factories sold for 54 percent of their original cost. Over the life of the program, however, war plants sold for only one-half to one-third of the cost to the government. Machine tools were temporarily leased and then sold. Gerald T. White, the historian of the Defense Plant Corporation, estimates that the cost recovery of facilities built under the program was not more than 35 percent.[45]

The effect of the surplus property disposal system was to create a buyer's market—the capital market was glutted. Since many defense plants had been built contiguous to existing privately owned factories, there was in practice often only one customer available. Although the details of each transaction may be disputed, there is little doubt that the decision to sell war plants quickly created a market in which a shrewd purchaser could obtain a highly favorable price. The decision to subsidize defense facilities lavishly was based on the assumption that after the war they would be useless.[46] But in practice, when the war ended there was a large pent-up demand for consumer products created by years of high wages and rationing. Good profits beckoned to the businessman who owned the facilities needed to service this demand.

Facilities built under "Certificates of Necessity" issued by army and navy officers were eligible to receive full tax amortization, and these proved to be the most lucrative postwar acquisitions. Through 1943 military officers routinely certified that 100 percent of the new facilities were required for defense production. Beginning in 1944, this figure was normally reduced to 35 percent. As long as a machine tool or factory addition was fully certified as a necessity of national defense, it could be written off against excess profits taxes. When the war ended, the residual value of the government-financed asset represented a windfall profit.[47]

Harry Truman denounced this form of gain as "legal profiteering." This was a novel definition of a very old concept, and it was unique to World War II. The full extent of these gains cannot be calculated precisely, but the best estimate is about 40 percent of the cost of the tax amortization program, or roughly $3 billion. Of this amount, two-thirds would have gone to the nation's largest corporations.[48] When added to the gains posted through the purchase of bargain-priced facilities constructed by the Defense Plant Corporation (only 35 percent of the cost was recovered), the great gains of the Good War were obtained in the postwar period, were paid in kind, and were entirely legal.

By war's end, many firms were also swollen with surplus cash. The average contingency reserve rose from 4.7 percent of profits before the war to 15.1 percent. By 1945 half the firms were subject to renegotiation of war contracts, and they retained another 4.0 percent of their profits to cover these costs. Although the War and Navy Departments recovered 68.2 percent of excess profits by renegotiating contracts during the war, contract renegotiation continued for years after V-J Day. In the postwar period, with public opinion angry at labor and grateful for the real contribution of war production to victory, businesses negotiated from a much stronger position than during the conflict. Each contract was reviewed by a business-dominated Price Adjustment Board, and its rulings were routinely accepted.[49]

The Warhog Survives

The government's antiprofiteering program was nothing if not comprehensive. It rested upon the inspiring words of a charismatic leader, upon a vast base of experience, upon the regulatory power of a modern state, upon the broadest possible exercise of the war powers of the chief executive, upon a system of price controls more comprehensive than any in American history, and finally upon a system of rationing that was broader than any Americans had experienced. Nevertheless, the program failed to reach the ambitious goal that Roosevelt repeatedly articulated—namely, of preventing the appearance of a new crop of war profiteers.

The nation's leading corporations weathered the national emergency quite comfortably. In the last four years before the war, 1936-39, the 2,230 largest American firms earned $10.2 billion after taxes. During the four war years, 1942-45, those firms reported earnings of $14.4 billion after taxes—an increase of 41 percent. The net worth of those firms grew from $31.6 billion to $38.2 billion. After taxes, reported returns on net worth increased from 8.1 percent before the war to 9.4 percent during the war. Most of the gain in profits (85.6 percent) came from the enormous rise in sales during the war, but not all of it. Price increases yielded 14.4 percent of the rise in profit; in other words, about one-seventh of the return came from price coercion. Firms that were directly engaged in war production and could therefore benefit from the generous tax amortization rules (29.4 percent) were able to understate their actual earnings by about 20 percent.[50]

A slightly different picture forms when all American firms are considered and a different base is employed. According to data assembled by Harold G. Vatter, aggregate corporate profits before taxes were $9.5 billion in 1940, ascending to a peak $28.0 billion in 1943. But net profits after taxes rose less rapidly, from $6.9 billion in 1940 to $12.2 billion in 1943. Thus while pretax profits nearly tripled during the war, after-tax profits rose by 77 percent in the first three years before stabilizing. In 1948, a good postwar year, after-tax profits were a robust $22.5 billion, nearly double the wartime level.[51]

Yet although the rise in retained earnings after taxes for all American businesses was restricted to between 41 and 77 percent, depending on the base year, the sample, and the reporting method, there is ample evidence that the success of the antiprofiteering campaign was limited in the sense of denying exceptional gains to all defense contractors. The aggregate data for corporations pertains to intangible entities and includes companies that were both directly and indirectly the beneficiaries of wartime prosperity. The first evidence of extraordinary gains by individual defense contractors was compiled by the Treasury Department to defend its record and to bring about compliance with its rules by bringing violations to public attention.

Testifying before the joint congressional committee on taxation in mid-1942, Treasury Secretary Henry Morgenthau Jr. described the improper practices of a number of companies holding war contracts. The most flagrant examples of profiteering came from the aircraft industry, which experienced explosive growth during the war (its output increased by thirty to fifty times) and which was the most profitable single industry when measured by after-tax return to net worth.[52] Morgenthau's examples of war profiteers (whom he did not name) included Robert J. Cannon, the president of the Cannon Manufacturing Company, a maker of spark plugs for aircraft. Cannon hired himself as the company's sales manager at an annual wage of $1,656,000. By this device he was attempting to evade corporate income taxes in the amount of $1,117,000. The head of the Switlik Parachute Company attempted to conceal income by renting property from his wife at an exorbitant rate—an annual rent of $31,104 for property which had originally cost only $45,412. The brother of Switlik's principal stockholder, who had no special training or ability, was hired at the then handsome salary of $15,000. The stockholder's son and daughter, recent graduates, were paid $7,500 each. Finally, two principal owners of Link Aviation Devices, which made pilot trainers, increased their salaries from $12,000 and $15,000 in 1939 to $72,000 and $90,000 in 1941. They jointly held a patent on the Link Trainer and increased their royalties from $87,000 in 1939 to $1,179,000 in 1941.[53]

Such scattered evidence is suggestive but not conclusive. To make a broader estimate of the effectiveness of the government's antiprofiteering program, corporate records of two dozen individual defense contractors have been examined for the wartime period.[54] The companies chosen represented a variety of defense-related industries. These included brass, steel, aluminum, and rubber products, small and large engines, machine tools, and shipbuilding. Except that the sample did not include aircraft construction, these firms were typical of American companies that produced for the war effort. The focus was on the individual who headed each firm, and the goal was to measure his economic well-being in terms of salary, dividends, and gain in net equity during the wartime period. It was not possible to estimate any gains that might have accrued from the postwar purchase of tools and facilities intended for war production.[55]

The data indicate that the average defense contractor prospered considerably during the war, and some prospered enormously. In 1939, the last peacetime year, the mean annual salary of the president of a corporation was $11,375. By 1945, the wartime peak, the mean annual salary had risen to $29,430—an increase of 159 percent (before taxes). One defense contractor reported that his salary climbed from $7,500 in 1939 to $82,635 in 1945. Another reported that during this time his salary soared from $7,500 to $73,041. In 1946, when defense contracts were cut back, these contractors saw their salaries fall to $72,500 and $24,307, respectively. In 1946 defense contractors suffered an average decrease in salary of 6.1 percent over the previous year (see table 2). Thus the government's attempt to restrict swollen executive salaries as a form of profiteering was but a qualified success.[56]

The data on dividends are less complete, but the pattern is similar. The mean dividend income for the eighteen corporation presidents doing defense work rose from $8,497 in 1939 to $15,739 in 1945, an increase of 85.2 percent (before taxes). In 1946 after the war had ended, dividends fell an average of 5.73 percent to $14,836. One defense contractor who had earned $5,083 in dividends in 1939 collected $82,522 in 1945. Another, who had received $2,854 in dividends in 1939, obtained $66,509 in 1946.

Income is an important measure of economic well-being, but wealth is even more important. An analysis of the eighteen corporation presidents reveals that their average net equity in their firms appreciated from $138,953 in 1936 to $355,084 in 1946—an increase of 155 percent. Again, very substantial gains were possible. One defense contractor enjoyed an expansion in his net equity from $610,209 to $2,156,449 between 1939 and 1945. Another saw his equity expand from $81,951 to $435,328. Most corporate presidents voluntarily complied with injunctions intended to restrict gains, but a substantial minority did not. Broadly speaking, the circumstance of most business leaders cooperating with the national effort while a few seized the opportunity to benefit unduly during the emergency replicated the home-front experience of previous American conflicts.

These data show that despite its best intentions, the Roosevelt administration was unable to fulfill its promise to restrict war profits. During the war the simplistic analysis of "war profiteering" gave way to a new reality. The task of restraining wartime profits proved to be vastly more difficult than had been assumed by those who railed against the "greedy munitions makers" during the heyday of isolationism. Restraint of war profits required more tenacious, single-minded determination than the Roosevelt administration possessed. In part, as the New Deal critic Rexford Tugwell understood, the reason for the limited success of the antiprofiteering program was that Roosevelt consciously compromised in the hope that more important production goals might be achieved. National unity would also be promoted by toning down the charges of war profiteering. Unity and productivity would in turn speed victory over fascism and pave the way for American entry into the United Nations. But the elimination of war profits was also a technical impossibility at a price that was acceptable to the American people. The construction of an antiprofiteering program that entirely eliminated the swollen profits of wartime without interfering with war production, discarding the democratic process, or dramatically expanding the police power of the state, exceeded the ingenuity of the nation's most seasoned and resourceful leaders.

Table 2. Salaries of Presidents of Large Wisconsin Defense Contracting Firms, 1939–1946

Industry	1939	1940	1941	1942	1943	1944	1945	1946
1. Brass	12,000	20,000	21,000	20,450	20,400	20,400	20,900	21,250
2. Brass	10,000	10,000	10,000	10,000	10,000	10,000	10,666	12,000
3. Engines	N/A	7,500	9,925	25,735	25,508	44,113	73,041	24,307
4. Aluminum	12,335	14,489	19,902	14,422	19,901	19,901	23,000	N/A
5. Brass	5,200	7,809	14,018	18,000	18,000	18,000	18,035	18,008
6. Brass	N/A	5,269	10,848	25,928	11,907	17,800	16,236	16,440
7. Aluminum	3,840	4,300	6,000	12,166	21,479	25,499*	14,852*	N/A
8. Aluminum	7,920	12,160	15,000	15,000	15,000	15,000	N/A	21,600
9. Steel Fabrication	4,079	2,099	5,109	9,088	21,741	17,396	18,578	6,000
10. Battery	12,647	11,897	12,874	13,093	13,440	14,040	23,040	15,040
11. Rubber	15,000	25,000	27,500	50,000	50,000	50,000	50,000	50,000
12. Aluminum	8,400	8,400	8,400	N/A	N/A	13,566	15,600	12,600
13. Rubber	7,500	7,500	18,261	34,018	39,106	62,815	82,635	72,500
14. Steel Fabrication	20,000	20,000	30,000	21,500	20,200	20,200	20,700	20,550
15. Aluminum	12,000	13,000	20,000	25,000	25,000	25,000	25,000	34,700
16. Shipbuilding	3,600	15,000	15,569	25,000	25,000	25,000	25,000	25,000
17. Iron Castings	N/A	20,066	30,000	33,500	33,500	33,576	33,510	35,550
18. Engines	33,000	33,000	33,000	33,000	33,000	33,000	33,000	48,583
19. Engines	5,625	15,025	15,905	31,708	29,750	33,750	33,750	42,500

(continued)

Table 2 continued.

Industry	1939	1940	1941	1942	1943	1944	1945	1946
20. Engines	10,000	12,000	12,000	15,000	17,250	18,250	28,512	31,000
21. Machine Tools	14,046	16,693	28,410	23,674	23,674	23,699	24,130	23,674
22. Machine Tools	14,082	26,468	18,850	34,550	38,800	33,800	33,800	33,800
23. Machine Tools	24,000	45,000	86,000	86,000	86,000	60,000	N/A	N/A
24. Machine Tools	3,600	3,600	8,600	12,326	11,663	12,537	18,484	15,068
Mean Salary	11,375	14,845	19,882	25,619	26,547	27,181	29,430	27,627
% Increase		30.51	33.9	28.9	3.6	2.4	8.3	-6.1

Source: Corporate Income Tax Records, Wisconsin State Archives.

Table 3. Dividend Income of Presidents of Eighteen Wisconsin Defense Contracting Firms

Year	1939	1940	1941	1942	1943	1944	1945	1946
Mean Income	8,497	10,932	11,248	10,076	12,033	10,466	15,739	14,834
% Increase	—	28.7	2.9	-10.4	19.4	-13.0	50.4	-5.7

Source: Corporate Income Tax Records, Wisconsin State Archives.

12

War Profits and Cold War Culture

I didn't know he was getting eight dollars a bag for it, Benjamin, a little bag.
Imagine taking money for other people's misery.
LAVINIA HUBBARD, IN LILLIAN HELLMAN'S
ANOTHER PART OF THE FOREST (1946)

Although the Second World War offered ample material to spark a new war profits controversy, circumstances after V-J Day precluded an early renewal of the old dispute. The nation rejoiced in the euphoria of victory, basked in relief from danger, and took great pride in the production miracle that had made those successes possible. But it also bore a deep anxiety that the hard days of the Depression would return, and few were interested in disturbing a prosperity that might turn out to be very fragile. The war profits controversy subsided, where it would sleep deep in the social memory. Yet while the profiteering issue has often lain dormant for considerable intervals, it has a penchant for awakening periodically to reveal again its characteristically stubborn resistance to defeat. What began as plunder and ransom, then changed to gouging and speculation, ultimately reappeared in thin disguise as a contestant in the cultural politics of cold war America.

There were sound reasons why the Spirit of '45 did not include a biting resentment of wartime gains. Of crucial importance was the war's fundamental popularity. In October 1939, two years before the United States entered World War II, 68 percent of the American public had considered participation in the Great War to have been a mistake. Yet in September 1947, two years after World War II ended, almost exactly an equal proportion judged that entry into the second war had *not* been an error.[1] A broad consensus held that the wartime sacrifices were necessary, and this left little room for complaints about which persons had borne them.

Although the New Deal's management of the war economy fell short of the standard it set for itself, most Americans realized that as measured by the standards of the Civil War or the Great War it was still a very respectable effort. Prices doubled during the Civil War and increased by 81 percent between 1916 and 1920, but rose by only 38 percent from 1941 to 1946. Taxes paid for 28 percent of the Civil War, 36 percent of the Great War, and 40 percent of World War II.[2] The tax structure, which featured sharply graduated personal income and excess profits taxes, was fundamentally progressive, persuading most that, broadly speaking, prosperous Americans had fairly borne their share of the economic sacrifice. In 1943 and again in 1945, the Gallup poll found that no less than 85 percent of the American public described the tax system as "fair." Significantly, there was no difference in this estimation between persons holding blue-collar and white-collar jobs. Indeed, there was evidence that the public preferred heavier taxes imposed on the lower income strata, less on the upper stratum.[3]

The remarkable rate of production achieved during the war was a proud national accomplishment, and businessmen rightfully claimed a major portion of the credit. Business prestige, which had been sharply weakened by the Great Depression, enjoyed a distinct (if incomplete) recovery. By contrast, labor's reputation waned. In early 1941, when production was perceived to be too slow, labor bore much of the blame. Fifty-two percent of the American public blamed strikes for the lag, and only 2 percent attributed it to "profit-seeking businessmen." As the war progressed, and after it ended, the antilabor mood persisted.[4]

Very significantly, the war years saw a dramatic leveling of shares of income and wealth in the United States. Although some business executives prospered greatly during the war, persons in lower economic strata did even better. In 1939 the share of income received by the top 5 percent of income recipients in the United States was 27.8 percent. By 1945, that share had fallen to 19.3 percent. For the top 1 percent of income recipients, the share fell from 11.8 percent to 8.8 percent. Persons who had been severely disadvantaged in the prewar period—farmers, blacks, southerners, and unskilled white men—reduced the income differential partially.[5] Although it can be argued that true economic well-being did not improve during the war, when hostilities ceased 64 percent of Americans reported that they had not made any real sacrifice for the war effort. Of those who said they had sacrificed, most said that their misfortune resulted from a relative entering military service, rather than from suffering a financial loss.[1] The flattening of the wealth curve served to reduce if not eliminate resentment of business gains. The languid stock market played a role as well. Since the Dow Jones industrial average did not return to its pre-Depression level until 1954, there were few if any conspicuously large stock market gains like those that marked the Great War.[7]

The popularity of World War II translated into a popularity for its warriors. During World War II the American people reached a sound decision to treat their soldiers and sailors more generously than those of previous conflicts. The minutemen of the Revolution had served for next to nothing, privates in the Grand Army of the Republic received $11 per month, and the doughboys pocketed but $30.[8] During the Depression, a private's pay actually declined to only $21 per month. Soon after Pearl Harbor, Congress wisely raised the minimum wage of a recruit to $50 monthly. This was the first of several actions designed to reduce the postwar anguish of sixteen million embittered veterans grousing about unfair wartime sacrifice. Although nothing could stop soldiers from grumbling, the fact remained that the United States forces were the best-fed, best-dressed, and ultimately the best-equipped troops in the war.[9]

When hostilities ceased, the government expressed its gratitude in a fresh version of the New Deal. Although Congress rejected Harold Ickes's idea of directly transferring government-owned tools and factories to the GIs, and John Maynard Keynes's concept of a capital levy with proceeds going to the veterans was never seriously considered, veterans of World War II received postwar compensation of great value. This included mustering-out pay, legal rights to old jobs, educational benefits, loan guarantees, preference for civil service jobs, unemployment compensation, and a priority right to purchase surplus property. By January 1947, former GIs had purchased $501 million worth of surplus property—22 percent of the total. In the first decade after the war, 7,800,000 veterans undertook educational training. When polled ten years after the war, 73.7 percent of the nondisabled veterans pronounced their separation benefits as "adequate."[10] This general satisfaction with postwar treatment eased the bitterness that was common among other generations of veterans. The comparative absence of anomie after World War II weakened the position of those who would complain about excessive war profits.

In the 1920s the political attack on profiteering had featured congressional investigations, a publicity campaign, and demands for a postwar bonus. During and after World War II, the political climate could not nourish such a campaign. The war was popular, but criticism was not. The restraint of the Truman committee contributed to the perception that the Roosevelt administration was successful in controlling profiteering. Indeed, an atmosphere existed under which serious criticism was informally proscribed. Paul Fussell has called this semiofficial policy "obligatory goodness," and with mocking irony Studs Terkel has memorably labeled the entire conflict "The Good War."[11] Since public opinion was unreceptive to direct criticism of World War II, during the cold war period profiteering was assailed primarily on a cultural level. The exception was a shrill and generally ignored attack by a disgruntled New Dealer, Bruce Catton (who later became a famed Civil War historian), in *The War Lords of Washington* (1948).[12]

Perhaps the earliest example of censure of war profiteering on a symbolic level was the first major motion picture whose subject matter centered on the war's legacy. Ironically entitled *The Best Years of Our Lives,* this touching film won the Academy Award as the best picture of 1946. Focusing on the difficulties faced by war veterans readjusting to civilian life, the film opens arrestingly with a homeward bound Air Corps captain arriving at an American airport, where he encounters a newly prosperous businessman. Stout but not obese, the executive appears to have been eating well and living comfortably. Equipped with a fine new set of clubs, he is leaving on a golfing vacation, and he conspicuously dismisses the extra charge for carrying a golf bag as excess baggage. Meanwhile, the highly decorated ex-bombardier pleads for a reservation on a flight home, only to find that the businessman has taken the last seat. As the pudgy businessman wings away to relax on the links, the war hero is obliged to endure yet another cramped and noisy military flight. Without much help, the weary veteran must make his own way back to normal civilian life. Golfing and air travel, powerful symbols of upper-class affluence, have become available to the merchant but remain closed to the soldier. In the race to enjoy life's amenities, the man of commerce has pulled well ahead of the returning serviceman, who has not yet even reached the starting line.[13]

In December 1946, war profiteering in another form made a striking appearance in Lillian Hellman's moving play, *Another Part of the Forest.* This story unveils the secret, bloody path to wealth of Marcus Hubbard, an Alabaman who has made a fortune in the illicit salt trade during the Civil War. To protect his favored trading position with the Yankee command, Hubbard has led Union troops to a camp where his neighbors' sons are in training for service in the Confederate army. Twenty-seven neighbor boys are massacred, and only by lying about his crime does Hubbard escape a lynching. But his wife, Lavinia, knows the truth, and she is despondent about the dead recruits and guilt-stricken about her husband's trade and treachery. "I didn't know he was getting eight dollars a bag for it, Benjamin, a little bag," she tells her son. "Imagine taking money for other people's misery." But mostly she cannot abide the luxury that Marcus's treason has provided, and she threatens to tear up her family's "evil money" to expiate their guilt.[14]

Hellman's play was about an earlier time and an earlier form of profiteering, thus addressing the issue circuitously, but it resonated with the postwar generation. A more direct censure appeared in 1947 when one of the most touching literary symbols of profiteering ever conceived emerged in the form of Joe Keller, the leading character in Arthur Miller's celebrated play, *All My Sons.* Keller is a crude and grasping owner of a company that has thrived during World War II from manufacturing engines for Air Corps pursuit planes. His conscience is not bothered in the slightest by the wealth he has acquired. "Jail? You want me to go to jail?" Keller chides defiantly. "Who worked for nothin' in that war? . . . When they work for nothin', I'll work for nothin'.

Did they ship a gun or a truck outa Detroit before they got their price? . . . Half the Goddam country is gotta go if I go!"

The play reveals that a number of the engines made by Keller's firm were known to have had cracked cylinder heads, but on Keller's orders they were delivered anyway. After twenty-one airmen are lost, Keller's pilot son learns of his father's crime and deliberately crashes his own plane. In the play's powerful conclusion, Keller learns of the circumstances of his son's suicide, and he too takes his own life. *All My Sons* assails contemporary profiteering directly, but the type of profiteering selected is the degraded quality form, which was uncommon during World War II. The thin factual underpinning weakens the play slightly, but nonetheless it remains a compelling statement. The play was also presented to the public as a motion picture, with Edward G. Robinson starring as Joe Keller.[15]

In 1955 Miller's achievement was surpassed by the appearance of Joseph Heller's *Catch-22*, which introduced Lieutenant Milo Minderbinder, perhaps the best known of all fictional profiteers. Minderbinder takes profiteering to the level of an ultimate absurdity. Milo obtains a personal squadron of bombers, which he leases to the Air Corps. After agreeing to bomb a bridge for a fee of cost plus 6 percent profit, Minderbinder is so colossally amoral that he also charges an equal fee from the enemy for defending the target. He even multiplies his gain by collecting a thousand-dollar commission for every friendly plane he shoots down. He defends his dealing with the Germans with transparent but engaging sophistry: "Maybe they did start the war, and maybe they are killing millions of people, but they pay their bills a lot more promptly than some allies of ours. . . . Can't you see it from my point of view?" Minderbinder also denies the charge that he started the war and argues: "I'm just trying to put it on a businesslike basis. Is anything wrong with that?"[16]

Heller's dark humor was a supremely effective challenge to the myth of World War II as "The Good War." The book's absurdist, deeply cynical pessimism surpasses in intensity the legendary disillusionment of the veterans of the Great War. A decade later the grim character of *Catch-22* and its use of profiteering as a vehicle of expressing it was carried forward by Kurt Vonnegut Jr. in *God Bless You, Mr. Rosewater*. Vonnegut described a symbolic Indiana family whose wealth had been built during the Civil War by converting a saw factory to make swords and bayonets for the Union army. The founder of the fortune, Noah Rosewater, had "hired a village idiot to fight in his place," wrote Vonnegut, who contended that "every grotesquely rich American represents property, privileges, and pleasures that have been denied to the many."[17]

Marcus Hubbard, Joe Keller, Milo Minderbinder, and Noah Rosewater are modern characters who followed by nearly a century the crooked Victorian path of Charles Holt, the first war profiteer in an American novel.[18] They are far more complex figures than the shallow Holt, and the fictional appearance of modern war profiteers in the early years after the allegedly Good War appealed to the American public, whose taste in culture, as in public affairs, had become increasingly sophisticated and mature.

The historian Robert E. Osgood has termed the great transformation of American thinking wrought by World War II as the New Realism. The most important element of New Realist thought was a basic understanding and acceptance of America's new role as a world leader. A major implication of that new status as a great power was that a vastly enhanced national defense establishment must become a permanent fixture of American life. A larger defense component would demand massive expenditures, and this implied that massive profits would accumulate.[19]

But how much defense profit would a New Realist accept? The unpleasant and unwelcome lesson taught by World War II was that the price of freedom included a percentage of profit, and that profit was legitimate.[20] Nevertheless, the cold war confrontation opened a new and very long ledger. Some Americans were ready to join William Shakespeare, a consummate Old Realist who four centuries earlier despaired in *Henry V* to "spend our vain command" in venturing to curtail profiteering.[21] But while there was no attempt to reimpose in the 1950s the profit restrictions of the Vinson-Trammell Act of the 1930s, most Americans of the postwar period yielded to profiteers grudgingly.

Like their ancestors, Americans of the cold war era remained deeply suspicious of wealth gained quickly and distrustful of wealth gained from arms. Americans might be willing to accept the internationalist foreign policy that their new position and overwhelming strength implied, but they found that the enormous price of that status was difficult to swallow. The cost, unfortunately, was exacted not only in economic terms (which were of an unprecedented scale) but also in other disturbing ways. The permanent existence of an immense, profit-seeking national defense industry implied an end to American exceptionalism. The American nation had been long spared the existence of a large standing army served by a colony of purveyors, but now those days were a thing of the past. Like the great powers of old, Americans would have to face the permanent and copious existence of war profit. Besides eroding American exceptionalism, large, ongoing defense expenditures would raise difficult new questions that would persist throughout the ethical twilight that was the cold war. Finally, these expenditures would dictate an indefinite continuation of much of the extensive procurement bureaucracy that had regulated defense profits during World War II.

Four decades of the cold war created new issues that redefined and veiled the meaning of profiteering once again. How could a nation call for patriotic waiver of "war" profits during a conflict, such as those in Korea or Vietnam, that was not a legally declared war? How could defense contractors be called to sacrifice if a conflict, like the cold war itself, was of indefinite duration? How could the claims of various sectors that are indirectly affected by large-scale defense spending—such as farm prices, labor rates, and regional growth—be balanced against direct defense needs and legitimate interests that are wholly unrelated to defense? The cold war blurred the difference between wartime and peacetime, and the rise of what some have called the National Security State obscured the difference between war profits and civilian profits.

Of the many numbing qualities of the cold war, one of the most desensitizing was its open-endedness. No private business can operate indefinitely if its profits are subnormal, and during the cold war the defense industry was profitable but not excessively so. During the 1950s investments in companies specializing in defense work were nearly twice as profitable as in nondefense firms, but during the 1960s defense-oriented companies lost ground, doing only about equally well as other New York Stock Exchange firms.[22] Nevertheless, the appetite of advanced military technology for money was so voracious that it inspired righteous wrath from claimants forced on a diet of restricted appropriations. Technological progress meant that a new war might be decided almost immediately. Having little need to prepare in advance for war had always made Americans different from Europeans, and now that distinction was gone. Americans struck out at the expensive shield that provided dubious security while eroding exceptionalism. In frustration, they denounced it angrily but emptily as the "military-industrial complex."

During their centuries of military experience, Americans had slowly made progress against the bane of war profits. One by one, several forms of profiteering had been eliminated or at least severely restricted. Scalp bounties, trading with the enemy, and plunder, types that had disgraced early American history, had thankfully fallen into disuse, declined to insignificance, or been isolated by legal and moral proscription (although the trading with the enemy thrived during the Civil War and plunder made a brief return during World War II).[23] The religious proscription of Puritan days—declaring profiteering a Provoking Evil—had been of limited effectiveness in curbing abuse, but it formed a basis for the Revolutionaries' secular appeal to shun war profits as poisonous to republican virtue.

Many of the practices that had outraged the mid-nineteenth century had also declined or been largely corrected. The custom of employing private citizens like W. W. Corcoran or Jay Cooke to fill the nation's war chest during the Mexican and Civil Wars (to their great profit) had been replaced by the employment of equally competent but far less well compensated public servants, such as William G. McAdoo and Henry Morgenthau Jr. The spectacle of unpatriotic stock and commodity speculation, which had tarnished the achievements of the Civil War generation, was moderated by the price control system of World Wars I and II. The salary control system of World War II shrank the bloated executive incomes that had disgraced the record of the Great War. The success of the wage and price control systems in turn rested upon a basis well established by the Civil War leadership: an appeal to fundamental patriotism in time of national danger.

Yet, resistant to these many remedies, profiteering persisted. As the twentieth century opened, war profiteering had become in the eyes of the Progressives a particularly ugly example of monopolistic greed. By the mid-1930s, profiteering had fallen to its deepest level of repugnance. From the pacifist isolationist perspective, its warmongering threatened the very future of civilization itself. Indeed, war profiteering offered evidence for those consummate pessimists who in despair had begun to believe in *l'histoire immobile,* the view that nothing ever changes or can be changed. Would a refusal to surrender war profits be the rock upon which the progress of civilization foundered?

But as a result of the Good War and the new menace of communist aggression, what had been a great issue in 1935 had become a matter of subordinate though still serious concern by 1955. Eisenhower's admonishment about a powerful military-industrial complex warned against a danger of fraud, distorted priorities, and the entrenched interest of defense contractors, but not about warmongering. The procurement issues of the cold war were significant but by no means new; for example, domination of military provisioning by contractors from large states was an issue as early as the Revolution and continued throughout the early national period. Defense spending during the cold war years clearly evidenced many flagrant examples of fraud and other abuses, but Eisenhower's warning was different from the Progressives' call for nationalization in order to avoid warmongering. The concept of a "military-industrial complex" suggested a mysterious, clandestine conspiracy against the public purse, but not a deliberate plot to stir up war to fatten profits. The alleged existence of a "military-industrial complex" was alarming, but it was not as menacing as the supposed plotting of the merchants-of-death.

The cold war devoured wealth in previously unimaginable amounts, but the methods of control worked out over the centuries were at last able to keep profiteering substantially in check. In the 1960s military profits finally had become "reasonable," in the sense that the gains of defense contractors were no more lavish than those of their counterparts who served the civilian economy. Profiteering had evolved into a matter of cultural interest rather than a leading political and economic problem, which was a salutary and long overdue development. After more than three centuries of effort, the American people had finally achieved an ability to manage economic sacrifice competently and evenhandedly.

Yet history shows that while profiteering has often gone into remission for long periods, like a resilient disease it has proven to be as persistent as it is pernicious. Only a foolhardy people would fail to remain vigilant against a reemergence of new forms of this ancient affliction.

Appendix A: Graft Convictions, Officers of the Continental Army, 1775–1781

Appendix A: Graft Convictions, Officers of the Continental Army, 1775–1781

	Name	Rank	Offense	Penalty
July 1775	Penuel Cheney	Surgeon's mate	Writing fraudulent drafts on the commissary account	Allowed to resign
August 1775	Oliver Parker	Captain	Defrauding troops of pay	Cashiered
August 1775	Jesse Saunders	Captain	Overdrawing provisions	Cashiered
October 1775	David Brewer	Colonel	Obtaining lieutenant's commission for absent son; overdrawing blankets	Cashiered
November 1775	—— Correy	Lieutenant	Defrauding troops of blanket money	Cashiered
October 1776	Cornelius Hardenbergh	Captain	Defrauding his men	Cashiered
April 1777	—— Carnes	Lieutenant	Converting public property (horse) to own use	Cashiered and ordered to make restitution
February 1778	Israel Davis	Captain	Withholding pay of five soldiers	Ordered to refund money from own wages
March 1778	Peter Vonk	Quartermaster	Misappropriating rum and soap	Cashiered
May 1778	Thomas Lucas	Captain	Overdrawing pay; selling discharge papers	Cashiered
June 1778	Neigal Gray	Lieutenant colonel	Defrauding troops	Cashiered, with name, (continued)

Appendix A continued

	Name	Rank	Offense	Penalty
			of supplies	crime, and place of abode published in newspapers
July 1778	James Davidson	Quartermaster	Defrauding soldiers of provisions and embezzling property	Cashiered
August 1778	Benjamin Flowers	Commissary general of stores (colonel)	Fraud by inflating prices of muskets	Cashiered
August 1778	Cornelius Sweers	Deputy commissary general	Accomplice to above	Cashiered
September 1778	Samuel Brewer	Colonel	Misapplying cloth	Cashiered
October 1778	Nathan Nutall	Quartermaster of stores	Embezzling public stores	Cashiered
February 1779	William Jenkins	Lieutenant	Embezzling and making false pay vouchers	Cashiered
April 1779	—— Lewes	Commissary sales to troops	Profiting on rum	Cashiered
August 1779	Jotham Loring	Lieutenant colonel	Embezzling flour	Cashiered
September 1779	William Godman	Captain-lieutenant	Embezzling a tent	Cashiered
October 1779	Benjamin Ballard	Assistant commissary	Selling public property	Cashiered and ordered

(continued)

Appendix A continued

	Name	Rank	Offense	Penalty
				to make restitution of $526.75
October 1779	Philip Gibbons	Lieutenant	Presenting fraudulent claims for $2,425	Cashiered
October 1779	Francis O'Neal	Surgeon stores	Selling hospital	Cashiered
March 1780	Theophilus Parke	Captain-lieutenant	Fraud and forgery	Dismissed "with infamy" (sword broken over head before troops, charge published in newspaper, prohibited from holding public office)
June 1780	Isaac Coren	Captain	Overdrawing provisions for three men	Cashiered
July 1780	—— Swain	Clothier general	Fraud	Cashiered
August 1780	Peter Manifold	Captain	Selling government horse	Reimbursement of horse of equal value
September 1780	Thomas Thomson	Forage master	Trading government oats for bridles	Cashiered and ordered to make restitution
October 1780	Albert Chapman	Major	Embezzlement	Forced to resign

(continued)

Appendix A continued

	Name	Rank	Offense	Penalty
December 1780	Thomas Dewees	Barrack master	Selling public wood	Cashiered, name to be published three times in English- and German-language newspapers
March 1781	John Collins	Deputy commissary of stores	Pilfering saltpeter; stealing saddle in escape	Forfeit pay; make restitution; name published; prohibited from holding public office; assigned to man-of-war without furlough
May 1781	John Bingham	Lieutenant	Misapplying bounty money	Cashiered

Source: Philander C. Chase, ed., *The Papers of George Washington: Revolutionary War Series,* vol. 1 (Charlottesville: University Press of Virginia, 1985); John C. Fitzpatrick, *The Writings of George Washington* (Washington, 1936).

Appendix B: Profits of Selected Defense Contractors, 1911-1920

1. Remington Arms Union Metallic Cartridge Company

	Gross Sales	Net Income	Income Tax	Invested Capital	Percentage of Net Income to Invested Capital	Dividends Paid
1911	$9,728,780.41	$892,346.03		$6,477,339.46	13.78	N/A
1912	10,539,442.67	799,615.19		7,094,014.37	11.27	N/A
1913	9,713,061.71	684,140.13		7,006,480.12	9.76	N/A
1914	13,033,493.16	1,694,847.03		8,754,136.40	19.36	N/A
1915	22,488,313.98	4,117,564.84		9,248,435.05	44.52	N/A
1916	37,636,881.71	(5,942,780.35)	-0-	43,231,816.03	-0-	N/A
1917	45,313,978.67	4,988,970.75	$708,031.55*	41,191,565.15	12.11	N/A
1918	117,051,153.41	5,646,803.14	663,874.25*	46,450,132.60	12.16	N/A
1919	32,531,954.65	(11,945,772.07)	-0-	38,604,841.23	-0-	N/A
1920	23,532,446.73	(116,377.41)	-0-	28,885,288.53	-0-	N/A
Total	321,569,507.10	819,357.28	1,371,905.80			N/A

* Includes excess profits tax.

2. Colt's Patent Fire Arms Company

	Gross Sales	Net Income	Income Tax	Invested Capital	Percentage of Net Income to Invested Capital	Dividends Paid
1911	$1,417,478.85	$322,930.38	$3,109.67	$1,473,394.26	21.92	$315,966.91
1912	1,902,936.92	569,458.76	3,148.21	1,341,976.20	42.43	200,000.00
1913	2,190,073.41	659,272.40	5,613.11	1,693,531.95	38.93	250,000.00
1914	2,238,421.66	794,646.47	7,946.46	N/A	N/A	262,500.00
1915	5,308,903.90	2,615,755.09	26,157.65	N/A	N/A	600,000.00
1916	10,034,508.52	5,799,586.38	115,991.73	9,628,756.46	60.23	1,575,000.00
1917	14,694,623.39	5,797,793.87	2,171,412.95*	8,932,106.15*	64.91	4,900,000.00#
1918	32,109,910.55	5,693,152.17	3,744,474.40*	11,113,685.68	51.23	1,800,000.00
1919	11,282,779.44	980,844.08	95,249.28	13,308,669.69	7.37	1,000,000.00
1920	4,859,271.33	843,599.08	80,506.56	10,516,615.22	8.02	1,000,000.00
Total	86,038,907.97	24,077,038.68	6,253,610.02*			11,903,466.91#

* Includes excess profits tax.

A stock dividend of $1,000,000 was paid in April 1917.

3. Winchester Repeating Arms Company

	Gross Sales	Net Income	Income Tax	Invested Capital	Percentage of Net Income to Invested Capital	Dividends Paid
1911	$2,787,029.11	$1,165,500.60	$11,605.01	N/A	N/A	N/A
1912	3,197,846.35	1,394,255.27	13,892.55	N/A	N/A	N/A
1913	10,590,816.53	1,258,132.92	12,581.33	N/A	N/A	N/A
1914	11,453,753.25	1,826,630.60	17,908.26	N/A	N/A	N/A
1915	19,436,869.14	3,538,980.50	35,389.81	N/A	N/A	N/A
1916	26,441,074.60	2,339,018.75	N/A	N/A	N/A	N/A
1917	28,471,861.62	(432,831.23)	-0-	$18,878,085.88	(2.29)	N/A
1918	61,257,937.95	4,934,032.21	2,523,149.46*	21,627,741.09	22.81	N/A
1919	25,185,940.56	3,828,848.44	934,429.28*	22,102,929.91	17.32	N/A
1920	18,263,224.16	(116,260.02)	-0-	30,238,258.19	(0.38)	N/A
Total	207,086,353.27	19,736,308.04	3,548,955.70*			N/A

* Includes excess profits tax.

4. Newport News Shipbuilding & Drydock Company

	Gross Sales	Net Income	Income Tax	Invested Capital	Percentage of Net Income to Invested Capital	Dividends Paid
1911	$6,781,853.78	$428,113.76	$4,281.13	$4,345,237.02	9.85	296,210.00
1912	7,835,634.35	501,632.76	5,016.33	4,454,506.88	11.26	290,210.00
1913	8,347,149.93	328,838.08	3,288.38	4,713,053.27	6.98	300,000.00
1914	7,503,640.30	431,208.24	4,312.08	4,735,165.45	9.11	300,000.00
1915	10,098,694.54	817,365.42	8,173.65	5,952,531.45	13.73	600,000.00
1916	11,506,201.54	814,891.04	16,297.82	4,222,557.45	19.30	600,000.00
1917	23,846,732.94	3,298,601.13	1,074,169.63	3,822,549.44	86.29	300,000.00
1918	39,744,100.00	3,990,311.37	2,493,369.67	5,500,091.76	72.55	300,000.00
1919	39,412,740.43	4,916,655.58	1,789,074.43	6,932,791.21	70.92	1,400,000.00
1920	37,962,853.52	5,764,089.69	1,576,211.05	7,667,134.10	75.18	2,300,000.00
Total	193,039,601.33	21,291,707.08	6,974,194.17			6,686,420.00

5. Jones & Laughlin Steel Company

	Gross Sales	Net Income	Income Tax	Invested Capital	Percentage of Net Income to Invested Capital	Dividends Paid
1911	N/A	$2,263,656.36	$22,586.36	$67,342,279.52	3.36	N/A
1912	N/A	1,705,296.63	17,002.97	68,863,729.19	2.48	N/A
1913	$8,527,477.82	4,629,033.80	46,290.34	69,625,512.57	6.65	N/A
1914	32,402,075.67	2,638,028.24	26,380.28	N/A	N/A	N/A
1915	44,431,034.71	6,593,394.24	65,933.94	N/A	N/A	N/A
1916	77,353,009.29	20,254,737.65	405,094.79	95,356,723.61	21.24	N/A
1917	132,570,498.83	48,869,577.22	19,636,202.90*	103,057,128.74	47.42	N/A
1918	148,943,706.65	29,654,130.20	17,090,598.92*	139,540,274.53	21.25	N/A
1919	119,129,200.58	16,970,731.57	3,011,235.38*	151,008,060.66	11.24	N/A
1920	175,567,789.58	32,288,372.27	8,935,689.02*	153,486,461.85	21.04	N/A
Total	738,924,793.13	165,866,958.18	49,257,014.90*			N/A

* Includes excess profits tax.

6. Semet Solvay Company

	Gross Sales	Net Income	Income Tax	Invested Capital	Percentage of Net Income to Invested Capital	Dividends Paid
1911	$1,062,557.52	$534,944.53	$5,349.74	$2,846,477.42	18.79	$176,000.00
1912	1,033,920.21	409,155.46	4,091.55	3,028,306.70	13.51	200,000.00
1913	7,953,464.29	604,329.59	6,043.29	3,736,372.52	16.17	200,000.00
1914	8,016,080.06	579,115.30	5,791.15	N/A	N/A	220,000.00
1915	16,455,590.40	5,157,254.98	51,572.55	N/A	N/A	600,000.00*
1916	36,971,464.90	10,476,198.56	209,523.97	22,997,497.29	45.55	1,900,000.00*
1917	38,087,860.64	6,783,122.36	1,370,628.24	23,910,684.21	28.37	3,779,996.00*
1918	44,179,486.37	1,362,528.78	158,692.26	13,799,332.45	9.87	1,305,583.98
1919	18,682,187.73	574,317.15	49,772.51	13,630,671.61	4.21	1,357,944.00
1920	28,515,693.91	1,475,621.06	169,721.60	16,456,504.58	8.97	1,358,310.88
Total	200,958,306.03	27,956,587.77	2,031,186.86	100,405,846.78	27.84	11,097,834.86*

Source: M. J. La Padula and H. C. Iller to Stephen C. Raushenbush, 16 August 1934, Records of the Special Committee Investigating the Munitions Industry (Nye committee), box 154, RG 46, National Archives. The information was taken from records of the Internal Revenue Service.

* In addition, the firm paid stock dividends in the amounts of $4,000,000 (1915), $319,996 (1916), and $652,791.99 (1917), for a total of $4,972,787.99.

NOTES

Introduction

1. Philip L. Barbour, ed., *The Jamestown Voyages under the First Charter, 1606-1609* (2 vols.; Cambridge: Cambridge University Press for the Hakluyt Society, 1969), 1:133–34, 170, 2:379; David Beers Quinn, *England and the Discovery of America, 1481–1620* (New York: Knopf, 1974), 455; J. Frederick Fausz, "An 'Abundance of Blood Shed on Both Sides': England's First Indian War, 1609–1614," *Virginia Magazine of History and Biography* 98 (January 1990): 11.

2. Peter D. McClelland, *The American Search for Economic Justice* (Cambridge, Mass.: Blackwell, 1990), 89; Hal R. Varian, "Distributive Justice, Welfare Economics, and the Theory of Fairness," *Philosophy and Public Affairs* 4 (winter 1975): 226–27.

3. Lester C. Thurow, "A Theory of Groups and Economic Distribution," *Philosophy and Public Affairs* 9, no. 1 (1979): 28.

4. Jeffrey G. Williamson and Peter H. Lindert, *American Inequality: A Macroeconomic History* (New York: Academic Press, 1990). See also Carole Shammas, "A New Look at Long-Term Trends in Wealth Inequality in the United States," *American Historical Review* 98 (April 1993): 412–31.

5. See McClelland, *American Search for Economic Justice*, 24–49, and James L. Huston, "The American Revolutionaries, the Political Economy of Aristocracy, and the American Concept of the Distribution of Wealth, 1765–1900," *American Historical Review* 98 (October 1993): 1094. In otherwise brilliant expositions, McClelland and Huston disregard the effect of military factors on wealth distribution.

6. Dwight D. Eisenhower, *At Ease: Stories I Tell to Friends* (Garden City, N.Y.: Doubleday, 1967), 349. See also Michael F. Noone, "The Military-Industrial Complex Revisited," *Air University Review* 30 (Nov.-Dec. 1978): 81–82, and Robert Griffith, "Dwight D. Eisenhower and the Corporate Commonwealth," *American Historical Review* 87 (February 1982): 90–93.

7. For the European story, readers should consult the monumental scholarship of the distinguished economic historian Fritz Redlich. See Redlich, "De Praeda Militari: Looting and Booty, 1500–1815," *Vierteljahrschrift für Sozial- und Wirtschaftsgeschichte* 39 (1956); Redlich, "The German Military Enterpriser and His Work Force: A Study in European Economic and Social History," *Vierteljahrshrift für Sozial- und Wirtschaftsgeschichte,* 47 and 48 (1964, 1965). See also Redlich, *Steeped in Two Cultures* (New York: Harper, 1971). An introduction to more recent European scholarship is Benjamin F. Cooling, ed., *War, Business, and World Military-Industrial Complexes* (Port Washington, N.Y.: Kennikat, 1981).

8. *Warhog, warsow,* and less frequently *liceman* were terms in general use in the United States throughout the Great War. Warhogs and warsows were male and female speculators who became wealthy as a consequence of the frenzied business activities and ample profits associated with war orders. A liceman searched out war contracts for a warhog or warsow. "Warhogs and War Millionaires [By a War Broker]," *Independent,* 9 July 1915, 80–81.

9. See Jonathon S. Boswell and Bruce R. Johns, "Patriots or Profiteers? British Businessmen and the First World War," *Journal of European Economic History* 2 (fall 1982): 423–45.

10. A precise economic definition of the economic "well-offness" of a consumer unit includes current income (both monetary and nonmonetary) and annual lifetime annuity value of its current net worth. See Burton A. Weisbrod and W. Lee Hansen, "An Income–Net Worth Approach to Measuring Economic Welfare," *American Economic Review* 58 (December 1968): 1315, and James N. Morgan and James D. Smith, "Measures of Economic Well-Offness and Their Correlates," *American Economic Review* 59 (May 1969): 454–55.

11. Fausz, "An 'Abundance of Blood Shed,'" 8n.

12. Boswell and Johns, "Patriots or Profiteers?" 426.

13. Hugh Rockoff, *Drastic Measures: A History of Wage and Price Controls in the United States* (Cambridge: Cambridge University Press, 1984), 20–22.

14. Elizabeth M. Nuxoll, "Congress and the Munitions Merchants: The Secret Committee of Trade during the American Revolution" (Ph.D. diss., City University of New York, 1979), 404. Even though this was a low percentage, several Revolutionary provisioners made enough sales at that rate so that it proved to be most lucrative.

15. John W. Hillje, "New York Progressives and the War Revenue Act of 1917," *New York History* 53 (October 1972): 441–43; U.S. House War Policies Commission, 71st Cong., 2d sess., 1931, 21; U.S. House Select Committee on Expenditures in the War Department, *Hearings,* 66th Cong., 1st sess., 1919, 1:485; "Should Wealth Be Conscripted?" *Nation,* 24 April 1935, 469; Jacob Vander Meulen, *The Politics of Aircraft: Building an American Military Industry* (Lawrence: University Press of Kansas, 1991), 140–43; transcript, meeting re Allied Purchasing, 26 May 1940, Henry Morgenthau Jr., Diaries, vol. 276, p. 38, Franklin D. Roosevelt Library, Hyde Park, N.Y.

16. Roosevelt quoted in Russell Samuel McMahan Jr., "The Protestant Churches during World War I: The Home Front, 1917–1918" (Ph.D. diss., St. Louis University, 1968), 140; Hoover to Wilson, 26 November 1917, Hoover Papers, Hoover Institution Archives, Stanford University, Stanford, Calif., box 7; "Patriotism and Profits: President Wilson's Message to Business," *Independent,* 21 July 1917, 112; "Patriotism and Prices," *Outlook,* 25 July 1917, 470–71.

17. Otis G. Hammond, ed., *Letters and Papers of Major-General John Sullivan, Continental Army* (Concord: New Hampshire Historical Society, 1931), 2:210, 351–52, 408–9.

18. Samuel Alleyne Otis to the president of Congress, 24 February 1778, in Edmund C. Burnett, *Letters of Members of the Continental Congress* (Washington: Carnegie, 1926), 3:98n.; U.S. Senate Special Committee Investigating the Munitions Industry, *Hearings,* 74th Cong., 2d sess., 1936, 791, 12011–12; F. D. Enfield, "Report from the Committee on Transactions," 11 January 1924, Records of the War Transactions Section, Department of Justice, box 49, RG 60, National Archives.

1. A Provoking Evil

1. Harold E. Seleskey, *War and Society in Colonial Connecticut* (New Haven: Yale University Press, 1990), 96.

2. C. G. Cruikshank, *Elizabeth's Army* (London: Oxford University Press, 1946), 5; George M. Trevelyan, *England under the Stuarts* (New York: Putnam, 1941), 146; Arthur H. Noyes, *The Military Obligation in Mediaeval England* (Columbus: Ohio State University

Press, 1930), 42–43, 57; Lois G. Schwoerer, *"No Standing Armies!" The Antiarmy Ideology in Seventeenth-Century England* (Baltimore: Johns Hopkins University Press, 1974), 8ff.

3. Sir Charles Oman, *A History of the Art of War in the Sixteenth Century* (New York: Dutton, 1937), 368, 373. There is some historical dispute in respect to the quality of British soldiers. In 1776 ex-convicts made up three regiments of the Royal Army, but evidently most troopers were drawn from the ranks of the unemployed. See Don Higginbotham, *The War of American Independence: Military Attitudes, Policies, and Practices, 1763–1789* (New York: Macmillan, 1971), 123, and Sylvia R. Frey, *The British Soldier in America: A Social History of Military Life in the Revolutionary Period* (Austin: University of Texas Press, 1981), 16.

4. Michael Powicke, *Military Obligation in Medieval England: A Study in Liberty and Duty* (Oxford: Clarendon Press, 1962), 1–23; Noyes, *Military Obligation,* 8.

5. William S. McKechnie, *Magna Carta: A Commentary on the Great Charter of King John* (2d ed.; Glasgow: Maclehose, 1914), 72–73, 260–61.

6. Powicke, *Military Obligation,* 223; David R. Millar, "The Militia, the Army, and Independency in Colonial Massachusetts" (Ph.D. diss., Cornell University, 1967), 7–12; Richard H. Marcus, "The Militia of Colonial Connecticut" (Ph.D. diss., University of Colorado, 1965), 36; Darrett Bruce Rutman, *A Militant New World, 1607–1640* (New York: Arno Press, 1979), 2, 16, 25; Douglas Edward Leach, "The Military System of Plymouth Colony," *New England Quarterly* 24 (September 1951): 343; Morrison Sharp, "Leadership and Democracy in the Early New England System of Defense," *American Historical Review* 50 (January 1945): 244–45; Samuel Eliot Morison, "Harvard in the Colonial Wars, 1675–1748," *Harvard Graduates' Magazine* 26 (June 1918): 555–56.

7. Noyes, *Military Obligation,* 56–57; Don Higginbotham, *War and Society in Revolutionary America: The Wider Dimensions of Conflict* (Columbia: University of South Carolina Press, 1988), 37; Fred Anderson, "The Colonial Background to the American Victory," in *The World Turned Upside Down,* ed. John Ferling (Westport, Conn.: Greenwood Press, 1988), 2; Adam J. Hirsch, "The Collision of Military Cultures in Seventeenth-Century New England," *Journal of American History* 74 (March 1988): 1188; Theodore H. Jabbs, "The South Carolina Militia, 1663–1733" (Ph.D. diss., University of North Carolina, 1973), 13; Leon de Valinger, *Colonial Military Organization in Delaware* (Wilmington: Delaware Tercentenary Commission, 1938), 20; Michael Roberts, *The Military Revolution, 1560–1660* (Belfast: Boyd, 1956), 18; Pennsylvania legislature, *Minutes of the Provincial Council of Pennsylvania* (Harrisburg, 1852), 6:534, 735.

8. Philip L. Barbour, *The Three Worlds of Captain John Smith* (London: Macmillan, 1964), 23; John Ferling, *Struggle for a Continent: The Wars of Early America* (Arlington Heights, Ill.: Davidson, 1993), 2; Rutman, *Militant New World,* 78, 457–58; Nathaniel B. Shurtleff, ed., *Records of the Governor and Company of the Massachusetts Bay in New England* (1628–86), 5 vols. (Boston: 1853–54), 1:29 (hereafter cited as *Mass. Records*).

9. Arthur H. Buffinton, "The Puritan View of War," *Publications of the Colonial Society of Massachusetts* 28 (April 1931): 71; Charles DeBenedetti, *The Peace Reform in American History* (Bloomington: Indiana University Press, 1980), 6; Jack S. Radabaugh, "The Militia of Colonial Massachusetts," *Military Affairs* 18 (spring 1954): 16–18; Millar, "Militia . . . in Colonial Massachusetts," 79–80; Increase Mather, *A Brief History of the War with the Indians in New-England* (1676; Albany, N.Y., 1862), 211–25; Sharp, "Leadership and Democracy," passim.

10. J. Franklin Jameson, ed., *Johnson's Wonder-Working Providence, 1628–1651* (New York: Scribner, 1910), 33; William Williams, *Martial Wisdom Recommended: A Sermon Preach'd at the Desire of the Honourable Artillery Company in Boston, June 6, 1737* (Boston: Henchman, 1737), 15–17; Moses Fiske, *The Duty of the Christian Soldier to Stand in the Warfare against his Spiritual Enemies: A Sermon Delivered before the Ancient and Honourable Artillery Company of Massachusetts, June 6, 1694* (Boston: Mudge, 1889), 6–7; DeBenedetti, *Peace Reform,* 6; Jabbs, "South Carolina Militia," 21.

11. Robert L. D. Davidson, *War Comes to Quaker Pennsylvania, 1682–1756* (New York: Columbia University Press, 1957), 11–12, 18, 28, 44, 111; William L. Shea, *The Virginia Militia*

in the Seventeenth Century (Baton Rouge: Louisiana State University Press, 1983), 79; Pennsylvania legislature, *Minutes of the Provincial Council of Pennsylvania* (Harrisburg, 1851), 4:435.

12. Higginbotham, *War and Society in Revolutionary America*, 25; J. Frederick Fausz, "An 'Abundance of Blood Shed on Both Sides': England's First Indian War, 1609–1614," *Virginia Magazine of History and Biography* 98 (January 1990): 37; Cruickshank, *Elizabeth's Army*, 60–61; Harold L. Peterson, "The Military Equipment of the Plymouth and Bay Colonies," *New England Quarterly* 20 (June 1947): 201.

13. Rutman, *Militant New World*, 291; Marcus, "Militia of Colonial Connecticut," 76–77; *Mass. Records*, 1:84, 2:222; John R. Bartlett, ed., *Records of the Colony of Rhode Island, 1664–1677*, 10 vols. (Providence, 1857), 2:114–15 (hereafter cited as *R.I. Records*).

14. Douglas E. Leach, *Arms for Empire: A Military History of the British Colonies in North America, 1607–1763* (New York: Macmillan, 1973), 13; James H. Trumbull and Charles J. Hoadly, eds., *The Public Records of the Colony of Connecticut*, 15 vols. (Hartford, 1850), 1:3–4 (hereafter cited as *Conn. Records*); *R.I. Records* 1:402–3, 2:567–72, 3:93; E. Milton Wheeler, "Development and Organization of the North Carolina Militia," *North Carolina Historical Review* 41 (July 1964): 307–23; William L. Saunders, ed., *The Colonial Records of North Carolina, 1765–68* (Raleigh: Daniels, 1890), 7:497; Louis D. Scisco, "Evolution of Colonial Militia in Maryland," *Maryland Historical Magazine* 53 (June 1940): 168.

15. *Mass. Records*, 1:84, 2:222; Marcus, "Militia of Colonial Connecticut," 37, 76–77; *R.I. Records*, 2:114–15; Wheeler, "North Carolina Militia," 318; Public Record Office, *Calendar of State Papers, Colonial Series, American and West Indies*, Reign of William III (1699) (London, 1908), 17:349.

16. *Conn. Records*, 1:3; Edward E. Atwater, *History of the Colony of New Haven* (New Haven: privately printed, 1881), 297; *Mass. Records*, 1:100, 137; Great Britain, *Acts of the Privy Council, Colonial Series*, vol. 3, 1720–45 (Hereford, 1910), 732–33.

17. Fausz, "An 'Abundance of Blood Shed,'" 16; Selesky, *War and Society in Colonial Connecticut*, 4; Harold L. Peterson, *Arms and Armor in Colonial America, 1526–1783* (New York: Bramhall, 1956), 12–20.

18. John K. Mahon, *History of the Militia and the National Guard* (New York: Macmillan, 1983), 17; Seleskey, *War and Society in Colonial Connecticut*, 15. One reason for the gunsmiths' slowness was the great variety of weapons utilized by colonial soldiers. The difficulty of repairing machines employing esoteric parts obtainable only at great distance has been the torment of American repairmen for centuries. See Richard B. Morris, *Government and Labor in Early America* (New York: Columbia University Press, 1946), 297n.

19. *Mass. Records*, 1:196; Hirsch, "Collision of Military Cultures," 1194–95.

20. Jabbs, "South Carolina Militia," 96–97; A. S. Salley, ed., *Journal of the Grand Council of South Carolina, August 25, 1671–June 24, 1680* (Columbia: Historical Commission of South Carolina, 1907), 11, 35, 39, 51.

21. Cruickshank, *Elizabeth's Army*, 66; *Conn. Records*, 1:12.

22. *Conn. Records*, 4:485, 9:341; Millar, "Militia . . . in Colonial Massachusetts," 67, 150; Public Record Office, *Calendar of State Papers, Colonial Series, American and West Indies*, Reign of William III (1701) (London, 1910), 19:389; Reign of Anne (1712–14) (London, 1926), 27:15, 128–29.

23. Higginbotham, *War of American Independence*, 124; Glenn Weaver, *Jonathon Trumbull: Connecticut's Merchant Magistrate, 1710–1785* (Hartford: Connecticut Historical Society, 1956), 5, 12; Stanley Pargellis, "The Four Independent Companies of New York," in *Essays in Colonial History Presented to Charles McLean Andrews by His Students* (New Haven: Yale, 1931), 111–12; Marcus, "Militia of Colonial Connecticut," 52–53.

24. Cruickshank, *Elizabeth's Army*, 101–103; Richard R. Johnson, "The Search for a Usable Indian: An Aspect of the Defense of Colonial New England," *Journal of American History* 64 (December 1977): 640; Seleskey, *War and Society in Colonial Connecticut*, 157.

25. Pargellis, "Four Independent Companies," 98–111; Public Record Office, *Calendar of State Papers, Colonial Series, American and West Indies*, Reign of Anne (1710–11) (London, 1924), 25:405–6.

26. Public Record Office, *Calendar of State Papers, Colonial Series, American and West Indies,* Reign of Anne (1704–1705) (London, 1916), 22:704; Reign of William III (1697–98) (London, 1905), 16:446–48; Pargellis, "Four Independent Companies," 118; *Conn. Records,* 11:237; *R.I. Records,* 5:227–28; Millar, "Militia . . . in Colonial Massachusetts," 209.

27. Fausz, "An 'Abundance of Blood Shed,'" 20; Jabbs, "South Carolina Militia," 98–100; Rutman, *Militant New World,* 207; *Conn. Records,* 2:500; James Axtell, "The White Indians of Colonial America," *William and Mary Quarterly,* 3d ser., 32 (January 1975): 59.

28. Johnson, "Search for a Usable Indian," 626; John Shy, *A People Numerous and Armed: Reflections on the Military Struggle for American Independence* (Ann Arbor: University of Michigan Press, 1990), 34; Richard I. Melvoin, *New England Outpost: War and Society in Colonial Deerfield* (New York: Norton, 1989), 96; Wheeler, "North Carolina Militia," 317; Selesky, *War and Society in Colonial Connecticut,* 19.

29. Leach, *Arms for Empire,* 107; Pennsylvania legislature, *Minutes of the Provincial Council of Pennsylvania* (Harrisburg, 1851-52), 7:84–85, 9:189; Melvoin, *New England Outpost,* 201; Mahon, *History of the Militia and the National Guard,* 22.

30. Marcus, "Militia of Colonial Connecticut," 132; Leach, *Arms for Empire,* 132; Samuel Penhallow, *The History of the Wars of New-England, With the Eastern Indians* (Boston, 1726; facsimile reprint, Freeport, N.Y. Books for Libraries, 1971), 39, 107, 111; Johnson, "Search for a Usable Indian," 630.

31. Indian scalps were not the only thing for which colonial legislators offered bounties—they did so for predators as well. This practice presents a grisly analogy which suggests the true attitude of whites toward Indians. Besides Indian scalps, Rhode Island (like other colonies) paid a handsome bounty for killing wolves. To collect his bounty, the wolf-killer was required to submit the head of the animal (rather than just the scalp) to a justice of the peace. In 1740 a hunter named Joseph Eady of Gloucester, Rhode Island, submitted an affidavit purporting to prove that he had killed a mother wolf and her seven pups, which would mean a considerable reward. But in a fit of doubt and penny-pinching, the Rhode Island leaders resolved "that the bounty on the old wolf's head be allowed, and no more; it being uncertain whether the young creatures were wolves or not." Eady lost out, and after a lusty exclamation of "God save the King," the Rhode Islanders adjourned, certain that they had not been defrauded. M. Roberts, *Military Revolution,* 31; Leach, *Arms for Empire,* 132; Richard Slotkin and James K. Folsom, *So Dreadful a Judgement: Puritan Responses to King Philip's War, 1676–1677* (Middletown, Conn.: Wesleyan, 1978), 28–29; Penhallow, *Wars of New-England,* 10; Jabbs, "South Carolina Militia," 286; Melvoin, *New England Outpost,* 201.

32. Joyce Appleby, "A Different Kind of Independence: The Postwar Restructuring of the Historical Process," *William and Mary Quarterly,* 3d ser., 50 (April 1993): 257; Powicke, *Military Obligation,* 211; Hirsch, "Collision of Military Cultures," 1205; Mahon, *History of the Militia and the National Guard,* 22, 26; Fausz, "An 'Abundance of Blood Shed,'" 22, 46.

33. Peterson, "Military Equipment," 198; Rutman, *Militant New World,* 583; Marcus, "Militia of Colonial Connecticut," 93; Seleskey, *War and Society in Colonial Connecticut,* 26; Johnson, "Search for a Usable Indian," 627; Jabbs, "South Carolina Militia," 17; John Shy, "A New Look at the Colonial Militia," *William and Mary Quarterly,* 3d ser., 20 (April 1963): 177–78, and *A People Numerous and Armed,* 33.

34. Jabbs, "South Carolina Militia," 240; Davidson, *War Comes to Quaker Pennsylvania,* 27, 53; Marcus, "Militia of Colonial Connecticut," 93; *Conn. Records,* 1:70–71; Fred Anderson, *A People's Army: Massachusetts Soldiers and Society in the Seven Years' War* (Chapel Hill: University of North Carolina Press, 1986), 157, 219, 221.

35. John Franklin Jameson, *Privateering and Piracy in the Colonial Period* (New York: Macmillan, 1923), ix.

36. Carl E. Swanson, "American Privateering and Imperial Warfare, 1739–1748," *William and Mary Quarterly,* 3d ser., 42 (July 1985): 372–82.

37. Charles C. Crittenden, *The Commerce of North Carolina, 1763–1787* (New Haven: Yale University Press, 1936), 148.

38. Fausz, "An 'Abundance of Blood Shed,'" 18; Shea, *Virginia Militia,* 57–58; Rutman, *Militant New World,* 408–9; Selesky, *War and Society in Colonial Connecticut,* 4.

39. *Mass. Records,* 1:48; *Conn. Records,* 1:2, 49; Philip S. Haffenden, *New England in the English Nation* (Oxford: Clarendon Press, 1974), 239–40.

40. Mather, *Brief History of the War,* 129, app. H; Public Record Office, *Calendar of State Papers, Colonial Series, American and West Indies,* Reign of Charles II (1675–76) (London, 1893), 9:466; Reign of Anne (1711–12) (London, 1925), 26:276. For other examples, see *Calendar of State Papers, Colonial Series, American and West Indies,* Reign of William and Mary (1789–92) (London, 1901), 13:46; George M. Waller, *Samuel Vetch: Colonial Enterpriser* (Chapel Hill: University of North Carolina Press, 1960), 83–99; John Winthrop to Fitz-John Winthrop, June 1706, *Massachusetts Historical Collections,* 6th ser., 3:334, Boston; William G. Godfrey, *Pursuit of Profit and Preferment in Colonial North America: John Bradstreet's Quest* (Waterloo, Ont.: Wilfred Laurier University, 1982), 152; and Seleskey, *War and Society in Colonial Connecticut,* 79.

41. Public Record Office, *Calendar of State Papers, Colonial Series, American and West Indies,* Reign of William III (1696–97) (London, 1904), 15:313–14, 557, 563; Reign of Anne (1704–1706) (London, 1916), 22:330; Reign of Anne (1708–1709) (London, 1922), 24:191; Reign of George I (1714-15) (London, 1928), 28:195; Pargellis, "Four Independent Companies," 52.

42. Millar, "Militia . . . in Colonial Massachusetts," 199; Great Britain, *Acts of the Privy Council of England: Colonial Series,* vol. 6, "The Unbound Papers," 1676–1783 (London, 1912), 330–31; Davidson, *War Comes to Quaker Pennsylvania,* 167, 226n; Public Record Office, *Calendar of State Papers, Colonial Series, American and West Indies,* Reign of Charles II (1681–85) (London, 1898), 11:514, 598; Reign of James II (1685–88) (London, 1899), 12:601–2; Douglas Edward Leach, Roots of Conflict: British Armed Forces and Colonial Americans, 1677–1763 (Chapel Hill: University of North Carolina Press, 1986), 159–60; Richard Pares, *War and Trade in the West Indies, 1739–1763* (Oxford: Oxford University Press, 1936), 457.

43. Public Record Office, *Calendar of State Papers, Colonial Series, American and West Indies,* Reign of William III (1696–97) (London, 1904), 15:212–14; Reign of William III (1702) (London, 1912), 20:495; Reign of Anne (1702–1703) (London, 1913), 21:135, 170, 181; Reign of Anne (1711–12) (London, 1925), 26:193.

44. Public Record Office, *Calendar of State Papers, Colonial Series, American and West Indies,* Reign of William III (1697–98) (London, 1905), 16:315, 508; Reign of William III (1702) (London, 1912), 20:136, 176; Reign of Anne (1708–1709) (London, 1922), 24:207.

45. *Mass. Records,* 1:55, 66, 225, 3:41.

46. O. A. Roberts, *History of . . . the Ancient and Honorable Artillery Company of Massachusetts, 1637–1888* (Boston: Mudge, 1895), 1:13; Bernard Bailyn, "The *Apologia* of Robert Keayne," *William and Mary Quarterly,* 3d ser., 6 (1950), passim.

47. On a similar occasion, Cotton offered his views on usury. Deriving much of his thinking from John Calvin, who believed that money must grow slowly and naturally, like a plant, Cotton distinguished between reasonable and excessive interest rates, and he appealed to the Puritan conscience to control them. See Jasper Resenmeier, "John Cotton on Usury," *William and Mary Quarterly,* 3d ser., 47 (October 1990): 553. Cf. O. A. Roberts, *Ancient and Honorable Artillery Company of Massachusetts,* 1:13, 19; Hugh Rockoff, *Drastic Measures: A History of Wage and Price Controls in the United States* (Cambridge: Cambridge University Press, 1984), 20–22.

48. Jameson, *Johnson's Wonder-Working Providence,* 33; Mahon, *History of the Militia and the National Guard,* 17.

49. Melvoin, *New England Outpost,* 105; Essex Institute, *Records and Files of the Quarterly Courts of Essex County, Massachusetts,* vol. 6, 1675–78 (Salem, Mass.: Essex Institute, 1917): 142; *Mass. Records,* 4:123.

50. Gerald S. Graham, ed., *The Walker Expedition to Quebec, 1711* (Toronto: Champlain, 1953), 23, 322, 341, 346; Public Record Office, *Calendar of State Papers, Colonial Series, American and West Indies,* Reign of Anne (1711–12) (London, 1925), 26:62.

51. Graham, *Walker Expedition,* 336; Public Record Office, *Calendar of State Papers, Colonial Series, American and West Indies,* Reign of Anne (1711–12) (London, 1925), 26:62; Leach, Roots of Conflict, 35–36.

52. Graham, *Walker Expedition,* 63–65, 379.

53. Ibid., 64–65.

54. Pargellis, "Four Independent Companies," 85, 271, 276–80; Leach, *Roots of Conflict,* 78, 80, 91.

55. *Conn. Records,* 9:216, 233, 488, 12:64; Seleskey, *War and Society in Colonial Connecticut,* 129.

56. Public Record Office, *Calendar of State Papers, Colonial Series, American and West Indies,* Reign of William III (1697–98) (London, 1905), 16:386, 415.

57. Great Britain, *Acts of the Privy Council, Colonial Series,* vol. 3, 1720–1745 (Hereford, 1910), 648–49, 701, 724–25; Great Britain, *Journal of the Commissioners for Trade and Plantations,* vol. 8, 1741–49 (London, 1931), 79.

58. Graham, *Walker Expedition,* 315.

59. Higginbotham, *War and Society in Revolutionary America,* 26–27.

2. Virtue Tested

1. Composed in part of an aristocratic officer corps and in part of elements of the British underclass, the army offered an unattractive, if not wholly accurate, model of British society. Friction developed sporadically between both of these contingents and American colonists. See Douglas Edward Leach, *Roots of Conflict: British Armed Forces and Colonial Americans, 1777–1763* (Chapel Hill: University of North Carolina Press, 1986), 163–66; John Philip Reid, *In Defiance of the Law: The Standing-Army Controversy, The Two Constitutions, and the Coming of the American Revolution* (Chapel Hill: University of North Carolina Press, 1981), 156–66, 216–17; John R. Alden, *A History of the American Revolution* (New York: Knopf, 1969), 48–49; John Shy, *Toward Lexington: the Role of the British Army in the Coming of the American Revolution* (Princeton: Princeton University Press, 1965), passim; J. G. A. Pocock, "Machiavelli, Harrington, and English Political Ideologies in the Eighteenth Century," *William and Mary Quarterly,* 3d ser., 22 (October 1965): 566; Lawrence Delbert Cress, *Citizens in Arms: The Army and the Militia in American Society to the War of* 1812 (Chapel Hill: University of North Carolina Press, 1982), 36–46.

2. J. G. A. Pocock, "Virtue and Commerce in the Eighteenth Century," *Journal of Interdisciplinary History* 3 (summer 1972): 120–25; Charles Royster, *A Revolutionary People at War: The Continental Army and the American Character,* 1775–1783 (Chapel Hill: University of North Carolina Press, 1979), 35; John T. White, "Standing Armies in Time of War: Republican Theory and Military Practice during the American Revolution" (Ph.D. diss., George Washington University, 1978), 37–40; Gordon S. Wood, *The Creation of the American Republic, 1776–1787* (Chapel Hill: University of North Carolina Press, 1969), 198.

3. Alexander Graydon, *Memoirs of His Own Time* (Philadelphia: Lindsay, 1846), 293.

4. In practice, this theory was only partially correct. Contrary to predictions, British and Hessian professional soldiers were almost invariably brave in battle, while the poorly trained American volunteers were not. But the remarkable ability of Americans to endure defeat and hardship, as at Long Island and Valley Forge, was not only inspiring but also ultimately decisive. Royster, *A Revolutionary People,* 46, 113; White, "Standing Armies," 95; Don Higginbotham, *George Washington and the American Military Tradition* (Athens: University of Georgia Press, 1985), 10.

5. White, "Standing Armies," 31; John E. Ferling, *A Wilderness of Miseries: War and Warriors in Early America* (Westport, Conn.: Greenwood Press, 1980), 80–81, 84; Higginbotham, *George Washington,* 54; Shy, *Toward Lexington,* 393; E. Wayne Carp, *To Starve the Army at Pleasure: Continental Army Administration and American Political Culture,* 1775–1783 (Chapel Hill: University of North Carolina Press, 1984), 13; Wood, *Creation of the American Republic,* 47–48; Edward Tabor Linenthal, *Changing Images of the Images of the Warrior Hero in America: A History of Popular Symbolism* (New York: Mellen, 1982), 51, 54.

6. Marvin Kitman, *George Washington's Expense Account* (New York: Simon and Schuster, 1970) is iconoclastic and ahistorical. It remains unclear how much Washington really received.

7. James Thomas Flexner, *The Traitor and the Spy: Benedict Arnold and John André* (Boston: Little, Brown, 1975), 357–58; James Thacher, *Military Journal during the American Revolutionary War: From 1775 to 1783* (Hartford: Andrus, 1854), 219.

8. Philander D. Chase, ed., *The Papers of George Washington: Revolutionary War Series* (Charlottesville: University Press of Virginia, 1985), 1:281; Cress, *Citizens in Arms*, 54–57.

9. John C. Fitzpatrick, *The Writings of George Washington* (Washington, 1936), 13:80. Cf. James Kirby Martin, "The Continental Army and the American Victory," in *The World Turned Upside Down: The American Victory in the War of Independence*, ed. John C. Ferling (Westport, Conn.: Greenwood Press, 1988), 32.

10. Edmund C. Burnett, ed., *Letters of Members of the Continental Congress* (Washington: Carnegie Institution, 1923), 2:356–57, 4:233, 255.

11. White, "Standing Armies," 102; Richard Buel Jr., *Dear Liberty: Connecticut's Mobilization for the Revolutionary War* (Middletown, Conn.: Wesleyan, 1980), 179.

12. John Laurens, *The Army Correspondence of Colonel John Laurens in the Years 1777–78* (1867; reprint, New York: Arno Press, 1969), 157; Cress, *Citizens in Arms*, 54–56.

13. John Adams, *Diary and Autobiography of John Adams* (Cambridge: Harvard University Press, 1961), 2:178–79; James T. Flexner, *George Washington: The Forge of Experience, 1732–1775* (Boston: Little, Brown, 1965), 137, 181.

14. Piers Mackesy, *The War for America, 1775–1783* (Cambridge: Harvard University Press, 1966), 369; Carp, *To Starve the Army at Pleasure*, 125; Edward E. Curtis, *The Organization of the British Army in the American Revolution* (New Haven: Yale University Press, 1926), 25; Ferling, *Wilderness of Miseries*, 73; Graydon, *Memoirs*, 148; Fitzpatrick, *Writings of George Washington*, 14:100.

15. Buel, *Dear Liberty*, 211; Higginbotham, *George Washington*, 52; Simeon Lyman, "Journal of Simeon Lyman of Sharon, Aug. 10 to Dec. 28, 1775," *Collections of the Connecticut Historical Society* 7 (1899): 121.

16. White, "Standing Armies," 117–18.

17. Burnett, *Letters of the Continental Congress*, 1:360.

18. White, "Standing Armies," 181; Higginbotham, *George Washington*, 89; Cress, *Citizens in Arms*, 56–60. Besides better pay, the new professional army included such features as long-term enlistments, a toughened military code, ouster of marginal officers, more uniform tactical movements, and implementation of French and German concepts. See Don Higginbotham, "The Military Institutions of Colonial America: The Rhetoric and the Reality," in *Tools of War: Instruments, Ideas, and Institutions of Warfare, 1445–1871*, ed. John A. Lynn (Urbana: University of Illinois Press, 1990), 148.

19. Royster, *Revolutionary People*, 65; Allen Bowman, *The Morale of the American Revolutionary Army* (Washington: American Council on Public Affairs, 1943), 13; Buel, *Dear Liberty*, 55–56, 102; William Heath, "The Heath Papers," *Collections of the Massachusetts Historical Society*, 7th ser., 5 (1905): 114–15.

20. William Coit, "Orderly Book of Captain William Coit's Company," *Collections of the Connecticut Historical Society* 7 (1899): 39; Fitzpatrick, *Writings of Washington*, 5:327, 368, 426, 468, 6:192, 7:111–12, 199, 428, 9:121, 15:163, 345, 21:452, 458n., 22:180, 400, 442, 486, 23:134, 455, 24:9, 31, 26:310; James Kirby Martin, "A 'Most Undiscipline, Profligate Crew': Protest and Defiance in the Continental Ranks, 1776–1783," in *Arms and Independence: The Military Character of the American Revolution*, ed. Ronald Hoffman and Peter J. Albert (Charlottesville: University Press of Virginia, 1984), 131.

21. Joseph Plumb Martin, *Private Yankee Doodle* (Boston: Little, Brown, 1962), 287; Royster, *Revolutionary People*, 195, 223; Albigence Waldo, "Valley Forge, 1777–1778: Diary of Surgeon Albigence Waldo, of the Connecticut Line," *Pennsylvania Magazine of History and Biography* 21, no. 3 (1897): 314.

22. Burnett, *Letters of the Continental Congress*, 4:7; J. P. Martin, *Private Yankee Doodle*, 172; Otis G. Hammond, ed., *Letters and Papers of Major-General John Sullivan, Continental Army* (Concord: New Hampshire Historical Society, 1931), 2:20; Heath Papers, 354; Waldo, *Diary*, 314.

23. Buel, *Dear Liberty*, 180; White, "Standing Armies," 272–77; Heath Papers, 33; E. James Ferguson, ed., *The Papers of Robert Morris* (Pittsburgh: University of Pittsburgh Press, 1973–1988), 5:521–24.

24. John Shy, "American Society and Its War for Independence," in *Reconsiderations on the Revolutionary War*, ed. Don Higginbotham (Westport, Conn.: Greenwood Press, 1976), 81–82; R. Arthur Bowler, *Logistics and the Failure of the British Army in America, 1775–1783* (Princeton: Princeton University Press, 1975), 167–211; Curtis, *British Army in the Revolution*, 98–100.

25. Mackesy, *War for America*, 369; Graydon, *Memoirs*, 214n.

26. Chase, *Papers of George Washington*, 1:56, 155n, 212, 229, 335, 375; Ferguson, *Papers of Robert Morris*, 2:75–76.

27. *Journals of the Continental Congress, 1774–1789*, ed. Worthington C. Ford et al. (Washington, D.C., 1904–37), 3:331.

28. Thacher, *Military Journal*, 196–97. Last-minute reprieves were standard military practice. They were designed to have the maximum psychological effect without actually going through with the execution. Used too frequently, reprieves could nullify the deterrent effect of the law, so some unlucky souls were hanged. See Fred Anderson, *A People's Army: Massachusetts Soldiers and Society in the Seven Years' War* (Chapel Hill: University of North Carolina Press, 1984), 122.

29. Don Higginbotham, "The American Militia: A Traditional Institution with Revolutionary Responsibilities," in his *Reconsiderations on the Revolutionary War*, 100; Higginbotham, *The War of American Independence: Military Attitudes, Policies, and Practices, 1763–1789* (New York: Macmillan, 1971), 133; Bowler, *Logistics*, 242; Ferling, *Wilderness of Miseries*, 99; Ira D. Gruber, *The Howe Brothers and the American Revolution*, (New York: Atheneum, 1972), 145–46; Victor Leroy Johnson, *The Administration of the American Commissariat during the Revolutionary War* (Philadelphia: University of Pennsylvania Press, 1941), 95; Charles Christopher Crittenden, *The Commerce of North Carolina, 1763–1789* (New Haven: Yale University Press, 1936), 153.

30. Richard K. Showman, ed., *The Papers of General Nathanael Greene* (Chapel Hill: University of North Carolina Press, 1976–91), 4:108 (hereafter cited as Showman, Greene Papers); Caleb Haskell, *Caleb Haskell's Diary, May 5, 1775–May 30, 1776* (Newburyport, Mass.: Huse, 1881), 13; S. Sydney Bradford, "Hunger Menaces the Revolution, December 1779–January, 1780," *Maryland Historical Magazine*, March 1966, 13; Robert Middlekauff, "Why Men Fought in the American Revolution," *Huntington Library Quarterly* 43 (spring 1980): 137; Fitzpatrick, *Writings of Washington*, 7:47, 9:432.

31. Chase, *Papers of George Washington*, 1:215; Coit, "Orderly Book," 80; Hammond, *Major-General John Sullivan*, 2:4–5; Fitzpatrick, *Writings of Washington*, 6:459; Middlekauff, "Why Men Fought in the American Revolution," 137.

32. Fitzpatrick, *Writings of Washington*, 7:57, 9:178–79. Cf. Higginbotham, *George Washington*, 94.

33. Fitzpatrick, *Writings of George Washington*, 8:452, 466; Robert C. Bray and Paul E. Bushnell, *Diary of a Common Soldier in the American Revolution, 1775–1783* (DeKalb: Northern Illinois University Press, 1978), 168.

34. Fitzpatrick, *Writings of Washington*, 8:452, 465–66, 9:432, 10:206–207; Showman, Greene Papers, 4:108.

35. General Sir William Howe whipped and executed plunderers, but to no avail. See Gruber, *Howe Brothers*, 242–44; Fitzpatrick, *Writings of Washington*, 7:47, 8:452; Bowler, *Logistics*, 61, 107; J. P. Martin, *Private Yankee Doodle*, 123–24; William Young, "Journal of Sergeant William Young, Written during the Jersey Campaign in the Winter of 1776-7," *Pennsylvania Magazine of History and Biography* 8, no. 3 (1884): 265–66, 268, 315; Thacher,

Military Journal, 71; Laurens, *Army Correspondence,* 94; Bray and Bushnell, *Diary of a Common Soldier,* 133; Buel, *Dear Liberty,* 274; Heath Papers, 148.

36. Chase, *Papers of George Washington,* 1:274.

37. Burnett, *Letters of the Continental Congress,* 6:373–74; Hammond, *Major-General John Sullivan,* 3:280; Buel, *Dear Liberty,* 283; Fitzpatrick, *Writings of Washington,* 16:285.

38. M. L. Robertson, "Scottish Commerce and the American War of Independence," *Economic History Review* 9, no. 1 (1956): 125; T. M. Devine, "A Glasgow Tobacco Merchant during the American War of Independence: Alexander Speirs of Elderslie, 1775 to 1781," *William and Mary Quarterly,* 3d ser., 33 (July 1976): 511–12; Bowler, *Logistics,* 66, 72–75; Mackesy, *War for America,* 37, 65–66.

39. Higginbotham, "The American Militia," 96–97; Royster, *Revolutionary People,* 272; Hammond, *Major-General John Sullivan,* 3:280; Burnett, *Letters of the Continental Congress,* 6:381; Buel, *Dear Liberty,* 257–67, 285; Curtis, *British Army in the Revolution,* 81–82; Willard O. Mishoff, "Business in Philadelphia during the British Occupation, 1777–78," *Pennsylvania Magazine of History and Biography* 41 (April 1937): 168, 170–73.

40. Richard K. MacMaster and David C. Skaggs, eds., "The Letterbook of Alexander Hamilton, Piscataway Factor," *Maryland Historical Magazine* 57 (1967): 148; Curtis, *British Army in the Revolution,* 119.

41. Robert L. Hildrup, "The Salt Supply of North Carolina during the Revolution," *North Carolina Historical Review* 22 (October 1945): 398, 413–17; Susie M. Ames, "A Typical Virginia Business Man of the Revolutionary Era: Nathaniel Littleton Savage and His Account Book," *Journal of Economic and Business History* 3 (May 1931): 411–13; Edward Countryman, *A People in Revolution: The American Revolution and Political Society in New York, 1760–1790* (Baltimore: Johns Hopkins University Press, 1981), 182–83; Burnett, *Letters of the Continental Congress,* 6:37, 170, 191; James A. Huston, *Logistics of Liberty: American Services of Supply in the Revolutionary War and After* (Newark: University of Delaware Press, 1991), 83; Crittenden, *Commerce of North Carolina,* 145.

42. James Blaine Hedges, *The Browns of Providence Plantations: Colonial Years* (Cambridge: Harvard University Press, 1952), 265–66; Pennsylvania legislature, *Minutes of the Provincial Council of Pennsylvania* (Harrisburg, 1852), 10:748; Hildrup, "Salt Supply of North Carolina," 416.

43. Laurens, *Army Correspondence,* 134.

44. Hedges, *Browns,* 257–58; Fitzpatrick, *Writings of Washington,* 21:182–83.

45. Ferling, *Wilderness of Miseries,* 22; Huston, *Logistics of Liberty,* 125; Ferguson, *Papers of Robert Morris,* 3:326; Hammond, *Major-General John Sullivan,* 1:206; Erna Risch, *Supplying Washington's Army* (Washington: U.S. Army, 1981), 82; Fitzpatrick, *Writings of George Washington,* 14:237, 16:222; Heath Papers, 170.

46. Carp, *To Starve the Army at Pleasure,* esp. 155–59.

47. During the French and Indian War, colonial legislatures sometimes paid their purchasing agents at even higher rates. In 1758 the Connecticut Assembly allowed a 12 percent markup on a lot of 1,000 shoes, stockings, and shirts, and a 50 percent markup on 5,000 pounds of sugar and 5,000 pounds of tobacco. Glenn Weaver, *Jonathon Trumbull: Connecticut's Merchant Magistrate, 1710–1785* (Hartford: Connecticut Historical Society, 1956), 35–36, 83; Virginia Bever Platt, "Tar, Staves, and New England Rum: The Trade of Aaron Lopez of Newport, Rhode Island, and Colonial North America," *North Carolina Historical Review* 48 (1971): 5, 13; Anderson, *A People's Army,* 181; Risch, *Supplying Washington's Army,* 61, and *Quartermaster Support of the Army: A History of the Corps, 1775–1939* (Washington: U.S. Army, 1988), 7, 27–28, 40–41.

48. Ferguson, *Papers of Robert Morris,* 3:370, 4:79–80, 355, 613; Huston, *Logistics of Liberty,* 109; Elizabeth M. Nuxoll, "Congress and the Munitions Merchants: The Secret Committee of Trade during the American Revolution" (Ph.D. diss., City University of New York, 1979), 207; Robert A. East, *Business Enterprise in the American Revolutionary Era* (New York: Columbia University Press, 1938), 211.

49. Jerome R. Garitee, *The Republic's Private Navy: The American Privateering Business as Practiced by Baltimore during the War of 1812* (Middletown, Conn.: Wesleyan, 1977), 7–9, 17; Gruber, *Howe Brothers,* 264–65; Higginbotham, *War of American Independence,* 345; Crittenden, *Commerce of North Carolina,* 125–26; Douglas Southall Freeman, *George Washington, a Biography,* vol. 4, *Leader of the Revolution* (New York: Scribner, 1951), 412n.

50. Showman, Greene Papers, 3:431n; Lawrence Shaw Mayo, *John Langdon of New Hampshire* (Concord, N.H.: Rumford, 1937), 134–37, 184–88; Adams, *Diary,* 4:4; Thomas P. Abernethy, "Commercial Activities of Silas Deane in France," *American Historical Review* 39 (April 1934): 48; Richard J. Morris, "Wealth Distribution in Salem, Massachusetts, 1759–1799: The Impact of Revolution and Independence," *Essex Institute Historical Collections* 114 (April 1978): 100–102.

51. Higginbotham, *War of American Independence,* 308–9.

52. Hedges, *Browns,* 270–277; Nuxoll, "Congress and the Munitions Merchants," 168.

53. Burnett, *Letters of the Continental Congress,* 1:195, 346; Donald E. Reynolds, "Ammunition Supply in Revolutionary Virginia," *Virginia Magazine of History and Biography,* January 1965, 66–67; Pennsylvania legislature, *Minutes of the Provincial Council,* 10:459; Kathleen Bruce, *Virginia Iron Manufacture in the Slave Era* (New York: Century, 1931), 40; John F. and Janet A. Stegeman, "The Cause Celebré of Caty Greene," *American History Illustrated* 12 (June 1977): 16; Freeman, *George Washington,* 5:508–9.

54. Heath Papers, 83.

55. East, *Business Enterprise,* 52–53; Ronald Hoffman, *A Spirit of Dissension: Economics, Politics, and the Revolution in Maryland* (Baltimore: Johns Hopkins University Press, 1973), 208–9; Buel, *Dear Liberty,* 170–71; Bernard Baruch, "Taking the Profit out of War," *Atlantic,* January 1926, 24; Burnett, *Letters of the Continental Congress,* 4:274, 293, 14:808.

56. Larry G. Bowman, "The Scarcity of Salt in Virginia during the American Revolution," *Virginia Magazine of History and Biography* 77 (October 1969): 469–70; Pennsylvania legislature, *Minutes of the Provincial Council* (Harrisburg, 1851–52), 11:13–14.

57. James A. Huston, *Logistics of Liberty,* 89; Crittenden, *Commerce of North Carolina,* 148; Hugh Rockoff, *Drastic Measures: A History of Wage and Price Controls in the United States* (Cambridge: Cambridge University Press, 1984), 24–41; Johnson, *Administration of the American Commissariat,* 55; E. Milton Wheeler, "Development and Organization of the North Carolina Militia," *North Carolina Historical Review* 41 (July 1964): 320; Pennsylvania legislature, *Minutes of the Provincial Council* (Harrisburg, 1851–52), 11:18.

58. Huston, *Logistics of Liberty,* 122, 127–28, 147; Fitzpatrick, *Writings of Washington,* 10:45–46.

59. Johnson, *Administration of the American Commissariat,* 24, 73–74; Risch, *Supplying Washington's Army,* 47–48; *Journals of the Continental Congress,* 5:627–68; Fitzpatrick, *Writings of Washington,* 7:160; Adams, *Diary,* 2:180.

60. Showman, Greene Papers, 6:248; Risch, *Supplying Washington's Army,* 417–38; Carp, *To Starve the Army at Pleasure,* 155–67.

61. Showman, Greene Papers, 1:228–29, 232, 3:378–79, 405n; Theodore Thayer, *Nathanael Greene, Strategist of the American Revolution* (New York: Twayne, 1960), 228–38; Huston, *Logistics of Liberty,* 67; Chester M. Destler, "Barnabas Deane and the Barnabas Deane & Company," *Connecticut Historical Society Bulletin,* January 1970, 14–17.

62. Higginbotham, *George Washington,* 82–83; Heath Papers, 44.

63. Graydon, *Memoirs,* 333; Fitzpatrick, *Writings of Washington,* 13:383.

64. Higginbotham, *War of American Independence,* 411–12; Buel, *Dear Liberty,* 300, 305–6; Larry R. Gerlach, "Connecticut and Commutation, 1778–1786," *Connecticut Historical Society Bulletin* 33 (April 1968): 56.

65. Gerlach, "Connecticut and Commutation," passim.

66. Buel, *Dear Liberty,* 296.

67. Hugh McLellan, ed., *Pliny Moore Papers: Captain Job Wright's Company* (Champlain, N.Y.: privately printed, 1928), 15.

3. Left-Handed Trade

1. John E. Ferling, *The First of Men: A Life of George Washington* (Knoxville: University of Tennessee Press, 1988), 403; Earl A. Molander, "Historical Antecedents of Military-Industrial Criticism," in *War, Business, and Society*, ed. Benjamin F. Cooling (Port Washington, N.Y.: Kennikat, 1977), 178.

2. U.S. House Committee on Military Affairs, *Universal Mobilization for War Purposes*, 68th Cong., 1st sess., 1924, H. Doc. 764, 238.

3. Ferling, *The First of Men*, 403; Richard H. Kohn, *Eagle and Sword: The Federalists and the Creation of the Military Establishment in America, 1783–1802* (New York: Free Press, 1975), 41; Donald R. Hickey, *The War of 1812: A Forgotten Conflict* (Urbana: University of Illinois Press, 1989), 6.

4. J. C. A. Stagg, *Mr. Madison's War: Politics, Diplomacy, and Warfare in the Early American Republic, 1783–1830* (Princeton: Princeton University Press, 1983), 127; Hickey, *The War of 1812*, 6; Merrit Roe Smith, "George Washington and the Establishment of the Harpers Ferry Armory," *Virginia Magazine of History and Biography* 81 (October 1973): 417, 435; James Blaine Hedges, *The Browns of Providence Plantations: Colonial Years* (Cambridge: Harvard University Press, 1952), 313; S. N. D. and Ralph H. North, *Simeon North, First Official Pistol Maker of the United States* (Concord, N.H.: Rumford, 1913), 43.

5. The Republicans spent comparatively more ($2.8 million) on coastal fortifications than did the Federalists. Hickey, *The War of 1812*, 8, 78; Kohn, *Eagle and Sword*, 260; Steven Watts, *The Republic Reborn: War and the Making of Liberal America, 1790–1820* (Baltimore: Johns Hopkins University Press, 1987), 131–32, 229–30; Stagg, *Mr. Madison's War*, 128.

6. Samuel Eliot Morison, "Dissent in the War of 1812," in Morison, Frederick Merk, and Frank Freidel, *Dissent in Three American Wars* (Cambridge: Harvard University Press, 1970), 3–4; Hickey, *The War of 1812*, 255–56.

7. *United States Gazette* [Philadelphia], 13, 16, and 20 July, 4 and 10 August, 3 September, 29 October 1812, 11 March, 1 and 12 April 1813; Watts, *The Republic Reborn*, 59–60, 85, 103.

8. *Niles' Weekly Register*, 25 December 1813, 280; 19 November 1814, 168; 28 January 1815, 350; 18 November 1815, 202; *United States Gazette*, 1 April 1813.

9. Hickey, *The War of 1812*, 85, 153; *United States Gazette*, 23 January 1814.

10. *Niles' Weekly Register*, 19 November 1814, 168; 28 January 1815, 350; *United States Gazette*, 1 April 1813, 7 January 1814; Hickey, *The War of 1812*, 215; Erna Risch, *Quartermaster Support of the Army: A History of the Corps, 1775–1939* (Washington: U.S. Army, 1989), 154.

11. *Niles' Weekly Register*, 18 November 1815, 202; Sarah McCulloh Lemmon, *Frustrated Patriots: North Carolina and the War of 1812* (Chapel Hill: University of North Carolina Press, 1973), 53.

12. Hickey, *The War of 1812*, 117.

13. Allan S. Everest, *The War of 1812 in the Champlain Valley* (Syracuse: Syracuse University Press, 1981), 151; John K. Mahon, *The War of 1812* (Gainesville: University of Florida Press, 1972), 281; *Niles' Weekly Register*, 3 July 1813, 288; 10 January 1813, 88; 2 October 1813, 76; 10 September 1814, 9; 19 November 1814, 168; Hickey, *The War of 1812*, 226.

14. Quoted in Hickey, *The War of 1812*, 216; Everest, *The War of 1812 in the Champlain Valley*, 152; *Niles' Weekly Register*, 19 November 1814, 168.

15. *Niles' Weekly Register*, 5 June 1813, 227; 27 November 1813, 214;, 1815 suppl., 175–76.

16. Hickey, *The War of 1812*, 224; *Niles' Weekly Register*, 10 September 1814, 9; 3 December 1814, 206.

17. The prize distribution was usually divided into twentieths. Ten shares went to officers according to an agreed-upon formula: captains, three-twentieths; lieutenants, two-

twentieths; junior officers, two-twentieths; and senior petty officers, three-twentieths. Ten shares went to the crew: three-twentieths to junior petty officers, seven-twentieths to seamen. If a warship in the U.S. Navy captured a prize, the crew was paid a bonus of twenty dollars for each person aboard the enemy ship. If more than one ship participated in the capture, all ships in sight of the seizure shared equally. If the enemy force was equal to or superior in strength to the American naval unit, the American sailors could keep the full value of the prize. If the American force was superior to the enemy, the U.S. government retained half the value of the prize. See *Niles' Weekly Register,* 22 August 1812, 411;, 9 January 1813, 298; *United States Gazette,* 13 January 1813; Jerome R. Garitee, *The Republic's Private Navy: The American Privateering Business as Practiced by Baltimore during the War of 1812* (Middletown, Conn.: Wesleyan, 1977), 67–74, 130–35, 168.

18. *United States Gazette,* 13 January 1813; Hickey, *The War of 1812,* 133.

19. *Niles' Weekly Register,* 29 November 1813, 208; Mahon, *The War of 1812,* 255.

20. *Niles' Weekly Register,* 11 September 1813, 29; 18 December 1813, 263; Hickey, *The War of 1812,* 173; Mahon, *The War of 1812,* 226; *United States Gazette,* 5 February 1814.

21. *United States Gazette,* 7 July 1812; Garitee, *The Republic's Private Navy,* 92–93; *Niles' Weekly Register,* 22 October 1814, 92–93; 27 October 1814, 111.

22. *Niles' Weekly Register,* 19 September 1812, 45; 13 February 1813, 383; William Henry Smith, ed., *The St. Clair Papers: The Life and Public Services of Arthur St. Clair* (Cincinnati: Clarke, 1882), 2:328–29.

23. Stagg, *Mr. Madison's War,* 148, 178; Hickey, *The War of 1812,* 76; Watts, *The Republic Reborn,* 12, 71–74; James Wallace Hammack Jr., *Kentucky and the Second American Revolution* (Lexington: University Press of Kentucky, 1976), 99–100.

24. *United States Gazette,* 6 January 1813, 1 October 1814.

25. Risch, *Quartermaster Support of the Army,* 146–77; William Crozier, *Ordnance and the World War: A Contribution to the History of American Preparedness* (New York: Scribner, 1920), 2–4; Keir B. Sterling, *Serving the Line with Excellence* (Aberdeen, Md.: U.S. Army, 1987), 16; Hickey, *War of 1812,* 78.

26. *United States Gazette,* 16 June 1813; Merrit Roe Smith, "Military Arsenals and Industry before World War I," in *War, Business, and Society: Historical Perspectives on the Military-Industrial Complex,* ed. Benjamin F. Cooling (Port Washington, N.Y.: Kennikat, 1977), 29–30; Merton J. Peck and Frederic M. Scherer, *The Weapons Acquisition Process: An Economic Analysis* (Boston: Graduate School of Business Administration, Harvard University, 1962), 3; *Niles' Weekly Register,* 19 September 1812, 60; 2 January 1813, 278; Hickey, *War of 1812,* 79; Richard Sullivan, *Report on the Merits of the Claim of the State of Massachusetts . . . for Expenses of the Militia during the Late War* (Boston: Russell, 1822), 27–29; Victor A. Sapio, *Pennsylvania and the War of 1812* (Lexington: University Press of Kentucky, 1970), 158–88; Hammack, *Kentucky and the Second American Revolution,* 27.

27. Stagg, *Mr. Madison's War,* 156; Risch, *Quartermaster Support of the Army,* 94, 119; Everest, *The War of 1812 in the Champlain Valley,* 147; Hickey, *War of 1812,* 78–79.

28. William Simmons to Congress, 10 October 1814, reprinted in *United States Gazette,* 16 November 1814. Condict's request appears in *United States Gazette,* 17 November 1814. See also *Biographical Directory of the American Congress, 1774–1971* (Washington, 1971), 772.

29. Hickey, *War of 1812,* 50, 248–49; *Niles' Weekly Register,* 7 March 1812, 14; 2 December 1813, 228–29; Joseph A. Hill, "The Civil War Income Tax," *Quarterly Journal of Economics* 8 (July 1894): 416.

30. Hickey, *War of 1812,* 122.

31. Speech of Congressman Thomas Jackson Oakley, reprinted in *United States Gazette,* 5 November 1814; Hickey, *War of 1812,* 167.

32. Donald R. Adams Jr., *Finance and Enterprise in Early America: A Study of Stephen Girard's Bank, 1812–1831* (Philadelphia: University of Pennsylvania Press, 1978), 25–27, 132.

33. Robert W. Johannsen, *To the Halls of the Montezumas: The Mexican War in the American Imagination* (New York: Oxford University Press, 1985), 57–59, 280, 303; Albert

Gallatin, "War Expenses," *Niles' National Register,* 21 February 1846, 396–97; Gallatin, *Expenses of the War* (Washington: Towers, 1848), passim.

34. Frederick Merk, "Dissent in the Mexican War," in Morison, Merk, and Freidel, *Dissent in Three American Wars,* 41; U.S. House, 29th Cong., 2d sess., 1847, H. Doc. 85, 2; *Niles' National Register,* 21 November 1846, 187; John H. Schroeder, *Mr. Polk's War: American Opposition and Dissent, 1846–1848* (Madison: University of Wisconsin Press, 1973), 51–62.

35. Merk, "Dissent in the Mexican War," 41; Thomas E. Thomas, *Covenant Breaking, and Its Consequences* (Rossville, Ohio: Christy, 1847), 72; Schroeder, *Mr. Polk's War,* 99–106.

36. Garrett Davis, *Speech of Mr. Garrett Davis of Kentucky . . . in the House of Representatives,* 14 May 1846, 13; D. D. Barnard, "The President's Message: The War," *American Review* 5 (January 1847): 4; "Executive Usurpation," *American Review* 5 (March 1847): 217ff.; "The Whigs and the War," *American Review* 6 (October 1847): 331–33. Cf. David Herbert Donald, *Lincoln* (New York: Simon and Schuster, 1995), 122–24; *American Review* 4 (August 1846): 172; K. Jack Bauer, *The Mexican War, 1846–1848* (New York: Macmillan, 1974), 358–59.

37. Alice Kessler Harris, ed., *Sermons on War by Theodore Parker* (New York: Garland, 1973), 9, 13, 43; John Weiss, *A Fast Day Sermon* (New Bedford, Mass.: Tilden, 1848), 6; Charles C. Shackford, *A Citizen's Appeal in Regard to the War with Mexico* (Boston: Andrews and Prentiss, 1848), 10, 16–17, 19.

38. By using a private mail service, speculators at New Orleans obtained news of the war declaration a day earlier than the general public, allowing them to purchase and stockpile commodities in anticipation of a war-related price increase. See Gene W. Boyett, "Money and Maritime Activities in New Orleans during the Mexican War," *Louisiana History* 17 (fall 1976): 420. James M. McCaffrey, *Army of Manifest Destiny: The American Soldier in the Mexican War, 1846–1848* (New York: New York University Press, 1992), 35; Thomas S. Jesup, *Report of the Quartermaster General,* 24 November 1847, in U.S. House, 30th Cong., 1st sess., 1848, Ex. Doc. 8, 544–49; *Niles' National Register,* 4 July 1846, 280.

39. Jesup, *Report of the Quartermaster General,* 1847, 548–4^: Ivor Debenham Spencer, *The Victor and the Spoils: A Life of William L. Marcy* (Providence, R.I.: Brown University Press, 1959), 164; *Niles' National Register,* 15 August 1846, 368; Risch, *Quartermaster Support of the Army,* 258–62.

40. Boyett, "Money and Maritime Activities," 423–27; Gurston Goldin, "Business Sentiment and the Mexican War with Particular Emphasis on the New York Businessman," *New York History* 33 (January 1952): 63–64.

41. Jesup, *Report of the Quartermaster General,* 1847, 550; George Gibson, *Report of the Commissary General of Subsistence,* 17 November 1846, in U.S. House, 29th Cong., 2d sess., 1847, Ex. Doc. 4, 190; Justin H. Smith, *The War with Mexico* (New York: Macmillan, 1919), 2:263; *Niles' National Register,* 8 November 1845, 147; 15 August 1846, 368; 3 October 1846, 67; Boyett, "Money and Maritime Activities," 416; Lewis E. Atherton, "Disorganizing Effects of the Mexican War on the Santa Fe Trade," *Kansas Historical Quarterly* 6 (May 1937): 120.

42. John Niven, *Gideon Welles, Lincoln's Secretary of the Navy* (New York: Oxford University Press, 1973), 210–17; Risch, *Quartermaster Support of the Army,* 237–40, 247–50; Sterling, *Serving the Line with Excellence,* 17–18.

43. Henry Stanton to Thomas S. Jesup, 22 November 1848, in Jesup, *Report of the Quartermaster General,* 18 November 1848, U.S. House, 30th Cong., 2d sess., 1848, Ex. Doc. 1, 240–43; *Niles' National Register,* 19 May 1847, 241; Risch, *Quartermaster Support of the Army,* 256–57.

44. Jesup, *Report of the Quartermaster General,* 24 November 1847, 548; "Shoeing the Army," *Scientific American,* 26 June 1847, 318.

45. *Niles' National Register,* 21 November 1846, 186.

46. James K. Polk, *Orders Respecting Military Contributions,* 3 March 1847, in U.S. House, 30th Cong., 1st sess., 1848, Ex. Doc. 8; William L. Marcy, *Report of the Secretary of War,* 1 December 1848, in U.S. House, 30th Cong., 2d sess., 1848, Ex. Doc. 1, 80.

47. Maj. O. Cross to Thomas S. Jesup, September 7, 1848, in Jesup, *Report of the Quartermaster General,* 18 November 1848, 208–12; Risch, *Quartermaster Support of the Army,* 298–99.

48. J. H. Smith, *War with Mexico,* 258–66; Henry Cohen, *Business and Politics in America from the Age of Jackson to the Civil War: The Career Biography of W. W. Corcoran* (Westport, Conn.: Greenwood Press, 1971), 26, 46; Henrietta M. Larson, *Jay Cooke, Private Banker* (Cambridge: Harvard University Press, 1936), 69.

49. Cohen, *Business and Politics in America,* 62; *Niles' National Register,* 11 September 1846, 293.

4. The "Shoddyocracy"

1. *New York Herald,* 9 January 1862.

2. Mark W. Summers, *The Plundering Generation: Corruption and the Crisis of the Union,* 1849–1816 (New York: Oxford University Press, 1987) observes that the decades before and after the Civil War were unusually corrupt.

3. Gustavus Myers, *History of the Great American Fortunes* (1909–17; New York: Modern Library, 1937), 291–92, 397, 400, 406.

4. Yale historian A. Howard Meneely told of "speculators, rogues, [and] conniving dishonest men." The definitive statement was the Pulitzer Prize–winning account by Fred A. Shannon. Shannon repeated a claim by Myers that the Civil War was the origin of the wealthy generation known as "Robber Barons." He charged that in the first months of the war "the contractors and officials . . . reveled in a saturnalia of graft." Without explaining his computations, Shannon declared that "it is fair to state that the army contractors handled at least a billion dollars of government money during the war, and from all the evidence, by conservative estimate, retained a half of it." Summarizing, Shannon observed that "profiteering even on this scale seems small to the present generation, so familiar with the pestilence of profiteers of 1917–1918 and since, but considering the wealth, population, industrial conditions, and available resources of the two periods even the World War does not present a larger degree of profiteering activities than did the Civil War." A. Howard Meneely, *The War Department,* 1861: *A Study in Mobilization and Administration* (New York: Columbia University Press, 1928), 269–71; Fred A. Shannon, *The Organization and Administration of the Union Army,* 1861–1865 (Cleveland: Clarke, 1928), 1:56, 71, 73. More recently, Ludwell H. Johnson argued that "favoritism, greed, dishonesty and treachery infected virtually every aspect of the Union war effort." See Johnson, "Northern Profit and Profiteers: The Cotton Rings of 1864–1865," *Civil War History* 12 (June 1966): 115.

The first tiny fissure in this version appeared in 1941. One of Myers's oft-repeated charges was that the great financier J. Pierpont Morgan had started his huge fortune by selling obsolete and unsafe rifles to the U.S. Army at extortionate prices. These overpriced weapons were allegedly inclined to blast off a soldier's thumb. Dozens of writers repeated Myers's account, known as the Hall carbine affair, including such authors as John Dos Passos, Carl Sandburg, and Bertrand Russell. In the 1930s the story received broad circulation in newspapers, over the radio, and in congressional testimony.

But R. Gordon Wasson, an executive in the Morgan banking firm who was disturbed by the bad publicity, wrote a brief monograph proving that Morgan played a minor role in the episode and that the rifles, while not state-of-the-art weapons, were neither obsolescent, unusually dangerous to use, nor sold above the market price. Morgan, moreover, was already prosperous, if not extremely wealthy, when the Civil War began. It is true that windfall profits were made in the Hall carbine deal (though not principally by Morgan), but only because the Army Bureau of Ordnance, thinking that the rebellion would be crushed in a few weeks, foolishly disposed of what it thought was an oversupply at fire-sale prices. The army sold five thousand Hall carbines to Arthur M. Eastman for $3.50 each in June 1861. Six weeks later, General John C. Frémont, the Union commander in Missouri, was desperate for arms in order to stop an imminent Confederate advance. He directed his agent in New York to repurchase the guns at $22 each. This was near the market price, and Morgan's only role was to provide financing for the deal. See R. Gordon Wasson, *The Hall Carbine Affair: A Study in Contemporary Folklore* (New York: Pendick, 1941, 1948; privately printed, 1971),

82–117, and S. N. D. North and Ralph H. North, *Simeon North: First Official Pistol Maker of the United States* (Concord, N. H.: Rumford, 1913), 167. Although Wasson's book appeared in three editions, only 1,116 copies were printed. For congressional testimony, see Thomas Hall Shastid, *How to Stop War-Time Profiteering* (2d ed.; Ann Arbor: Wahr, 1937), 12–13.

Wasson had demonstrated a deficiency in Myers's scholarship, but the mistake went largely uncorrected and was repeated as recently as 1990. See Ernest A. McKay, *The Civil War and New York City* (Syracuse: Syracuse University Press, 1990), 98–99, and Richard F. Kaufman, *The War Profiteers* (Garden City, N.Y.: Anchor, 1972), 9–10. This failure was mostly because of the obscurity of the book, but also because Wasson was obviously not a disinterested investigator. In the most comprehensive account of the Civil War mobilization, Allan Nevins absolved Morgan of discredit in the Hall carbine affair but continued to claim that the Civil War was the origin of his wealth and that of others, such as John D. Rockefeller and Andrew Carnegie. The war probably delayed Carnegie's rise to great wealth, and in any case all three men gained the bulk of their money later and would certainly have done so without the war. See Allan Nevins, *The War for the Union,* vol. 2, *War Becomes Revolution* (New York: Scribner, 1960), 510; vol. 3, *The Organized War, 1863–1864* (New York: Scribner, 1971), 329; Joseph Frazier Wall, *Andrew Carnegie* (New York: Oxford University Press, 1970), 189–90.

A second significant case study also undercut the conventional account of Civil War profiteering. In 1959 Russell F. Weigley published a careful biography of Montgomery C. Meigs, the quartermaster general of the Union army. Although Weigley agreed that corruption existed in the early months of the war, he showed that Meigs was honest and efficient. He termed the progressives' criticism of the army supply system "extravagant" and argued that, over the course of the war, the system was competently managed. Some writers supported his findings, but others did not. See Russell F. Weigley, *Quartermaster General of the Union Army: A Biography of M. C. Meigs* (New York: Columbia University Press, 1959).

In 1990 J. Matthew Gallman examined the experience of wartime Philadelphia. He found that while some Philadelphians prospered mightily during the war, only a handful did so as a direct result of the conflict. See J. Matthew Gallman, *Mastering Wartime: A Social History of Philadelphia during the Civil War* (Cambridge: Cambridge University Press, 1990), 324–26.

For those who do not share this view, see James A. Frost, "The Home Front in New York during the Civil War," *New York History* 42 (July 1961): 297; Maury Klein, "The War and Economic Expansion," *Civil War Times Illustrated* 8 (January 1970): 37; Earl K. Molander, "Historical Antecedents of Military-Industrial Criticism," in *War, Business, and Society,* ed. Benjamin F. Cooling (Port Washington, N.Y.: Kennikat, 1977), 178; Samuel Richey Kamm, *The Civil War Career of Thomas A. Scott* (Philadelphia: University of Pennsylvania Press, 1940), 77; George Edgar Turner, *Victory Rode the Rails: The Strategic Place of the Railroads in the Civil War* (Indianapolis: Bobbs-Merrill, 1953), 111–12, and Erwin Stanley Bradley, *Simon Cameron, Lincoln's Secretary of War* (Philadelphia: University of Pennsylvania Press, 1966), 195.

5. Howard K. Beale, ed., *Diary of Gideon Welles: Secretary of the Navy under Lincoln and Johnson* (3 vols.; New York: Norton, 1960), 1:549.

6. Frank E. Vandiver, "The Civil War as an Institutionalizing Force," in William F. Holmes and Harold M. Hollingsworth, eds., *Essays on the American Civil War* (Austin: University of Texas Press, 1968), 82; Richard Franklin Bensel, *Yankee Leviathan: The Origins of Central State Authority in America, 1859–1877* (Cambridge: Cambridge University Press, 1990), 125–35.

7. Keir B. Sterling, *Serving the Line with Excellence: The Development of the U.S. Army Ordnance Corps* (Aberdeen, Md.: U.S. Army, 1987), 28.

8. Beale, *Gideon Welles,* 1:247–49; Marvin R. Cain, *Lincoln's Attorney General: Edward Bates of Missouri* (Columbia: University of Missouri Press, 1965), 253–54. During the first year of the war, the Confederacy employed privateers. See Raymond H. Robinson, *The Boston Economy during the Civil War* (New York: Garland, 1988), 178–79.

9. Morford, *The Days of Shoddy: A Novel of the Great Rebellion of* 1861 (Philadelphia: Peterson, 1863), 85, 90–91, 192–93. See also Morford, *Shoulder Straps: A Novel of New York and the Army*, 1862 (Philadelphia: Peterson, 1863), 421.

10. Abraham Lincoln to Congress, 27 May 1862, in *New York Times*, 28 May 1862.

11. Allan Nevins, *The War for the Union*, vol. 1, *The Improvised War, 1861–1862* (New York: Scribner, 1959), 351, 365; Wasson, *Hall Carbine Affair*, 10.

12. *New York Herald*, 16 April 1861, 6.

13. Nevins, *War for the Union*, 1:351, 365.

14. "Our Finances," *Harper's Weekly*, 21 December 1861, 802; *New York Herald*, 26 September 1863; Norman B. Wilkinson, *Lammot du Pont and the American Explosives Industry, 1850–1884* (Charlottesville: University Press of Virginia, 1984), 76.

15. *New York Tribune*, 15 June 1861; Lincoln to Congress, 27 May 1862, reprinted in *New York Times*, 28 May 1862; Charles F. Stone, "Washington on the Eve of the War," in *Secret Missions of the Civil War*, ed. Philip Van Doren Stern (Chicago: Rand McNally, 1959), 37–53, passim.

16. *New York Herald*, 4 May 1861.

17. New York Assembly, *Report of State Officers*, 85th sess., 1862, Rept. 15, 3–38.

18. Fred A. Shannon repeated a newspaper allegation that shoddy cloth disintegrated in the rain, but this cannot be corroborated. For the charge that shoddy cloth disintegrated, see *New York Herald*, 27 November 1861, and Shannon, *Organization of the Union Army*, 1:94–95. See also New York Assembly, Rept. 15, 187, 193; New York Assembly, Rept. 194, 22, 35–37; *New York Herald*, 30 August 1861; *New York Tribune*, 11 June 1861; *New York Times*, 5 September 1861.

19. New York Assembly, Rept. 15, 188; New York Assembly, Rept. 194, 30, 34.

20. Ohio Quartermaster General, *Annual Report of the Quartermaster General*, 1861, Exec. Docs., pt. 1, 1861, p. 594.

21. *New York Herald*, 21 August 1861, 4. The *Oxford English Dictionary* incorrectly dates the first use of this term as 1862 (see vol. 9, p. 722), and the *Dictionary of Americanisms* repeats the mistake. Douglas Fermer identifies the error in his book *James Gordon Bennett and the New York Herald: A Study of Editorial Opinion in the Civil War Era, 1854–1867* (Woodbridge, U.K.: Boydell, 1986), 256n 38, although he dates the first use three days later.

22. *New York Herald*, 13 January 1862, 9 December 1862, 13 June 1864, 14 June 1864.

23. *New York Herald*, 24, 25, and 30 August 1861, 21 June 1864, 14 October 1864.

24. *New York Herald*, 30 August, 6 and 16 November 1861, 3 April 1862, 15 April 1863.

25. New York Assembly, Rept. 194, 2; Oliver Willcox Norton, *Army Letters, 1861–1865* (New York: privately published, 1903), 169; "At Mrs. Shoddy's," *Harper's Weekly* 29 January 1865, 475. Cf. Robert Tomes, "The Fortunes of War," *Harper's Monthly*, July 1864, 227–31; Gallman, *Mastering Wartime*, 286–88.

26. The distinction was that local rates were for traffic carried entirely upon a railroad's own track, whereas through rates were for shipments that went from one railroad to another.

27. Kamm, *Civil War Career of Thomas A. Scott*, 69; U.S. House, *Government Contracts*, 37th Cong., 2d sess., 1861, H. Rept. 2, 2:631.

28. *New York Times*, 9 August 1862.

29. Hermon K. Murphey, "The Northern Railroads and the Civil War," *Mississippi Valley Historical Review* 5 (December 1918): 331; Bradley, *Simon Cameron*, 201; Richard Ray Duncan, "The Social and Economic Impact of the Civil War on Maryland" (Ph.D. diss., Ohio State University, 1963), 29; Thomas Weber, *The Northern Railroads in the Civil War* (Westport, Conn.: Greenwood Press, 1970), 128.

30. Meneely, *War Department*, 244; U.S. House, *Government Contracts*, 2:490, 558–59, 818–19. There were some exceptions to the rate schedule. The Illinois Central, a land-grant railroad, was allowed only a small markup above cost. The Baltimore and Ohio, which was often ravaged by Confederate raids, was allowed to charge 10 percent more for freight than

other companies in order to compensate it for the damage to its roadbed. The Louisville and Nashville received a similar dispensation. See Weber, *Northern Railroads,* 130; Nevins, *War for the Union,* 2:460–61, 3:249; Murphey, "Northern Railroads," 330.

31. U.S. House, *Government Contracts,* 2:636, 726.

32. Ibid., 2:708–10.

33. Ibid., 2:72–73, 100, 111–12, 119, 330–37, 413–16, 780; William DeLoss Love, *Wisconsin in the War of the Rebellion* (Chicago: Church, 1866), 293.

34. U.S. House, *Government Contracts,* 2:xviii, 557, 610–11.

35. Murphey, "Northern Railroads," 332; Nevins, *War for the Union,* 2:46–61; Weigley, *Meigs,* 238.

36. One historian even went so far as to title a chapter "The Itching Palm of Simon Cameron." George Edgar Turner, *Victory Rode the Rails: The Strategic Place of the Railroads in the Civil War* (Indianapolis: Bobbs-Merrill, 1953), 45–61.

37. "Extraordinary Prosperity of Our Railroads," *Scientific American,* 10 October 1861, 219; U.S. House, *Government Contracts,* 2:633; Duncan, "Impact of the Civil War on Maryland," 29–30; Gallman, *Mastering Wartime,* 279–80; Emma Lou Thornbrough, *Indiana in the Civil War,* (Indianapolis: Indiana Historical Bureau, 1965), 337; Murphey, "Northern Railroads," 328; Edward G. Everett, "Pennsylvania's Mobilization for War" (Ph.D. diss., University of Pittsburgh, 1954), 239–41; John F. Stover, *History of the Baltimore and Ohio Railroad* (West Lafayette, Ind.: Purdue University Press, 1987), 115–17; Emerson D. Fite, *Social and Industrial Conditions in the North during the Civil War* (New York: Macmillan, 1910), 44, 162n.

38. John Niven, *Gideon Welles: Lincoln's Secretary of the Navy* (New York: Oxford University Press, 1973), 362–64; *Dictionary of American Biography,* 13:168–69.

39. Niven, *Welles,* 362–64; Richard S. West Jr., "The Morgan Purchases," *United States Naval Institute Proceedings* 66 (January 1940): 76–77; U.S. House, *Government Contracts,* 1:31–32. Morgan's own estimate of his commissions was only $70,000. See Fred Nicklason, "The Civil War Committee," *Civil War History* 17 (September 1971): 233.

40. John Tucker, *Reply to the Report of the Select Committee of the Senate on Transports for the War Department* (Philadelphia: Moss, 1863), 27, 50–55.

41. U.S. House, *Government Contracts,* 2:1214–16, 1225–26, 1228, 1235–36, 1248–52, 1256–60, 1270; U.S. Senate Select Committee to Inquire into the Chartering of Transport Vessels for the Banks Expedition, *Report in Part,* 37th Cong., 3d sess., 1862, S. Rept. 75, 19–20, 44 (cited hereafter as U.S., Banks Expedition); "Prosperity of the Shipping Interest," *Scientific American,* 14 September 1861, 171.

42. U.S. House, *Government Contracts,* 2:1235–36, 1248–52, 1436–37; Ludwell H. Johnson, "The Butler Expedition of 1861–1862," *Civil War History* 11 (September 1965): 234–35.

43. U.S. House, *Government Contracts,* 1:19–20; West, "Morgan Purchases," 74.

44. U.S. House, *Government Contracts,* 1:7–9, 12, 2:227–28; *New York Herald,* 11 and 14 October 1861, 18 December 1862.

45. G. V. Fox to W. L. Hudson, 11 November 1861, in *Confidential Correspondence of Gustavus Vasa Fox, Assistant Secretary of the Navy, 1861–1865,* ed. Robert Means Thompson and Richard Wainwright (New York: Naval History Society, 1920), 1:400–401.

46. U.S., Banks Expedition, 16–20, 26–27, 108–12.

47. Ibid., 75–76, 123; Edwin P. Hoyt, *The Vanderbilts and Their Fortunes* (Garden City, N.Y.: Doubleday, 1962), 153–54.

48. U.S. Senate, *Message of the President of the United States,* 39th Cong., 1st sess., 1866, Ex. Doc. 46; W. A. Croffut, *The Vanderbilts and the Story of Their Fortune* (Chicago: Belford, 1886), 54–56.

49. In July 1862, Marshall O. Roberts leased the *Coatzatcoalcon* with the understanding that it was to carry 1,200 soldiers for the Port Royal expedition. It was loaded with 1,250 men, plus an extra 30,000 gallons of water. The ship sustained $60,000 in damages. See *New York Herald,* 20 July 1862.

50. Tucker, *Reply to . . . the Select Committee,* 54.

51. U.S. Conduct Committee, U.S. Senate, Joint Committee on the Conduct of the War, 37 Cong., 3d Sess., Rept. 108, pt. 3, 1862, 250, 272.

52. Allan Nevins, *Frémont: Pathmarker of the West* (New York: Appleton Century, 1939), 626–27; Weigley, *Meigs,* 187–99.

53. Justus B. McKinstry, *Vindication of Brig. Gen. J. McKinstry, Formerly Quarter-master Western Department* ([St. Louis], [1862]), 32ff.

54. Ibid., 15, 58; U.S. Senate, *Proceedings of the Commission on Ordnance Contracts and Claims—1862,* 37th Cong., 2d sess., 1862, Ex. Doc. 72, 313–14 (cited hereafter as U.S. Senate, *Commission on Ordnance Contracts*).

55. U.S. War Department, *The War of the Rebellion, . . . Official Records of the Union and Confederate Armies,* ser. 3, 2:107 (cited hereafter as *Official Records*).

56. U.S. Conduct Committee, pt. 3, pp. 5, 250, 272, 276–79; U.S. House, *Government Contracts,* 2:757–58, 794; David Herbert Donald, *Lincoln* (New York: Simon and Schuster, 1995), 316; Nevins, *Frémont,* 538; Shannon, *Organization of the Union Army,* 1:63; *New York Herald,* 16 October 1861, 28 December 1861, 9 April 1862, 27 July 1862; *New York Times,* 27 and 28 May 1862; U.S. Senate, 38th Cong., 2d sess., 25 February 1865, Rept. 134, passim.

57. During World War I, the army tripled in about the same period of time. During World War II, the pace of mobilization, while by no means slow, was not as meteoric. The army quadrupled in twelve months, and by then the industrial capacity of the nation was much greater. Weigley, *Meigs,* 204; Brocks M. Kelley, "Fossildom, Old Fogeyism, and Red Tape," *Pennsylvania Magazine of History and Biography* 90 (Jan. 1966), 112.

58. John Francis Mitchell, "Springfield, Massachusetts, and the Civil War" (Ph.D. diss., Boston University, 1960), 50, 196–97; *New York Herald,* 14 July 1861.

59. "Armory and Arms," *Scientific American,* 27 July 1861, 53; Ripley to Stanton, 19 June 1861, *Official Records,* ser. 3, 1:279; Edward M. Morin, "Springfield during the Civil War Years, 1861–1865," *Historical Journal of Western Massachusetts* 3 (fall 1974): 35–36.

60. New York Assembly, *Report of the Commander-in-Chief . . . Relative to Arms and Equipments of the Militia,* 1 January 1862, Doc. 21, 4–5.

61. U.S. Senate, *Commission on Ordnance Contracts,* 141.

62. U.S. War Department, *Official Records,* ser. 3, 1:675–76, 2:563, 578, 588.

63. U.S. Senate, *Commission on Ordnance Contracts,* 32.

64. Freight to the United States added about fifty-eight cents to the price.

65. Half a century later, Britain would complain bitterly about American price gouging during its own national emergency. *Marcellus Hartley: A Brief Memoir* (New York: privately printed, 1903), 30–32, 146, 148.

66. U.S. Senate, *Commission on Ordnance Contracts,* 141.

67. U.S. War Department, *Official Records,* ser. 3, 2:112–13.

68. Ibid.

69. U.S. Senate, *Commission on Ordnance Contracts,* 23, 103–4; Meneely, *War Department,* 285.

70. U.S. Senate, *Commission on Ordnance Contracts,* 164, 171, 248, 250, 285; *New York Herald,* 4 January 1865.

71. U.S. Senate, *Commission on Ordnance Contracts,* 1:138; Nevins, *War for the Union,* 3:307.

72. "A New Rifle Manufactory," *Scientific American,* 6 September 1862, 148.

73. *New York Herald,* 15, 16, and 22 December 1864. The rate of depreciation is based on two estimates sworn to under oath. The intermediate producer was Eli Whitney and Company, which valued similar guns at $13.86 each for tax purposes in 1862. For the Whitney data, see U.S. Senate, *Commission on Ordnance Contracts,* 1:348, 386, and "Manufacturing Returns, October, 1862," div. 2, collection dist. 2, p. 4, *Internal Revenue Assessment Lists for Connecticut, 1862–1866,* roll 10, National Archives.

74. U.S. Senate, *Commission on Ordnance Contracts,* 263–66, 312; *Senate Journal,* 37th Cong., 2d sess., 26 June 1862, 715.

75. U.S. Senate, *Commission on Ordnance Contracts*, 8–9, 13, 16, 357, 380, 500.

76. U.S. War Department, *Official Records*, ser. 3, 4:804–5.

77. Mitchell, "Springfield, Massachusetts," 213n., 221; John Niven, *Connecticut for the Union: The Role of the State in the Civil War* (New Haven: Yale University Press, 1965), 358.

78. "Colt's Armory—The Colonel on the Side of the Government," *Scientific American*, 25 May 1861, 333.

79. William B. Edwards, *The Story of Colt's Revolver* (Harrisburg: Stackpole, 1953), 279, appendix; U.S. Senate, *Commission on Ordnance Contracts*, 133, 138, 315.

80. Insurance coverage was only $660,000, so the Colt heirs lost heavily. Edwards, *Colt*, 342; "Colt's Armory," *Scientific American*, 5 July 1862, 5; "Conflagration at Colt's Armory," *Scientific American*, 20 February 1864, 117–18; U.S. Senate, *Commission on Ordnance Contracts*, 61, 65; Nevins, *War for the Union*, 1:357.

81. Charles T. Haven and Frank A. Belden, *A History of the Colt Revolver* (New York: Bonanza, 1940), 383.

82. U.S. War Department, *Official Records*, ser. 3, 1:423; Nevins, *War for the Union*, 1:363, 2:469; Niven, *Connecticut*, 358.

83. *New York Herald*, 20 December 1864.

84. U.S. Senate Joint Committee on the Conduct of the War, *Heavy Ordnance*, 38th Cong., 2d sess., 1865, S. Rept. 121, 6, 9, 184.

85. U.S. Senate, *Commission on Ordnance Contracts*, 549–71.

86. U.S. Senate Committee on the Conduct of the War, 17–19, 86, 100; U.S. Senate Joint Committee on the Conduct of the War, *Heavy Ordnance*, 38th Cong., 2d sess., 1865, S. Rept. 121, 7–8, 24, 85, 100, 108; U.S. War Department, *Official Records*, ser. 3, 4:627. In 1865, "all large guns" were forged by the Rodman process, including both Parrott and Dahlgren guns, but evidently Rodman did not receive royalties on these designs. See U.S. Senate, *Heavy Ordnance*, Rept. 121, 24, 109.

87. Colin T. Naylor Jr., *Civil War Days in a Country Village* (Peekskill, N.Y.: Highland, 1961), 14–15.

88. U.S. Senate, *Heavy Ordnance*, Rept. 121, p. 140. In 1863 the wealthiest munitions executive, Henry du Pont, who was the wealthiest man in Delaware, received an income of $117,985. Wilkinson, *Lammot du Pont*, 106.

89. U.S. War Department, *Official Records*, ser. 3, 1:676–77.

90. *New York Herald*, 13 January 1862. Cotton inventories appreciated by $480,000 at the Amoskeag Company, by $250,000 at the Stark Mills, and by $165,000 at the Manchester Mills. See "Great Advance in Cotton," *Scientific American*, 5 October 1861, 210.

91. Theodore Collier, "Providence in Civil War Days," *Rhode Island Historical Society Collections*, 27 (July 1934), 103–4; Fite, *Social and Industrial Conditions in the North*, 87; "The Internal Revenue Tax," *Scientific American*, 4 April 1864, 243; Robert P. Sharkey, *Money, Class, and Party: An Economic Study of the Civil War and Reconstruction* (Baltimore: Johns Hopkins University Press, 1959), 141–43.

92. Nevins, *War for the Union*, 2:475; *New York Herald*, 26 September 1863; *New York Times*, 7 May 1864.

93. "Our Special Correspondence," *Scientific American*, 12 December 1864, 373; *New York Herald*, 3 March, 21 May 1864; U.S. Senate, *Report of the Select Committee on Naval Supplies*, 38th Cong., 1st sess., 1864, S. Rept. 99, pp. 178, 193.

94. "The National Resources, and Their Relation to Foreign Commerce and the Price of Gold," *North American Review* 100 (January 1865): 138–39.

95. Amasa Walker, "The National Finances," *Merchants Magazine and Commercial Review* 12 (January 1865): 26; R. H. Stanley, *Eastern Maine and the Rebellion* (Bangor, Me.: Stanley, 1887), 185; Marian V. Sears, "Gold and the Local Stock Exchanges of the 1860s," *Explorations in Entrepreneurial History*, 2d ser., 6 (winter 1969): 204–9; Robert Tomes, "The Fortunes of War," *Harper's Monthly*, July 1864, 230; "Wall Street in War Time," *Harper's Monthly*, April 1865, 616; *New York Herald*, 20 May 1863; Henrietta M. Larson, *Jay Cooke, Private Banker* (Cambridge: Harvard University Press, 1936), 154–60; Ellis Paxson Oberholtzer, *Jay Cooke, Financier of the Civil War* (Philadelphia: Jacobs, 1907), 1:497.

96. Mitchell, "Springfield, Massachusetts," 218–19.

97. A. S. Roberts, "The Federal Government and Confederate Cotton," *American Historical Review* 32 (January 1927): 264n; Robert Frank Futrell, "Federal Trade with the Confederate States, 1861–1865: A Study of Governmental Policy" (Ph.D. diss., Vanderbilt University, 1950), 119; Nevins, *War for the Union,* 3:352; Fred Harvey Harrington, *Fighting Politician: Major General N. P. Banks* (Philadelphia: University of Pennsylvania Press, 1948), 135; Thomas W. O'Connor, "Lincoln and the Cotton Trade," *Civil War History* 7 (March 1961): 27–28.

98. Confederate blockade runners could find no better prices in Europe. The price advantage obtained by blockade runners selling in European markets ranged from a low of three times cost to a high of ten times cost. See Louise B. Hill, "State Socialism in the Confederate States of America," *Southern Sketches,* 1st ser., 9 (1936): 7; Nevins, *War for the Union,* 3:341, 372–73; O'Connor, "Lincoln and the Cotton Trade," 32.

99. Allan G. Bogue, *The Congressman's Civil War* (Cambridge: Cambridge University Press, 1989), 43; Nevins, *War for the Union,* 2:475; Futrell, "Federal Trade," 11; E. Merton Coulter, "Commercial Intercourse with the Confederacy in the Mississippi Valley," *Mississippi Valley Historical Review* 5 (March 1919): 378; U.S. House Joint Committee on the Conduct of the War, *Trade in Military Districts,* 37th Cong., 3d sess., 1863, 561–65.

100. Futrell, "Federal Trade," 102–3; Benjamin P. Thomas and Harold M. Hyman, *Stanton: The Life and Times of Lincoln's Secretary of War* (New York: Alfred A. Knopf, 1962), 364.

101. Coulter, "Commercial Intercourse," 379; *New York Herald,* 23 and 25 April 1861; U.S. House Committee on the Conduct of the War, *Trade in Military Districts,* 565; Howard K. Beale, ed., *The Diary of Edward Bates, 1859–1866* (Washington, 1933), 414.

102. Everett, "Pennsylvania's Mobilization," 173, 176; Futrell, "Federal Trade," 4–5; W. H. H. Terrell, *Indiana in the War of the Rebellion: Report of the Adjutant General* (Indianapolis: Indiana Historical Bureau, 1960), 497–98; John D. Barnhart, *The Impact of the Civil War on Indiana* (Indianapolis: Indiana Civil War Commission, 1962), 11.

103. Ludwell H. Johnson, "Contraband Trade during the Last Year of the Civil War," *Mississippi Valley Historical Review* 49 (March 1963): 639; Harrington, *Fighting Politician,* 135; Niven, *Gideon Welles,* 464; Ludwell H. Johnson, "The Louis A. Welton Affair: A Confederate Attempt to Buy Supplies in the North," *Civil War History* 15 (March 1969): 33–34.

104. O'Connor, "Lincoln and the Cotton Trade," 33; Harrington, *Fighting Politician,* 135; E. L. Erikson, ed., "Hunting for Cotton in Dixie: From the Diary of Captain Charles E. Wilcox," *Journal of Southern History* 4 (November 1938): 509; Futrell, "Federal Trade," 92–93, 108; *New York Herald,* 17 January 1864; Ella Lonn, *Salt as a Factor in the Confederacy* (University: University of Alabama Press, 1965), 160–70. Soldiers customarily traded Southern tobacco for Northern coffee. See Francis A. Lord, "That 'Wonderful Solace': Virginia Tobacco and the Civil War," *Virginia Cavalcade* 20 (spring 1971): 46.

105. Futrell, "Federal Trade," 422–24, 446.

106. Ibid., 85, 91, 209.

107. Ibid., 136; Roberts, "The Federal Government and Confederate Cotton," 263; Coulter, "Commercial Intercourse," 385; U.S. Senate Select Committee, *Alleged Traffic with Rebels in Texas,* 41st Cong., 3d sess., 1870, S. Rept. 377, passim; Ludwell H. Johnson, "Northern Profit and Profiteers: The Cotton Rings of 1864–1865," *Civil War History* 12 (June 1966): 103–14; Johnson, "Blockade or Trade Monopoly? John A. Dix and the Union Occupation of Norfolk," *Virginia Magazine of History and Biography,* January 1985, 77.

108. Ludwell H. Johnson, *Red River Campaign: Politics and Cotton in the Civil War* (Baltimore: Johns Hopkins University Press, 1958), 102–3; David D. Porter to Gustavus Vasa Fox, 24 May 1862, *Confidential Correspondence of Gustavus Vasa Fox,* 2:110–11.

109. Futrell, "Federal Trade," 122; Johnson, *Red River Campaign,* 102–3; Niven, *Gideon Welles,* 464–65; Porter to Gustavus Vasa Fox, 19 February 1864, *Confidential Correspondence of Gustavus Vasa Fox,* 2:201; Coulter, "Commercial Intercourse," 393.

110. Roberts, "The Federal Government and Confederate Cotton," 269; James G. Randall, *The Confiscation of Property during the Civil War* (Indianapolis: Mutual, 1913), 41–43.

111. *National Cyclopaedia of American Biography* (New York: White, 1893), 3:440.

112. Norton, *Army Letters*, 26; Albert Castel, *William Clarke Quantrill: His Life and Times* (New York: Fell, 1862), 127, 134; Castel, *A Frontier State at War: Kansas, 1861–1865* (Ithaca: Cornell University Press, 1958), 44–63, 130; Futrell, "Federal Trade," 69; *New York Herald*, 26 June, 19 July 1862; *New York Times*, 4 January 1863; U.S. Senate Committee on the Conduct of the War, Rept. 108, 630–31; Frank L. Byrne, "'A Terrible Machine': General Neal Dow's Military Government on the Gulf Coast," *Civil War History* 12 (March 1966): 5–13, and passim.

113. Elisabeth Joan Doyle, "'Rottenness in Every Direction': The Stokes Investigation in Civil War New Orleans," *Civil War History* 18 (March 1972): 33, 38–39; U.S. House, *Government Contracts*, 2:299–301, 833, 1000–1006, 1020–26, 1035–37, 1077–78, 1090–1136; Barnhart, *Impact of the Civil War on Indiana*, 12; *New York Herald*, 22 September 1863. Harold M. Hyman, "Deceit in Dixie," *Civil War History* 3 (1957): 73, describes bribery in the provost marshal's office.

114. Fred A. Shannon, "The Mercenary Factor in the Creation of the Union Army," *Mississippi Valley Historical Review* 12 (March 1926): 536; Shannon, *Organization of the Union Army*, 2:55–60, 71; James Barnett, "The Bounty Jumpers of Indiana," *Civil War History* 4 (December 1958): 431–33; Thornbrough, *Indiana in the Civil War*, 140. Manhattan was not the only haven for bounty jumpers. The practice was widespread in Brooklyn as well. See *New York Times*, 27 December 1864.

115. Shannon, *Organization of the Union Army*, 2:11–46. For an opposite view see Eugene C. Murdock, "Was It a 'Poor Man's Fight'?" *Civil War History* 10 (September 1964): 241–45, and Hugh G. Earnhart, "Commutation: Democratic or Undemocratic?" *Civil War History* 12 (June 1966): 132–42. Evidence that the progressive historians may have been repeating a partisan charge levied by Democrats against Republicans may be found in the *New York Leader*, 29 August 1863, quoted in James A. Frost, "The Home Front in New York during the Civil War," *New York History* 42 (July 1961): 289–90.

116. Eugene C. Murdock, "New York's Civil War Bounty Brokers," *Journal of American History* 53 (September 1966): 265, 268n.; Will Plank, *Banners and Bugles: A Record of Ulster County, New York and the Mid-Hudson Region in the Civil War* (Marlborough, N.Y.: Centennial, 1963), 104.

117. Blake McKelvey, ed., *Rochester in the Civil War* (Rochester, N.Y.: Society, 1944), 56; *New York Herald*, 25 December 1864.

118. Shannon, *Organization of the Union Army*, 2:75; U.S. War Department, *Official Records*, ser. 3, 4:1225.

119. *New York Herald*, 26 May 1861, 27 December 1862, 19, 20, and 22 May 1864; *New York Times*, 27 December 1862, 22 and 23 May 1864. For a somewhat similar episode in Pennsylvania, see Everett, "Pennsylvania's Mobilization," 252–53.

120. Mack Walker, "The Mercenaries," *New England Quarterly* 39 (September 1966): 393; Robert L. Peterson and John A. Hudson, "A Foreign Recruitment for Union Forces," *Civil War History* 7 (June 1961), passim.

121. Bogue, *Congressman's Civil War*, 63; *New York Herald*, 9 April 1863, 4.

122. For example, see essays on the issues of the war in the deeply moralistic *Christian Examiner* that discuss slavery and the preservation of the Union but that ignore profiteering: "The War," *Christian Examiner* 71 (July 1861): 95–115; "The Relation of War to Human Nature," *Christian Examiner* 71 (November 1861): 313–40; "Our War Policy, and How It Deals with Slavery," *Christian Examiner* 73 (September 1862): 246–48.

123. *Congressional Globe*, 37th Cong., 1st sess., 2 August 1861, 405; *New York Herald*, 4 February 1862, 15 March 1862, 25 February 1863; "Inspectors of Supplies," *Scientific American*, 15 June 1861, 377.

124. Larson, *Jay Cooke*, 148, 165, 177; Oberholtzer, *Jay Cooke*, 1:144, 255, 325, 512–13; Everett, "Pennsylvania's Mobilization," 219.

125. Weigley, *Montgomery Meigs*, 252; U.S. War Department, *Official Records*, ser. 3, 2:1–2.

126. Weigley, *Montgomery Meigs*, 275; *House Journal*, 37th Cong., 2d sess., 1862, 255; *New York Herald*, 25 July 1862, 23 February 1863.

127. *Congressional Globe*, 26 March 1863; George P. Sanger, *Public Laws of the United States of America* (Boston: Little, Brown, 1863), 12:698; *New York Times*, 24 November 1863. As of 27 June 1994, the record award was $22.5 million, paid to Douglas R. Keeth, a vice president of Sikorsky Aircraft who helped the Justice Department obtain a judgment of $150 million against his company. Now known officially as the False Claims Act of 1986, the law is also dubbed informally the "Lincoln Law." See *New York Times*, 10 July 1989, 27 June 1994. On the Contracts Committee, see Bogue, *Congressman's Civil War*, 80–88.

128. However, an attempt to force an end to speculation in gold by act of Congress proved a failure. See Donald, *Lincoln*, 507; U.S. Senate, *Committee on Naval Supplies*, 68; *New York Herald*, 12 May 1862, 13 December 1862, 22 September 1863.

129. "The Capacities of Government and Private Armories," *Scientific American*, 17 October 1863, 241; Nevins, *War for the Union*, 3:305–6; George Winston Smith and Charles Judah, *Life in the North during the Civil War* (Albuquerque: University of New Mexico Press, 1966), 183–86; "The Government Bakery," *Scientific American*, 12 September 1863, 167; *New York Herald*, 24 October 1864.

130. U.S. Senate, *Commission on Ordnance Contracts*, 316; Kamm, *Civil War Career of Thomas A. Scott*, 106n; U.S. Senate, *Committee on Naval Supplies*, 204–8; Meneely, *War Department*, 1861, 263.

131. "Immorality in Politics," *North American Review* 98 (January 1864): 105–6. Cf. "Democracy on Trial," *Christian Examiner* 235 (March 1863): 276–77.

132. "Army Peculators," *New York Tribune*, 25 May 1861, 4; *New York Times*, 24 November 1863.

133. *New York Herald*, 14 April 1861, 19 and 30 October 1861.

134. The demand for silk was so great that Connecticut farmers experimented unsuccessfully with growing silkworms. "War and Increasing Wealth," *Scientific American*, 13 June 1863, 377; "Unseemly Extravagance," *Scientific American*, 21 May 1864, 326–27; *New York Herald*, 16 September 1862, 12 September 1864 (quote); *New York Times*, 2 April 1864; Nevins, *War for the Union*, 3:216; Jarlath R. Lane, *A Political History of Connecticut during the Civil War* (Washington: Catholic University Press, 1941), 202; Niven, *Connecticut for the Union*, 383–84; Kenneth Nolan Metcalf and Lewis Beeson, *Effects of the Civil War on Manufacturing in Michigan* (Lansing: Michigan Civil War Centennial Observance Commission, 1966), 12; Fite, *Social and Industrial Conditions*, 260; Castel, *Frontier State*, 205.

135. Women also devoted great attention to the larger issues. When the National Convention of Loyal Ladies of the North met in New York in 1863, the main question addressed was whether or not freed slaves should have the vote (they should). John B. McMaster, *Our House Divided* (Greenwich, Conn.: Fawcett, 1961), 533–34, 536–37; Collier, "Providence in Civil War Days," 105; *New York Herald*, 15 May 1863, 16, 17, and 27 May 1864; *New York Times*, 4 and 7 May 1864.

136. "The Striking Contrasts of the War," *New York Herald*, 27 June 1863, 6. See also *New York Herald*, 6 November 1861, 14 January 1862, 15 April, 11 June, 16 September, 13 October 1863, 5 September 1864; Norton, *Army Letters*, 169.

137. *New York Times*, 2 April 1864; E. L. Godkin, "Immorality and Political Corruption," *North American Review* 107 (1868): 248–66; "Democracy on Trial," *Christian Examiner* 235 (March 1863): 262–94; "The Age of Extortion," *Scientific American*, 11 July 1863, 25.

138. Joseph A. Hill, "The Civil War Income Tax," *Quarterly Journal of Economics* 8 (July 1894): 418–21, 424n; "The New Tax Law," *Scientific American*, 17 August 1861, 98; *New York Herald*, 16 June 1862.

139. Hill, "Civil War Income Tax," 418; Paul Beers, "Civil War Income Tax," *Civil War Times Illustrated* 9 (April 1970): 25; *New York Herald*, 16 June 1862; Stephen J. DeCanio and Joel Mokyr, "Inflation and the Wage Lag during the American Civil War," *Explorations in Economic History* 14 (1977): 325.

140. "Taxation in America," *New York Times,* 18 February 1864, 4; *New York Herald,* 15 October 1861, 16 June 1862; "Manufacturers' Opposition to Income Tax," *Merchants Magazine and Commercial Review* 49 (August 1863): 151–52. Cf. "The Tax on Manufacturers," *Scientific American,* 20 June 1863, 393, Wilkinson, *Lammot du Pont,* 86, and Bensel, *Yankee Leviathan,* 168–72.

141. Beers, "Civil War Income Tax," 25; DeCanio and Mokyr, "Inflation and the Wage Lag," 325–27; Leonard P. Curry, *Blueprint for Modern America: Nonmilitary Legislation of the First Civil War Congress* (Nashville: Vanderbilt University Press, 1968), 179–80.

142. *New York Herald,* 4, 13, 17, and 20 March 1862, 26 September 1862; "The New Tax Law," *Scientific American,* 17 August 1861, 98; Murphey, "Northern Railroads," 334.

143. *New York Times,* 20 January 1865; *New York Herald,* 14 January 1865; Nevins, *War for the Union,* 3:228; "Rich Monopolists Shirking Their Tax," *Scientific American,* 20 September 1862, 186; U.S. Commissioner of Internal Revenue, *Laws of the United States Relating to Internal Revenue* (Washington, 1866), 122; Sanger, *Public Laws of the United States of America,* 12:483; P. J. Staudenraus, ed., *Mr. Lincoln's Washington: Selections from the Writings of Noah Brooks, Civil War Correspondent* (New York: Yoseloff, 1967), 35, 402.

144. Ludwell H. Johnson, "The Plundering Generation: Uneasy Reflections on the Civil War," *Continuity* 9 (1985): 118–19.

145. U.S. House, *Government Contracts,* 2:117; Everett, "Pennsylvania's Mobilization," 36; New York Assembly, *Report of State Officers,* 1862, Rept. 15, 25.

146. *New York Herald,* 10 June 1862; U.S. Senate, *Message of the President of the United States,* 39th Cong., 1st sess., 1866, Ex. Doc. 46.

147. U.S. House, *Government Contracts,* 1861, 2:1109; *New York Herald,* 2 May 1861; Sharkey, *Money, Class, and Party,* 143; U.S. Senate, *Commission on Ordnance Contracts,* 246.

148. Nevins, *War for the Union,* 1:350–54; Harriet C. Owsley, "Henry Shelton Sanford and Federal Surveillance Abroad," *Mississippi Valley Historical Review* 48 (September 1961): 223; Douglas H. Maynard, "The Forbes-Aspinwall Mission," *Mississippi Valley Historical Review* 45 (June 1958): 88. Ernest A. McKay argues that Aspinwall's contribution was trivial. See McKay, *Civil War and New York City,* 99–100.

149. *New York Herald,* 24, 26, 27, and 29 April 1861; Victor Hicken, *Illinois in the Civil War* (Urbana: University of Illinois Press, 1966), 322; Niven, *Connecticut for the Union,* 381.

150. "The Shoddy Aristocracy in the Parish Mansion," *New York Herald,* 6 September 1863, 4. Cf. "The New Union League Club," *New York Herald,* 1 September 1863.

151. Allan Nevins, ed., *Diary of the Civil War, 1860–1865* (New York: Macmillan, 1962), 302–7, esp. 305n, 306n; Guy James Gibson, "Lincoln's League: The Union League Movement during the Civil War" (Ph.D. diss., University of Illinois, 1957), 77, 204ff.

152. Thomas and Hyman, *Stanton,* 363.

153. The best examples are Johnson, "The Plundering Generation," McKay, *The Civil War and New York City,* and David Herbert Donald, *Lincoln* (New York: Simon and Schuster, 1995), 325. See also Stephen B. Oates, *A Woman of Valor: Clara Barton and the Civil War* (New York: Free Press, 1994), 277.

5. Toward the Great War

1. U.S. Senate, *Sale of Arms by Ordnance Department,* 42d Cong., 2d sess., 1872, S. Rept. 183, 29–30, 218, 229, 529, 531; U.S. Senate, *Sale of Ordnance Stores,* 42d Cong., 2d sess., 1872, S. Doc. 73, 1; Allan Nevins, *Hamilton Fish: The Inner History of the Grant Administration* (New York: Ungar, 1936, 1957), 403; Merrit Roe Smith, "Military Arsenals and Industry before World War I," in *War, Business, and Society: Historical Perspectives on the Military-Industrial Complex,* ed. Benjamin F. Cooling (Port Washington, N.Y.: Kennikat, 1977), 35–39.

2. The War Department was also willing to sell to a German agent, H. Boker & Co., but German chancellor Otto von Bismarck turned down the purchase as unneeded. Nevins, *Fish,* 400–404; U.S. Senate, *Sale of Arms by Ordnance Department,* 201, 528; Edward L. Pierce,

Memoir and Letters of Charles Sumner, vol. 4, *1860–1874* (Boston: Roberts, 1894), 504–506; John Gerow Gazley, *American Opinion of German Unification, 1848–1871* (New York: Columbia University Press, 1926), 322–58.

3. Hans Trefousse, *Carl Schurz: A Biography* (Knoxville: University of Tennessee Press, 1982), 15–27, 58–72.

4. Owing to the size of the purchase this was later reduced to 2.5 percent. *Congressional Globe,* 14 February 1872, 1009; U.S. Senate, *Sale of Arms by the Ordnance Department,* 227, 316.

5. Pierce, *Sumner,* 509; Trefousse, *Schurz,* 79.

6. The French consul general in New York City, Victor Place, was tried in France but not convicted. U.S. Senate, *Sale of Ordnance Stores,* 1; U.S. Senate, *Sale of Arms by Ordnance Department,* xxi, lxix–lxxi.

7. Nevins, *Fish,* 404.

8. Hiram S. Maxim, *My Life* (New York: McBride, 1915), 154–210; David A. Armstrong, *Bullets and Bureaucrats: The Machine Gun and the United States Army* (Westport, Conn.: Greenwood Press, 1982), 83–85; John Ellis, *The Social History of the Machine Gun* (Baltimore: Johns Hopkins University Press, 1975), 33–37, 74; J. D. Scott, *Vickers: A History* (London: Weidenfeld, 1962), 36–39; William Manchester, *The Arms of Krupp, 1587–1968* (Boston: Little, Brown, 1964), 185, 207; Benjamin F. Cooling, *Gray Steel and Blue Water Navy: The Formative Years of America's Military-Industrial Complex* (Hamden, Conn.: Archon, 1979), 63, 66, 82, 93.

9. James L. Abrahamson, *America Arms for a New Century: The Making of a Great Military Power* (New York: Free Press, 1981), 10–15, 19–21; Cooling, *Gray Steel,* 21–50.

10. Abrahamson, *America Arms,* 137–39; Cooling, *Gray Steel,* 59–60, 76, 110.

11. Robert Hessen, *Steel Titan: The Life of Charles M. Schwab* (New York: Oxford University Press, 1975), 164–66; Cooling, *Gray Steel,* 72–73, 95–96. Samuel P. Huntington coined the term *business pacifist.* See Abrahamson, *America Arms,* 137–78, esp. 145.

12. George Weiss, "What the War Has Done for Steel," *Forum* 57 (January 1917): 118. During periods of business depression, the price would be considerably less. See Hessen, *Steel Titan,* 79–80.

13. For example, see Clyde H. Tavenner, "How the War Trust Is Robbing the Government While Driving Us on Toward the Brink of War," *Congressional Record* 15 February 1915, 418.

14. Cooling, *Gray Steel,* 117–18.

15. Hessen, *Steel Titan,* 50–57; Cooling, *Gray Steel,* 118.

16. Hessen, *Steel Titan,* 56–57.

17. This genre would include such books as Gustavus Myers's *The History of the Great American Fortunes* (1907), John Winkler's *Incredible Carnegie* (1931), Matthew Josephson's *The Robber Barons* (1934), George Seldes's *Iron, Blood, and Profits* (1934), Helmuth C. Engelbrecht and Frank C. Hanighen's *Merchants of Death* (1934), and Philip Noel-Baker's *The Private Manufacture of Armaments* (1936).

18. On the charges against Morgan, see chap. 4, note 4. See also Hessen, *Steel Titan,* 307–10. In his 1977 book, Benjamin F. Cooling accepted the committee's report without qualification. See Cooling, *Gray Steel,* 118.

19. Scott, *Vickers,* 42–43, 86–87.

20. Hessen, *Steel Titan,* 91–92; Cooling, *Gray Steel,* 121–29.

21. Hessen, *Steel Titan,* 98; Manchester, *Krupp,* 168–69, 90–99; Francis Butler Simkins, *Pitchfork Ben Tillman, South Carolinian* (Baton Rouge: Louisiana State Press, 1944), 311, 348–49; Cooling, *Gray Steel,* 122.

22. Hessen, *Steel Titan,* 91–101; Manchester, *Krupp,* 221. Cooling cites somewhat different prices. See *Gray Steel,* 226–31.

23. Cooling, *Gray Steel,* 136; Abrahamson, *America Arms,* 19–20, 29–39; Graham Cosmas, *An Army for Empire: The United States Army in the Spanish-American War* (Columbia: University of Missouri Press, 1971), 14–15, 33.

24. Cosmas, *An Army for Empire,* 54.

25. Ibid., 163–64, 286–94, 303.

26. Ibid., 156; Tavenner, "How the War Trust Is Robbing the Government," 420; James A. Frear, "Frauds in War Contracts Are Uncovered," *Eau Claire (Wis.) Leader*, 13 June 1917, clipping in Henry A. Frear Papers, State Historical Society of Wisconsin, Madison, box 4. Cf. Edward Robb Ellis, *Echoes of Distant Thunder: Life in the United States, 1914–1918* (New York: Coward, McCann, 1975), 385–87.

27. Kansas populists generally supported the war as a humanitarian crusade and denounced war opponents as servants of "heartless commercialism." John M. Cooper Jr., *The Vanity of Power: American Isolationism and the First World War, 1914–1917* (Westport, Conn.: Greenwood Press, 1969), 265; O. Gene Clanton, *Kansas Populism: Ideas and Men* (Lawrence: University Press of Kansas, 1969), 212; Bruce Palmer, "*Man over Money": The Southern Populist Critique of American Capitalism* (Chapel Hill: University of North Carolina Press, 1980), passim; Stuart Creighton Miller, "*Benevolent Assimilation": The American Conquest of the Philippines, 1899–1903* (New Haven: Yale University Press, 1982), 29.

28. The wage was per month. U.S. House Select Committee on Expenditures in the War Department, 66th Cong., 3d sess., 1921, H. Rept. 1400, 2:2698; Newton D. Baker, *Why We Went to War* (New York: Council on Foreign Relations, 1936), 118; William F. Holmes, *The White Chief: James Kimble Vardaman* (Baton Rouge: Louisiana State University Press, 1970), 297; Andrew S. Hicks to John M. Nelson, 10 January 1929, John M. Nelson Papers, State Historical Society of Wisconsin, Madison, box 2.

29. Abrahamson, *America Arms*, 15–16, 66, 90–91; Cooling, *Gray Steel*, 166, 169; John W. Adams, "The Influences Affecting Naval Shipbuilding Legislation, 1910–1916," *Naval War College Review* 22 (December 1969): 41–42; John Morgan Gates, *Schoolbooks and Krags: The United States Army in the Philippines, 1898–1902* (Westport, Conn.: Greenwood Press, 1973), 22–25.

30. Lincoln quoted in Manfred Jonas, *Isolationism in America, 1935–1941* (Ithaca: Cornell University Press, 1966), 122; Sherman in Abrahamson, *America Arms*, 6.

31. Quoted in Karen Falk, "Public Opinion in Wisconsin during World War I," *Wisconsin Magazine of History* 25 (June 1942): 392; Alexander M. Arnett, *Claude Kitchin and the Wilson War Policies* (1937; reprint, New York: Russell, 1971), 104.

32. Abrahamson, *America Arms*, 43, 55; Adams, "Naval Shipbuilding Legislation," 46–49; David Starr Jordan, *War and Waste: A Series of Discussions of War and War Accessories* (1914; reprint, New York: Garland, 1972), 60–67, quote on 241.

33. Hamilton Holt, "Straining an Historic Friendship," *Independent*, 1 May 1913, 977; Holt, "Armament Octopus," *Independent*, 13 April 1914, 80; Holt, "Is it Necessary?" *Independent*, 25 October 1915, 120–21; Claude Kitchin, "The Nation's Preparedness," in *Leading Opinions Both for and against National Defense*, comp. Hudson Maxim (New York: Hearst, 1916), 84ff.; James Parker Martin, "The American Peace Movement and the Progressive Era" (Ph.D. diss., Rice University, 1975), 67, 281; Jonas, *Isolationism in America*, 122–23, 273; Jordan, *War and Waste*, 195; Arnett, *Kitchin and the Wilson War Policies*, 64; John M. Craig, "Lucia True Ames Mead: American Publicist for Peace and Internationalism," in *Women and American Foreign Policy: Lobbyists, Critics, and Insiders*, ed. Edward P. Crapol (Westport, Conn.: Greenwood Press, 1987), 75. Jordan quoted in Abrahamson, *America Arms*, 77.

34. Quoted in Armin Rappaport, *The Navy League of the United States* (Detroit: Wayne State University Press, 1962), 22.

35. Holt, "Armament Scandals," *Independent*, 1 May 1913, 946; Holt, "Straining an Historic Friendship," 976; "Preparedness," *Independent*, 20 September 1915, 379; David Starr Jordan to editor, *Army and Navy Journal*, 21 September 1912, David Starr Jordan Collection, Hoover Institution, Stanford University, Stanford, Calif., box 13 (cited hereafter as JC); Clyde Henry Tavenner, "The War Trust: How Can We Beat It?" *La Follette's Weekly Magazine*, 22 August 1914, 4; Jordan, MSS for "The Standing Incentives for War," *Unpopular Review* 1 (January–June 1914), JC, box 52; Barton J. Bernstein and Franklin A. Leib, "Progressive Republican Senators and American Imperialism, 1898–1916: A Reappraisal," *Mid-America* 50

(July 1968): 203. See also Tavenner, "At the Mercy of the Armor Ring," *La Follette's Weekly Magazine,* 28 February 1914, 4, and "War Preparations Are a Matter of Profit to the Armor Ring," *La Follette's Weekly Magazine,* 1 August 1914, 4.

36. Jordan, *War and Waste,* 52–53, 123, 173–75; Henry Noel Brailsford, *The War of Steel and Gold: A Study of the Armed Peace* (5th ed.; London: G. Bell, 1915), 162–63; Charles V. Genthe, *American War Narratives,* 1917–1918 (New York: Lewis, 1969), 5–6; Arthur Marwick, *War and Social Change in the Twentieth Century: A Comparative Study of Britain, France, Germany, Russia, and the United States* (London: Macmillan, 1974), 3.

37. Jane Addams quoted in Marie Louise Degen, *The History of the Women's Peace Party* (1939; New York: Franklin, 1974), 20–21; Jordan, "The Crime of the Century," unpublished MSS [August 1914], JC, box 43.

38. David Starr Jordan, "From Feudalism to Capitalism," unpublished MSS [1872?], JC, box 44; Jordan, "Exercises in History: Story of the Civil War," unpublished notes, 10 September 1912, JC, box 52; Jordan, "The People and the War Castes," *Stanford Sequoia* 27 (December 1917): 83; Jordan, *The Human Harvest: A Study of the Decay or Races through the Survival of the Unfit* (1907; New York: Garland, 1972), 3; Jordan, "The Farmer and War," *Farm and Fireside,* 5 December 1914, clipping, JC, box 44; Jordan to B. G. M. Baskett, 2 February 1912, JC, box 17; Jordan to Eliot, 10 August and 30 September 1912, and Eliot to Jordan, 23 August and 11 September 1912, JC, box 19; Jordan, "War and Waste," *Peace* 1 (September 1912), passim; Jordan, letters to editor, *Army and Navy Journal,* 12 September and 18 November 1912, JC, box 13; Jordan, "Does Human Nature Change?" MSS [n.d.], JC, box 44. Jordan was not an absolute pacifist. He served as vice chairman, commander of the District of Columbia, Society of American Wars. See its membership list, State Historical Society of Wisconsin, Madison, and Charles Chatfield, *For Peace and Justice: Pacifism in America, 1914–41* (Knoxville: University of Tennessee Press, 1971), 27.

39. David M. Kennedy, *Over Here: The First World War and American Society* (New York: Oxford University Press, 1980), 23, 178–79; Cooper, *Vanity of Power,* 93, 100–101. Further evidence of the power of the social memory of the Civil War may be found in Theda Skocpol, *Protecting Soldiers and Mothers: The Political Origins of Social Policy in the United States* (Cambridge: Harvard University Press, 1992), 130–51.

40. Warren G. Harding to D. John Markey, chairman, military affairs committee, American Legion, 6 October 1922, reprinted in Marquis James, "The Profiteer Hunt: The Quest Begins and the Trail Grows Hot," *American Legion Weekly,* 23 March 1923, 7.

41. A new charge that began to appear was that Judge Thomas Mellon had founded his family's great fortune on sales to the War Department during the Civil War. See George Seldes, *Iron, Blood, and Profits: An Exposure of the World-Wide Munitions Racket* (New York: Harper, 1934), 226.

42. This was precisely the point made by Senator Robert M. La Follette Sr. of Wisconsin, when the Senate considered expelling him for opposing the war. The Senate quashed the resolution. U.S. Senate Committee on Privileges and Elections, *Senator from Wisconsin,* 65th Cong., 3d sess., 1918, Rept. 614, 16. Cf. U.S. House Committee on Military Affairs, *Universal Mobilization for War Purposes,* 68th Cong., 1st sess., 1924, H. Doc. 764, 149–51.

43. U.S. House, *Universal Mobilization,* 145, 148; James Ford Rhodes, *History of the United States from the Compromise of 1850,* vol. 5, *1864–66* (New York: Macmillan, 1904): 219.

44. Warren I. Cohen, *The American Revisionists: The Lessons of Intervention in World War I* (Chicago: University of Chicago Press, 1967), 141; Melvin I. Urofsky, "Josephus Daniels and the Armor Trust," *North Carolina Historical Review* 65 (July 1968): 239; Jordan, *War and Waste,* 37.

45. Prominent intellectuals who shared this view included Randolph Bourne, Crystal Eastman, Frederick C. Howe, David Starr Jordan, Paul U. Kellogg, and Lillian D. Wald. See Michael Wreszin, *Oswald Garrison Villard, Pacifist at War* (Bloomington: Indiana University Press, 1965), v, 53; Anthony Gronowicz, ed., *Oswald Garrison Villard: The Dilemma of the Absolute Pacifist in Two World Wars* (New York: Garland, 1983), 152–53; Agnes Anne Trotter,

"The Development of the Merchants of Death Theory of American Intervention in World War I, 1914–1939" (Ph.D. diss., Duke University, 1966), 109; John A. Thompson, *Reformers and War: American Progressive Publicists and the First World War* (New York: Cambridge University Press, 1987), 137, 156; Arthur S. Link, *Wilson* (6 vols.; Princeton: Princeton University Press, 1947–1966), 5:418; Bernstein and Leib, "Progressive Republican Senators and American Imperialism," 204; Maxim, *Leading Opinions*, 44; Frederick C. Howe, *Why War* (1916; New York: Garland, 1972), 314.

46. Trotter, "Merchants of Death Theory," 38; Wreszin, *Villard*, v, 53; C. Roland Marchand, *The American Peace Movement and Social Reform* (Princeton: Princeton University Press, 1972), 176; Thompson, *Reformers and War*, 137; Ida M. Tarbell, *All in the Day's Work: An Autobiography* (New York: Macmillan, 1939), 322.

47. Howe, *Why War*, 314–15; Craig, "Lucia True Ames Mead," 72–73.

48. U.S. Senate Special Committee Investigating the Munitions Industry, *Hearings*, 74th Cong., 2d sess., 1936, 7587 (cited hereafter as the Nye committee).

49. Hessen, *Steel Titan*, 7.

50. David S. Patterson, *Towards a Warless World: The Travail of the American Peace Movement* (Bloomington: Indiana University Press, 1976), passim, esp. 68–70; Walter I. Trattner, "Progressivism and World War I: A Re-appraisal," *Mid-America* 44 (1962): 136; David Starr Jordan, *The Root of the Evil* (New York: Friends of German Democracy, 1918), 4; Katherine Anthony, *Susan B. Anthony: Her Personal History and Her Era* (Garden City, N.Y.: Doubleday, 1954), 204.

51. "The Armament Makers of Europe," *New Republic*, 23 May 1934, 33; Geoffrey Jones and Clive Trebilcock, "Russian Industry and British Business, 1910–1930: Oil and Armaments," *Journal of European Economic History* 11 (spring 1982): 72–81, 87; Earl A. Molander, "Historical Antecedents of Military-Industrial Criticism," in *War, Business, and Society*, ed. Benjamin F. Cooling (Port Washington, N.Y.: Kennikat, 1977), 180–81.

52. Manchester, *Krupp*, 244–45.

53. Perris and Wells made the most direct connection between war and profits. The relevant titles are John A. Hobson, *Imperialism: A Study* (New York: Pott, 1902), Norman Angell, *The Great Illusion* (1912; New York: Arno, 1972), Brailsford, *War of Steel and Gold*, George Herbert Perris, *The War Traders* (London: National Peace Council, 1914), and H. G. Wells, *The End of the Armament Rings* (Boston: World Peace Foundation [1914]).

54. Thompson, *Reformers and War*, 104; Alexander Arnett, "Claude Kitchin versus the Patrioteers," *North Carolina Historical Review* 14 (1937): 93–94, and Arnett, *Kitchin and the Wilson War Policies*, 93–94; Anthony Sampson, *The Arms Bazaar: From Lebanon to Lockheed* (New York: Viking, 1977), 38–39n, 57–59; Clive Trebilcock, *The Vickers Brothers: Armaments and Enterprise, 1854–1914* (London: Europa, 1977); Peter Filene, "The World Peace Foundation and Progressivism: 1910–1918," *New England Quarterly* 36 (December 1963): 487; Angell, *Great Illusion*; Brailsford, *War of Steel and Gold*; Perris, *War Traders*; Wells, *End of the Armament Rings*. The flow of ideas was not entirely one-way. In 1901 the leading American pacifists Lucia and Edwin Mead visited the prominent English anti-imperialist John A. Hobson, who was then working on his most important treatise, *Imperialism*. In 1911 Hobson visited the American pacifist David Starr Jordan in the latter's home. Jordan also met with George Perris in the National Liberal Club in London. See Craig, "Lucia True Ames Mead," 73–74; Jordan to Hobson, 15 March 1911, and Hobson to Jordan, 14 March 1919, JC, box 23; "Builders of Democracy," speech MSS [1918?], JC, box 43; Lucia Ames Mead, *Outline of Lessons on War and Peace* (Boston: World Peace Foundation, pam. ser., 5, no. 1 [February 1915]), lesson 4. Hobson's influence is evident in Scott Nearing, *War: Organized Destruction and Mass Murder by Civilized Nations* (New York: Vanguard, 1931), 43–44.

55. On May Day 1916, Liebknecht participated in a peace demonstration in Berlin, charging that German profiteers wanted war with the United States. He was immediately arrested and sentenced to thirty months in prison. Karl Liebknecht, *Militarism* (New York: Huebsch, 1917), and *Militarism and Anti-Militarism* ([1917]; New York: Fertig, 1969), xi, 50, 86; Manchester, *Krupp*, 271–72, 278–79; Hamilton Holt, "Germany's Military Scandal,"

Independent, 1 May 1913, 1004; Felice A. Bonadio, "The Failure of German-American Propaganda in the United States, 1914–1917," *Mid-America* 41 (January 1959): 48–49; Clifton J. Child, "German-American Attempts to Prevent the Exportation of Munitions of War, 1914–15," *Mississippi Valley Historical Review* 25 (December 1938): 356, 360; Merle Curti, *Peace or War: The American Struggle, 1636–1936* (New York: Norton, 1936), 218–19.

56. Martin, "American Peace Movement and the Progressive Era," 150, 159; *La Follette's Weekly Magazine,* 9 September 1914, 15; 26 September 1914, 6; 17 October 1914, 1; Jordan, "The Aim of the War," speech MSS [December 1917], JC, box 39. In 1915 an obscure, pro-German journalist alleged the existence of a plot to engage the United States in a war in order to benefit Wall Street speculators, international bankers, major newspapers, and defense contractors. See Charles A. Collman, *The War Plotters of Wall Street* (New York: Fatherland, 1915), 131–30.

57. Hamilton Holt, "A Wall Street Craze," *Independent,* 18 October 1915, 86; Link, *Wilson,* 4:iv, 23; John Patrick Finnegan, *Against the Specter of a Dragon: The Campaign against Military Preparedness, 1914–1917* (Westport, Conn.: Greenwood Press, 1974), 124ff.; Palmer, *"Man over Money,"* 3–5, 28–38; Louis Galambos, *The Public Image of Big Business in America: 1880–1940* (Baltimore: Johns Hopkins University Press, 1975), 175; John A. Thompson, *Reformers and War: American Progressive Publicists and the First World War* (New York: Cambridge University Press, 1987), 42–43, 46, 161.

58. Crystal Eastman to David Starr Jordan, 26 February 1916, and Jordan to Eastman, 16 March 1916, JC, box 19; Degen, *Women's Peace Party,* 48, 72–73; Frederick C. Howe, *Why War,* 10; Finnegan, *Against the Specter of a Dragon,* 133. Mrs. Amos Pinchot was chairwoman of the New York City branch of the Women's Peace Party. In the 1930s, Amos Pinchot became a prominent member of the isolationist bloc. See Wayne S. Cole, *Roosevelt and the Isolationists, 1932–45* (Lincoln: University of Nebraska Press, 1983), 26, 385.

59. Marchand, *American Peace Movement,* 244–46, 382; Amos Pinchot to conference committee in respect to the Council of National Defense bill, 18 September 1917, Records of the Special Committee Investigating the Munitions Industry (Nye committee), RG 154, National Archives; Blanche Wiesen Cook, "Woodrow Wilson and the Anti-militarists, 1914–1917" (Ph.D. diss., Johns Hopkins University, 1970), 3–4; Jane Addams, *Peace and Bread in Time of War* (Morningside Heights, N.Y.: King's Crown, 1945), 2.

60. In the early twentieth century, Howe, then in her eighties, became a supporter of Philippine annexation. See Miller, *"Benevolent Assimilation,"* 117. On her role during the Franco-Prussian War, see Edwin D. Mead, *Women and War: Julia Ward Howe's Peace Crusade* (Boston: World Peace Foundation, pam. ser., 4, no. 6 [October 1914]): 6–7; Degen, *Women's Peace Party,* 14.

61. Marchand, *American Peace Movement,* 244, 247; Dorothy Detzer, "Record of the Women's International League on the Question of Munitions" [June 1934], and Records of the Madison Chapter, Women's International League of Peace and Freedom, State Historical Society of Wisconsin, Madison, box 1; Degen, *Women's Peace Party,* 18, 25–26, 35, 41, 159–60; Craig, "Lucia True Ames Mead," 72–73; Robert H. Ferrell, "The Merchants of Death, Then and Now," *Journal of International Affairs* 26, no. 1 (1972): 29; Jane Addams, *Peace and Bread,* 7.

62. Nineteenth-century American pacifism had a strong religious flavor, and much of the feeble opposition to rearmament in the early 1890s came from congressmen who served districts with numerous Quaker constituents, as in Pennsylvania, Rhode Island, or Iowa. In the aftermath of the Spanish War, discontent with defense spending broadened and became more secular. See Curti, *Peace or War,* 139, 144–45.

63. Tobacco and cotton farmers of the rural South resented the British blockade against sales of their products to the Continental markets. Arnett, *Claude Kitchin and the Wilson War Policies,* 142–43.

64. "American Shellmakers under Fire," *Literary Digest,* 3 February 1917, 236; John M. Cooper Jr., "Progressivism and American Foreign Policy," *Mid-America* 51 (1979): 265, 268; Henry C. Ferrell Jr., "Regional Rivalries, Congress, and MIC: The Norfolk and Charleston Navy Yards, 1913–1920," in Cooling, *War, Business, and Society,* 60–72; Richard Lowitt, *George*

W. Norris: The Making of a Progressive (Syracuse: Syracuse University Press, 1963), 149; Holmes, White Chief, 304; Ben Baack and Edward Ray, "Special Interests and the Adoption of the Income Tax in the United States," Journal of Economic History 45 (September 1985): 615; Robert P. Wilkins, "Middle Western Isolationism: A Re-examination," North Dakota Quarterly 25 (summer 1957): 70; Wilkins, "The Non-Partisan League and Upper Midwest Isolationism," Agricultural History 39 (April 1965): 108–9; Trotter, "Merchants of Death Theory," 61n.

65. "An Advertising-Crusade against Our Traffic in Arms," Literary Digest, 17 April 1915, 861; Wilkins, "Middle Western Isolationism," 70–71; Wilkins, "Non-Partisan League," 103; Wayne S. Cole, Senator Gerald P. Nye and American Foreign Relations (Minneapolis: University of Minnesota Press, 1962), 131; Trattner, "Progressivism and World War I," 135; Larry Remele, "The Tragedy of Idealism: The National Nonpartisan League and American Foreign Policy," North Dakota Quarterly 42 (autumn 1974): 83; Thomas N. Guinsburg, The Pursuit of Isolationism in the United States from Versailles to Pearl Harbor (New York: Garland, 1982), 33; Bonadio, "Failure of German-American Propaganda," 42; David Starr Jordan, "Emotionalism and Foreign Policy in America" [n.d.], JC, box 44. Cf. Jordan, "An Argument for War" [1916?], JC, box 42. Not all midwesterners and southerners were so inclined. Clyde H. Tavenner, who represented a largely rural district that included Rock Island, Illinois, advocated expansion of the large federal arsenal there. Tavenner was also a longtime critic of private defense contractors. He opposed American engagement in the European war and lost a bid for reelection. Senator Claude A. Swanson, a strong free-silver Bryanite from Virginia, worked to divert shipbuilding from New York and Boston to Norfolk. Even Claude Kitchin of North Carolina, generally a strong opponent of military intervention, demanded that the United States protest Britain's decision to declare as contraband of war the products of his state, tobacco and cotton. He was joined by other southerners. See James A. Frear, Forty Years of Progressive Public Service, Reasonably Filled with Thorns and Flowers (Washington: Associated Writers, 1937), 122; Trotter, "Merchants of Death Theory," 16–20, 92–93, Henry Ferrell, "Regional Rivalries," 61; Arnett, "Claude Kitchin," 142–43; and Holmes, White Chief, 299–302.

6. Warhogs and Warsows

1. "Buying Here for the Armies," Independent, 7 December 1914, 357–58; Merle Curti, Peace or War: The American Struggle, 1636–1936 (New York: Norton, 1936), 191–93; Paolo E. Coletta, William Jennings Bryan: II. Progressive Politician and Moral Statesman, 1909–1915 (Lincoln: University of Nebraska Press, 1969), 264–65, 269, 272–74; William F. Holmes, The White Chief: James Kimble Vardaman (Baton Rouge: Louisiana State University Press, 1970), 302.

2. Merle Curti, Bryan and World Peace, (1931; New York: Octagon Books, 1969), 223–26; Blanche Weisen Cook, "Woodrow Wilson and the Anti-militarists, 1914–1917" (Ph.D. diss., Johns Hopkins University, 1970), 182–84.

3. Similar organizations of lesser importance were the Army League and the Association for National Service. Progressive suspicion of these organizations became so acute that on 17 August 1917 Secretary of the Navy Josephus Daniels ordered that no officer or agent of the Navy League was to be admitted to any American naval vessel or naval station. See U.S. Senate Special Committee Investigating the Munitions Industry, Hearings, 74th Cong., 1st sess., 1935, 12019 (cited hereafter as Nye committee). See also James L. Abrahamson, America Arms for a New Century: The Making of a Great Military Power (New York: Free Press), 134; Larry Wayne Ward, The Motion Picture Goes to War: The U.S. Film Effort during World War I (Ann Arbor: UMI Research Press), 34–36; U.S. House Select Committee, National Security League, 65th Cong., 3d Sess., 1919, 883; "To organize a people's National Liberal League," speech MSS, [1918?], John M. Nelson Papers, State Historical Society of Wisconsin, Madison, box 12; David Starr Jordan to C. H. Bailey, Army and Navy

News, 22 April 1915, David Starr Jordan Collection, Hoover Institution Archives, Stanford University, Stanford, Calif., box 13; James A. Frear, *Forty Years of Progressive Public Service, Reasonably Filled with Thorns and Flowers* (Washington: Associated Writers, 1937), 108ff.; Cook, "Woodrow Wilson and the Anti-militarists," 46; Armin Rappaport, *The Navy League of the United States* (Detroit: Wayne State University Press, 1962), 42; C. Roland Marchand, *The American Peace Movement and Social Reform* (Princeton: Princeton University Press, 1972), 10.

4. Hudson Maxim, *Defenseless America* (New York: Hearst, 1915).

5. Oswald Garrison Villard, *Fighting Years: Memoirs of a Liberal Editor* (New York: Harcourt Brace, 1939), 300; Ward, *Motion Picture Goes to War,* 39; Alexander M. Arnett, *Claude Kitchin and the Wilson War Policies* (1937; New York: Russell, 1971), 200.

6. Alfred D. Chandler Jr. and Stephen Salsbury, *Pierre S. du Pont and the Making of the Modern Corporation* (New York: Harper, 1971), 365.

7. Robert D. Cuff, "Private Success, Public Problems: The Du Pont Corporation and World War I," *Canadian Review of American Studies* 20 (1989): 188, n. 10.

8. Most of the NSL men were supporters of former president and fellow New Yorker Theodore Roosevelt.

9. Chandler and Salsbury, *du Pont,* 324, 366, 396; Charles A. Collman, *The War Plotters of Wall Street* (New York: Fatherland, 1915), 27, 71–72; U.S. House, *National Security League,* 883, 909, 913–34; Frear, *Forty Years,* 108ff.; U.S. Senate, Nye committee, 12320–28; Robert D. Ward, "The Origin and Activities of the National Security League, 1914–1919," *Mississippi Valley Historical Review* 47 (July 1960): 53–54.

10. Arthur S. Link, *Wilson* (6 vols.; Princeton: Princeton University Press, 1947–66), 4:334, 337; Hudson Maxim, comp., *Leading Opinions Both for and against National Defense* (New York: Hearst, 1916), 118–20; Theodore S. Woolsey, "The Case for the Munitions Trade," *Leslie's Illustrated Weekly,* 29 January 1915, 106.

11. For example, see the anonymous poem "To Arms for Peace": "Their gestured menace bids us be aware / Lest we would be slaves, prepare, prepare! . . . Let all the vulcan furnaces be driven— / Forge thunderbolts, out-thundering the heaven! . . . / Go, fortify the earth, the sea, the air, / And fortify our hearts—Prepare, prepare!" Quoted in Maxim, *Leading Opinions,* 121.

12. Norman Angell, *America and the European War* (Boston: World Peace Foundation, pam. ser., 5, no. 1 [February 1915]), passim, and Angell, *The World's Highway* (New York: Doran, 1915), 36–37; H. G. Wells, *The End of the Armament Rings* (Boston: World Peace Foundation, [1914]), 1; Charles V. Genthe, *American War Narratives, 1917–1918* (New York: Lewis, 1969), 5–6; Herbert Hoover, "German Practices in Belgium," Hoover Papers, Hoover Institution Archives, Stanford, Calif.

13. Wilson may have privately advocated naval expansion as early as August 1912, but in deference to the Bryanite little navy wing of his party he did not make his views public. Wilson had been a member of the American Peace Society as early as 1908, and he was reputed to favor government manufacture of munitions. See Ruhl J. Bartlett, *The League to Enforce Peace* (Chapel Hill: University of North Carolina Press, 1944), 52; *New York Times,* 4 February 1924; Robert D. Cuff, *The War Industries Board: Business-Government Relations during World War I* (Baltimore: Johns Hopkins University Press, 1973), 11, 243–44; Cuff, "We Band of Brothers—Woodrow Wilson's War Managers," *Canadian Review of American Studies* 5 (fall 1974): 137; and Cook, "Woodrow Wilson and the Anti-militarists," 24–34.

14. John M. Cooper Jr., "Progressivism and Foreign Policy," *Mid-America* 51 (1979): 270–75.

15. Wartime prosperity was achieving what the progressives could not, however. Between 1914 and 1919, real wages of urban unskilled workers rose by 10 percent, while those of skilled workmen fell by 11 percent. Ben Baack and Edward Ray, "Special Interests and the Adoption of the Income Tax in the United States," *Journal of Economic History* 45 (September 1985): 612; Jeffrey G. Williamson and Peter H. Lindert, *American Inequality: A Macroeconomic History* (New York: Academic Press, 1980), 79, 110. Cf. Carole Shammas, "A

New Look at Long-Term Trends in Wealth Inequality in the United States," *American Historical Review* 98 (April 1993): 427. The Spanish-American War tax is briefly discussed in W. Elliot Brownlee Jr., "Wilson and Financing the Modern State: The Revenue Act of 1916," *Proceedings of the American Philosophical Society* 129, no. 2 (1985): 192.

16. Allen F. Davis, "The Flowering of Progressivism," in Arthur S. Link, ed., *The Impact of World War I* (New York: Harper, 1969), 46–47; Ida M. Tarbell, *All in the Day's Work: An Autobiography* (New York: Macmillan, 1939), 322; Cooper, "Progressivism and Foreign Policy," 274; Cook, "Woodrow Wilson and the Anti-militarists," 24–26; Marchand, *American Peace Movement*, 249; John A. Thompson, *Reformers and War: American Progressive Publicists and the First World War* (New York: Cambridge University Press, 1987), 140–41; John W. Hillje, "New York Progressives and the War Revenue Act of 1917," *New York History* 53 (October 1972): 446, 450; Brownlee, "Wilson and Financing the Modern State," 192. For other attempts to convert the campaign for military preparedness to progressive ends, see Penn Borden, *Civilian Indoctrination of the Military: World War I and Future Implications for the Military-Industrial Complex* (Westport, Conn.: Greenwood Press, 1989), passim.

17. Fifteen Unitarian Ministers, *The Soul of America in Time of War* (Boston: Beacon, 1918), passim; Len G. Broughton, *Is Preparedness for War Unchristian?* (New York: Doran, 1916), passim.

18. "The 'War Orders' and American Industry," *Review of Reviews* 52 (August 1915): 223; "America's New Industries," *Review of Reviews* 54 (July 1916): 93–94; "Russia's Market for America's Goods," *Review of Reviews* 54 (July 1916): 22–23; Ward, *Motion Picture Goes to War*, 5; Richard Lewinsohn, *The Profits of War through the Ages* (London: Routledge, 1936), 241; Harold C. Syrett, "The Business Press and American Neutrality, 1914–1917," *Mississippi Valley Historical Review* 32 (September 1945): 217; Ronald Shaffer, *America in the Great War: The Rise of the War Welfare State* (New York: Oxford University Press, 1991), 55–57.

19. George Weiss, "What the War Has Done for Steel," *Forum* 57 (January 1917): 116; "Warhogs and War Millionaires [By a War Broker]," *Independent*, 19 July 1915, 80.

20. U.S. Senate, Nye committee, 2092–2102, 2220.

21. "The Truth about Our Munitions-Making," *Forum* 56 (July 1915): 34–36.

22. Brownlee, "Wilson and Financing the Modern State," 180. A bill to tax the profits of munitions manufacturers, introduced by Senator William Kenyon of Iowa on 13 December 1915, was tabled. See Agnes Anne Trotter, "The Development of the Merchants of Death Theory of American Intervention in World War I, 1914–1939" (Ph.D. diss., Duke University, 1966), 81.

23. The terms of the trust agreement excluded Europeans from entering the American military market and vice versa, unless the price maintained by the domestic supplier was not undercut. William S. Stevens, "The Powder Trust, 1872–1912," *Quarterly Journal of Economics* 26 (May 1912): 466, 478, and Stevens, "The Dissolution of the Powder Trust," *Quarterly Journal of Economics* 28 (November 1912): 202; Chandler and Salsbury, *du Pont*, chap. 10; Steven J. McNamee, "Du Pont–State Relations," *Social Problems* 34 (February 1987): 4.

24. Chandler and Salsbury, *du Pont*, 393.

25. U.S. House Select Committee on Expenditures in the War Department, 66th Cong., 1st sess., 1919, 2:1616 (cited hereafter as Graham committee); William Crozier, *Ordnance and the World War: A Contribution to the History of American Preparedness* (New York: Scribner, 1920), 247.

26. Chandler and Salsbury, *du Pont*, 325, 368; Trotter, "Merchants of Death Theory," 132–33; U.S. House, Graham committee, 4:4101–2; U.S. Senate, Nye committee, 1042.

27. Chandler and Salsbury, *du Pont*, 325, 359; U.S. Senate, Nye committee, 1021–25, 1030–31, 1038; Cuff, "Private Success, Public Problems," 173–74.

28. Kitchin's former law partner, Edward L. Travis, was jailed for attempting to peddle influence to obtain a Navy Department contract. Kitchin intervened with Navy Secretary Josephus Daniels in hopes of securing Travis's release. See E. David Cronon, ed., *The Cabinet Diaries of Josephus Daniels* (Lincoln: University of Nebraska Press, 1963), 327.

29. Brownlee, "Wilson and Financing the Modern State," 180; Arnett, *Kitchin and the Wilson War Policies,* 105, 110–11, 200; David A. Armstrong, *Bullets and Bureaucrats: The Machine Gun and the United States Army, 1861–1916* (Westport, Conn.: Greenwood Press, 1982), 198–99.

30. Brownlee, "Wilson and Financing the Modern State," 192.

31. Ibid., 191–92, 201–2; Link, *Wilson,* 5:63–64; Chandler and Salsbury, *du Pont,* 396; Robert M. La Follette, *War Profits Tax: Is It Disloyal to Advocate Just Taxation of War Profits and Surplus Incomes?* (Washington, D.C., 1917), 28.

32. Brownlee, "Wilson and Financing the Modern State," 191–92, 201–2; Chandler and Salsbury, *du Pont,* 396; Link, *Wilson,* 5:65; Arnett, *Kitchin and the Wilson War Policies,* 108–11; Holmes, *White Chief,* 306; letter of Otto H. Kahn, Kuhn, Loeb and Company, in *New Republic,* 29 December 1917, 252; Arthur Marwick, *War and Social Change in the Twentieth Century: A Comparative Study of Britain, France, Germany, Russia, and the United States* (London: Macmillan, 1974), 58.

33. The elementary level of planning is revealed in John A. Topping, "How Industrial Leaders Face the War: Iron and Steel," *World's Work* 34 (May 1917): 26.

34. Lewis K. Morse, "The Price Fixing of Copper," *Quarterly Journal of Economics* 33 (November 1918): 73; Abraham Berglund, "Price Fixing in the Iron and Steel Industries," *Quarterly Journal of Economics* 32 (August 1918): 599; Weiss, "What the War Has Done for Steel," 118–21; William Hard, "Prices and Patriotism," *New Republic,* 4 August 1917, 12; "The 'War Munitions Companies,'" *Nation,* 27 January 1916, 117; "Warhogs and War Millionaires [By a War Broker]," *Independent,* 9 July 1915, 80; George Soule, *Prosperity Decade* (London: Pilot, 1947), 77–78.

35. S. S. Huebner, "The American Security Market during the War," *Annals of the American Academy of Political and Social Science* 68 (November 1916): 96, 99; Robert B. Barsky and J. Bradford De Long, "Bull and Bear Markets in the Twentieth Century," *Journal of Economic History* 50 (June 1990): 273; Weiss, "What the War Has Done for Steel," 115; Cuff, "Private Success, Public Problems," 177; Harry Richard Kuniansky, *A Business History of Atlantic Steel Company, 1901–1968* (New York: Arno, 1976), 106, 131; Albert W. Atwood, "Americans Made Rich and Powerful by the War," *American Magazine* 81 (February 1916): 17; W. Elliot Brownlee Jr., "Economists and the Formation of the Modern Tax System in the United States: The World War I Crisis," in Mary O. Furner and Barry Supple, eds., *The State and Economic Knowledge: The American and British Experiences* (Cambridge: Cambridge University Press, 1989), 407; Thomas N. Guinsburg, *The Pursuit of Isolationism in the United States Senate from Versailles to Pearl Harbor* (New York: Garland, 1982), 25.

36. Warren I. Cohen, *The American Revisionists: The Lessons of Intervention in World War I* (Chicago: University of Chicago Press, 1967), 144; Daniel R. Beaver, *Newton D. Baker and the American War Effort* (Lincoln: University of Nebraska Press, 1966), 106.

37. Trotter, "Merchants of Death Theory," 50; Lewinsohn, *Profits of War,* 237; Kathleen Burk, *Britain, America, and the Sinews of War* (Boston: Allen and Unwin, 1985), 14–19; U.S. Senate, Nye committee, 7484. A more recent estimate places the Morgan earnings much higher. British spending may have reached $18 billion, leaving the Morgan Company with a gross commission of $180 million. See Kathleen Burk, "The Mobilization of Anglo-American Finance during World War I," in *Mobilization for Total War: The Canadian, American and British Experience, 1914–1918, 1939–1945,* ed. N. F. Dreisziger (Waterloo, Ont.: Wilfrid Laurier University Press, 1981), 27n, 34.

38. Trotter, "Merchants of Death Theory," 58; U.S. Senate, Nye committee, 5337, 7812; "Press Poll on Prohibiting the Export of Arms," *Literary Digest,* 6 February 1915, 225–26, 275–76, 280; "Negro Invasion Is Not Feared," unidentified clipping, 11 January 1915, Frear Papers. Cf. "Justifying Munitions-Exports," *Literary Digest,* 28 August 1915, 389–90.

39. "Buying Here for the Armies," 357–58; "The Business of Selling Death, by One of the Salesmen," *Independent,* 26 April 1915, 142; "Orders for War Supplies," *Independent,* 26 April 1915, 172–73.

40. Charles Noble Gregory, "The Sale of Munitions of War by Neutrals to Belligerents," *Annals of the American Academy of Political and Social Science* 60 (July 1915):

186; Syrett, "Business Press and American Neutrality," 218–19; "Justifying Munitions-Exports," 389–90; "The Case for the Munitions Trade," *Review of Reviews* 52 (September 1915): 350; Robert Lansing, *War Memoirs of Robert Lansing* (Indianapolis: Bobbs-Merrill, 1935), 54–58; Edmund von Mach, "An Argument against the Exportation of Arms," *Annals of the American Academy of Political and Social Science* 60 (July 1915): 192–94; Clifton J. Child, "German-American Attempts to Prevent the Exportation of Munitions, 1914–15," *Mississippi Valley Historical Review* 25 (December 1938): 354.

41. "To Organize a People's National Liberal League," unpublished speech MSS [1918?], Nelson Papers, box 12. Cf. U.S. Senate, Nye committee, 9051.

42. For recent evidence on this point, see Reinhard R. Doerries, *Imperial Challenge: Ambassador Count Bernstorff and German-American Relations, 1908–1917*, trans. Christa D. Shannon (Chapel Hill: University of North Carolina Press, 1989).

43. Cohen, *American Revisionists,* 12–14; Cooper, "Progressivism and Foreign Policy," 271–72; "How End the High Cost of Living," unpublished speech MSS [1918?], Nelson Papers, box 12; "Mammon," unpublished, undated speech MSS, Nelson Papers, box 12; U.S. House Committee on Military Affairs, *Universal Mobilization for War Purposes,* 68th Cong., 1st sess., 1924, Doc. 764, 189.

44. Quoted in Robert L. Morlan, *Political Prairie Fire: The Nonpartisan League, 1915–1922* (Minneapolis: University of Minnesota Press, 1955), 138.

45. Robert P. Wilkins, "The Non-Partisan League and Upper Midwest Isolationism," *Agricultural History* 39 (April 1965): 103.

46. Robert M. La Follette, "Dollars or Bullets?" *La Follette's Weekly Magazine,* 14 December 1912, 3; "War Profits," *La Follette's Weekly Magazine,* 1 August 1914, 3; Padraic Colum Kennedy, "La Follette's Foreign Policy: From Imperialism to Anti-Imperialism," *Wisconsin Magazine of History* 46 (summer 1965): 288–91; Barton J. Bernstein and Franklin A. Leib, "Progressive Republican Senators and American Imperialism, 1898–1916: A Reappraisal," *Mid-America* 50 (July 1968): 203; John M. Cooper Jr., *The Vanity of Power: American Isolationism and the First World War, 1914–1917* (Westport, Conn.: Greenwood Press, 1969), 183; *Congressional Record,* 21 April 1914, 7020; 29 May 1914, 9459.

47. Cooper, *Vanity of Power,* 42; U.S. Senate Committee on Privileges and Elections, 65th Cong., 3d sess., 1918, Rept. 614, 42.

48. Quoted in Norman L. Zucker, *George W. Norris, Gentle Knight of American Democracy* (Urbana: University of Illinois Press, 1966), 128. The tone of Norris's rhetoric belies the title of Zucker's book. Almost thirty years later, Norris reflected, "As I now look back over the years, I still feel that I was right." But, he added inconsistently, "no single group and no single industry could be charged with being wholly guilty or wholly innocent." See Norris, *Fighting Liberal: The Autobiography of George W. Norris* (New York: Macmillan, 1945), 192, 196–97.

49. Marvin Robert Bendinger, "Corruption in the World War," *American Mercury* 34 (February 1935): 225.

7. Supplying the Doughboys

1. "The Truth about Our Munitions-Making," *Forum* 56 (July 1916): 36.

2. Jeffrey G. Williamson and Peter H. Lindert, *American Inequality: A Macroeconomic History* (New York: Academic Press, 1980), 79, 110. Cf. Carole Shammas, "A New Look at Long-Term Trends in Wealth Inequality in the United States," *American Historical Review* 98 (April 1993): 412–31.

3. Michael S. Sherry, *The Rise of American Air Power: The Creation of Armageddon* (New Haven: Yale University Press, 1987), 3, 11.

4. Robert D. Cuff, "American Mobilization for War, 1917–1945: Political Culture vs. Bureaucratic Administration," in *Mobilization for Total War: The Canadian, American and*

British Experience, 1914–1918, 1939–1945, ed. N. F. Dreisziger (Waterloo, Ont.: Wilfrid Laurier University Press, 1981), 76–79; U.S. House Select Committee on Expenditures in the War Department, *Hearings,* 66th Cong., 1st sess., 1919, 1:992 (cited hereafter as Graham committee).

5. James L. Abrahamson, *America Arms for a New Century: The Making of a Great Military Power* (New York: Free Press, 1981), 158–171; Stephen Skowronek, *Building a New American State: The Expansion of National Administrative Capacities, 1877–1920* (Cambridge: Cambridge University Press, 1981), 219–22, 227, 237; Phyllis A. Zimmerman, *The Neck of the Bottle: George W. Goethals and the Reorganization of the U.S. Army Supply System, 1917–1918* (College Station: Texas A&M University Press, 1992), 24–25; John A. Topping, "How Industrial Leaders Face the War: Iron and Steel," *World's Work* 34 (May 1917): 26.

6. U.S. House, Graham committee, *Hearings,* 1:522, 990–91; John K. Ohl, "The Navy, the War Industries Board, and Industrial Mobilization for War, 1917–1918," *Military Affairs* 40 (February 1976): 17; Skowronek, *Building a New American State,* 236–37.

7. U.S. House, Graham committee, *Hearings,* 1:522, 990–91; William Crozier, *Ordnance and the World War: A Contribution to the History of American Preparedness* (New York: Scribner, 1920), 20.

8. Robert D. Cuff, *The War Industries Board: Business-Government Relations during World War I* (Baltimore: Johns Hopkins University Press, 1973), 64, 244; U.S. House, Graham committee, subcommittee no. 5, *Ordnance,* 1:452–59; Skowronek, *Building A New American State,* 238; Zimmerman, *Neck of the Bottle,* 25.

9. U.S. House, Graham committee, *Hearings,* 1:452–59; Clive Trebilcock, "War and the industrial mobilisation: 1899 and 1914," in *War and Economic Development: Essays in Memory of David Joslin,* ed. J. M. Winter (Cambridge: Cambridge University Press, 1975), 155; Porter Emerson Browne, *The Uncivil War* (New York: Doran, 1918), 117.

10. Cuff, *War Industries Board,* 11, 62, 244; Crozier, *Ordnance and the World War,* 42; U.S. House, Graham committee, *Hearings,* 1:522; Zimmerman, *Neck of the Bottle,* 87.

11. Cuff, *War Industries Board,* 16–17; U.S. Senate Special Committee Investigating the Munitions Industry, *Minutes of the Advisory Commission of the Council of National Defense,* 74th Cong., 2d sess., 8 January, 23 March 1917, Pt. 8, 1936, 10, 33 (cited hereafter as Nye committee).

12. Abrahamson, *America Arms,* 175; Cuff, *War Industries Board,* 11, 62, 72; Grosvenor B. Clarkson, *Industrial America in the World War: The Strategy behind the Line* (Boston: Houghton Mifflin, 1923), 134; U.S. House, Graham committee, *Ordnance,* 3:4026.

13. U.S. Senate, Nye committee, *Minutes,* 24 March 1917, pt. no. 7, 1936, pp. 46–50.

14. "The 'War Orders' and American Industry," *Review of Reviews* 52 (August 1915): 223–24; Clarkson, *Industrial America,* 452–53; George Soule, *Prosperity Decade* (London: Pilot, 1947), 65.

15. "Warhogs and War Millionaires [By a War Broker]," *Independent,* 9 July 1915," 80; U.S. House, Graham committee, *Majority Report,* 1919, H. Rept. 1400, 66; Marquis James, "The Profiteer Hunt: The Quest Begins and the Trail Grows Hot," *American Legion Weekly,* 23 March 1923, 8; U.S. House, Graham committee, *Hearings,* 1:421.

16. U.S. House, Graham committee, *Leather Goods,* 1921, H. Rept. 1307, 3; U.S. House, Graham committee, *Ordnance,* 3:3570, 3641–42.

17. Crozier, *Ordnance and the World War,* 18–19, 27.

18. U.S. House, Graham committee, *Hearings,* 1:480.

19. U.S. Senate, Nye committee, *Hearings,* 6274; U.S. House, Graham committee, *Hearings,* 1:421.

20. "On Taking the Profit Out of Munitions-Making," *Nation,* 27 January 1916, 93.

21. U.S. House, Graham committee, *Hearings,* 1:335; Graham Committee, subcommittee 2, *Camps,* 2:2797.

22. U.S. House, Graham committee, *Hearings,* 1:404; Graham committee, *Camps,* 1:993–1002, 1082, 1095.

23. U.S. House, Graham committee, *Camps,* 2:2527, 2533–35, 2807–9.

24. U.S. Senate, Nye committee, *Minutes of the General Munitions Board,* 16 May 1917, pt. 6, 1936, p. 91; U.S. House, Graham committee, *Camps,* 2:2525, 2389–91, 2699, 2712–20.

25. U.S. House, Graham committee, *Camps,* 2:2440.

26. Ibid., 1:904, 1077–78, 1145, 1152, 2:2144–45, 2355, 2426.

27. Ibid., 1:493, 2: 2357, 2698–98.

28. Ibid., 1:1125, 1294–95; 2:2300–2301, 2527; testimony of Bernard M. Baruch, U.S. House Committee on Military Affairs, *Universal Mobilization for War Purposes,* 68th Cong., 1st sess., 1924, H. Doc. 764, 129.

29. Careful survey was imperative in order to obtain a good supply of water and also to detect any drainage problems. In both the Civil War and the Spanish-American War more soldiers died from disease than from wounds, so good sanitation in the training camps could not be taken lightly. U.S. House, Graham committee, *Camps,* 2:2357, 2697–99.

30. U.S. House, Graham committee, *Majority Report,* 66th Cong., 2d sess., 1921, 3:379; Graham committee, *Camps,* 1921, H. Rept. 816, 428.

31. U.S. House, Graham committee, *Camps,* 1:115–16, 906–9, 918, 1047–48, 1127; 2:2527.

32. Ibid., 2:2558.

33. Ibid., 2:1822, 1930, 1945, 1966–68, 2558–59, 2779–83, 2795, 2904.

34. Ibid., 2:1930, 1945, 1966–68.

35. Parkhurst Whitney, "Next Time—*Everybody's* War," *American Legion Weekly,* 29 September 1922, 7–8.

36. U.S. House, Graham committee, *Camps,* 1:1517–18, 2:1660–61.

37. Ibid., 1:1260–67, 1332, 1383, 1399; 2:1948, 1982–83, 1992–98, 2559.

38. Ibid., 1:1260–67, 1292–93, 1338, 1517–18, 2:2070, 2365, 2559, 2564–65.

39. Ibid., 2:2907.

40. Ibid., 1:1324.

41. Ibid., 1:1228, 1438–45, 2:1605–12, 1651–57, 2116–24, 2563, 2796.

42. Ibid., 1:2608–9.

43. Ibid., 2:2063–80, 2620–50, 2943.

44. Ibid., 2:2553.

45. The investigation was conducted by the former Republican presidential candidate and associate justice of the Supreme Court, Charles Evans Hughes. The findings are recorded in U.S. House, Graham committee, subcommittee 1, *Aviation,* 3:3862–68.

46. U.S. Senate Committee on Military Affairs, *Aircraft Production,* 65th Cong., 2d sess., 1918, S. Rept. 380 (cited hereafter as Thomas committee).

47. U.S. House, Graham Committee, *Aviation,* 1920.

48. U.S. Senate, Nye committee, *Hearings,* 1935–36.

49. U.S. House, Graham committee, *Aviation,* 1:278, 281, 509.

50. Ibid., 1:186–87, 190–91, 367–69, 556b, 3:2973; Marquis James, "Who Got the Money?" *American Legion Weekly,* 8 September 1922, 5, and James, "Who Got the Money? II: The Airplane Production Mess," *American Legion Weekly,* 15 September 1922, 5–8.

51. Bolling was a civilian who was commissioned a major and later promoted to colonel. U.S. House, Graham committee, *Aviation,* 1:165–66, 3:2973, 2987.

52. Ibid., 1:66, 68, 242, 246, 270.

53. Minutes of the joint meeting of the Council of National Defense and the Advisory Commission, 24 March 1917, in U.S. Senate, Nye committee, *Minutes of the Council of National Defense,* pt. 7, 1936, p. 55.

54. U.S. House, Graham committee, *Aviation,* 1:377–79; Jordan A. Schwarz, *The Speculator: Bernard M. Baruch in Washington, 1917–1965* (Chapel Hill: University of North Carolina Press, 1981), 71.

55. U.S. House, Graham committee, *Aviation,* 1:50–51, 58–59; 3:3862–68; James, "Who Got the Money? The Aircraft Mess," 7.

56. U.S. House, Graham committee, *Aviation,* 3:2655, 3881–98.

57. U.S. Senate, Thomas committee, pt. 2, p. 2.

58. U.S. House, Graham committee, *Aviation,* 2:1371–89; U.S. Senate, Thomas committee, pt. 2, p. 2.

59. U.S. House, Graham committee, *Aviation,* 1:831, 2:1893–94, 1899–1904, 3:3519–20.

60. Ibid., 2:1371–89, 2738, 3:3537.

61. U.S. House, Graham committee, *Hearings,* 1:87–113, 831, 2:969.

62. U.S. House, Graham committee, *Aviation,* 1:87–113.

63. Ibid., 1:87–113, 2:1469–84, 1532.

64. Ibid., 1:699–700; 2:1503–4, 1650.

65. U.S. Senate, Thomas committee, 1:194, 2:1157–63.

66. Ibid., 2:1182.

67. U.S. Senate, Thomas committee, 1:513–19; Marquis James, "The Profiteer Hunt, VI. The Case of the Packard Motor Car Company," *American Legion Weekly,* 27 April 1923, 5. Slightly different figures for the cost of Liberty motors are given in James, "Who Got the Money? II. The Airplane Production Mess," 8.

68. U.S. Senate, Thomas committee, 1:2–5, 2:1157–63.

69. U.S. House, Graham committee, *Aviation,* 1:182, 197–98, 203–5, 3:3835–37, 3617–17, 3803.

70. Ibid., 3:3803; Edwin C. Parsons, *I Flew with the Lafayette Escadrille* (1937; New York: Arno, 1972), 139.

71. U.S. House, Graham committee, *Aviation,* 3:3461–76.

72. Clarkson, *Industrial America,* 347; Richard Lewinsohn, *The Profits of War through the Ages* (London: Routledge, 1936), 165; E. David Cronon, ed., *The Cabinet Diaries of Josephus Daniels* (Lincoln: University of Nebraska Press, 1963), 115n.

73. U.S. House, Graham committee, *Ordnance,* 1:353; Clarkson, *Industrial America,* 347; Cuff, *War Industries Board,* 58–60; Hugh Rockoff, *Drastic Measures: A History of Wage and Price Controls in the United States* (Cambridge: Cambridge University Press, 1984), 44; David M. Kennedy, "Rallying Americans for War, 1917–1918," in *The Home Front and War in the Twentieth Century: The American Experience in Historical Perspective,* ed. James Titus (Washington, 1984), 50–51; Jacob Vander Meulen, *The Politics of Aircraft: Building an American Military Industry* (Lawrence: University Press of Kansas, 1991), 24, 63–64.

74. U.S. Senate, Nye committee, *Minutes of the War Industries Board,* pt. 4, 1936, 12–13; Nye committee, *Minutes of the Price Fixing Committee of the War Industries Board,* 1936, pt. 5, p. 399; U.S. House, Graham committee, *Ordnance,* 1:321–29, 335, 352–53, 460; Clarkson, *Industrial America,* 347.

75. Melvin I. Urofsky, *Big Steel and the Wilson Administration: A Study in Business-Government Relations* (Columbus: Ohio State University Press, 1969), 116; Cuff, *War Industries Board,* 125. Daniels also had strong words for the Bethlehem Steel Company. See "American Shell Makers under Fire," *Literary Digest,* 3 February 1917, 236.

76. "War Stock Speculation," *Independent,* 9 August 1915, 176; "Efforts to Halt Arms-Exports," *Literary Digest,* 26 June 1915, 1520; Lewinsohn, *Profits of War,* 162–66, 272–73; Agnes Anne Trotter, "The Development of the Merchants of Death Theory of American Intervention in World War I, 1914–1939" (Ph.D. diss., Duke University, 1966), 47–48, 121; Clarkson, *Industrial America,* 319.

77. Robert H. Montgomery, "Summarized Evidence," 10 March 1932, Records of the War Policies Commission, box 175, RG 107, National Archives; Cronon, *Josephus Daniels,* 176; Clarkson, *Industrial America,* 319; Urofsky, *Big Steel,* 207–23; U.S. Federal Trade Commission, *War-Time Profits and Costs of the Steel Industry* (Washington: Government Printing Office, 1925), 29.

78. U.S. Senate, Nye committee, *Minutes of the Price Fixing Committee of the War Industries Board,* 20 March 1918, pt. 5, 1936, 48, 62.

79. U.S. House, Graham committee, *Hearings,* 1:414, 418, 475.

80. U.S. House, Graham committee, *Camps,* 2:2485–94; Cuff, *War Industries Board,* 60, 230–31; Trotter, "Merchants of Death Theory," 101, 135; U.S. Federal Trade Commission, *War-Time Profits,* 38.

81. "Patriotism and Profits: President Wilson's Message to Business," *Independent,* 21 July 1917, 112.

82. U.S. House, Graham committee, *Hearings,* 1:1890–96, 2017, and passim; Anthony Gronowicz, ed., *Oswald Garrison Villard: The Dilemma of the Absolute Pacifist in Two World Wars* (New York: Garland, 1983), 152–53.

83. U.S. House, Graham committee, *Hearings,* 1:1894–96; U.S. Senate, Nye committee, *Hearings,* 5804, 5813. There were also very costly expenses incurred in renting wharfage for shipments and repairs. See Marquis James, "The Profiteer Hunt, VIII. The President Hears about It," *American Legion Weekly,* 25 May 1923, 17.

84. U.S. House, Graham committee, *Hearings,* 1:1937–40, 2021.

85. U.S. Senate, Thomas committee, 2:876, 880; James, "The Profiteer Hunt: VIII. The President Hears about It," 28.

86. Cronon, *Josephus Daniels,* 291; Marquis James, "The Profiteer Hunt. The Cost-Plus Contract—A Product of Haste," *American Legion Weekly,* 30 March 1923, 8, 29.

87. U.S. House, Graham committee, *Ordnance,* 1:1253–56, 2:1505, 1514; U.S. Senate, Thomas committee, 1:514–15; U.S. Senate, Nye committee, *Hearings,* 4548, 5757–60, 6617. Grace paid income taxes in 1917–18 in the amount of $1,810,000, or about 66 percent of his income.

88. James, "The Profiteer Hunt, VIII. The President Hears about It," 28; Marquis James, "The Profiteer Hunt, V. Sought from Two Contractors, $3,500,000," *American Legion Weekly,* 20 April 1923, 28; U.S. Senate, Thomas committee, 1:500.

89. U.S. House, Graham committee, *Ordnance,* 1:794, 2:1687, 1729, 1970. A fire destroyed many of the records of the Thompson-Starrett Company. Although there allegations of arson, this charge has never been substantiated.

90. U.S. House, Graham committee, *Aviation,* 3:2759, *Ordnance,* 3:3210–31, 3256, 3576.

91. U.S. Senate, Nye committee, *Hearings,* 4284; J. Franklin Crowell, *Government War Contracts* (New York: Oxford University Press, 1920), 43.

92. U.S. Senate, Nye committee, *Hearings,* 6357, 6271.

93. Crowell, *Government War Contracts,* 43; U.S. Senate, Nye committee, *Hearings,* 4270; U.S. Senate, Nye committee, *Minutes of the War Industries Board,* 19 March 1918, pt. 4, 1936, p. 231. Cf. pp. 24, 57, 71.

94. U.S. House Committee on Military Affairs, *Universal Mobilization for War Purposes,* 47.

95. Ibid., 50, 55; U.S. House, Graham committee, *Hearings,* 1:1854.

96. Crowell, *Government War Contracts,* 340–41; U.S. Senate, Nye committee, *Hearings,* 4259, 4428; "Hoover on Profiteering and Taxes," *World's Work* 36 (August 1918): 453–54; "Profiteering: Federal Trade Commission Report," *Nation,* 13 July 1918, 345; Urofsky, *Big Steel,* 223–34.

97. U.S. House Committee on Military Affairs, *Universal Mobilization for War Purposes,* 56; U.S. House, Graham committee, *Camps,* 2:2693–94; Stuart D. Brandes, *American Welfare Capitalism* (Chicago: University of Chicago Press, 1976), 26–27.

98. Alexander D. Noyes, *The War Period of American Finance* (New York: Putnam, 1926), 195.

99. "The Money Power," speech manuscript [1920?], John M. Nelson Papers, State Historical Society of Wisconsin, Madison, box 12.

100. U.S. House Committee on Military Affairs, *Universal Mobilization for War Service,* 40, 78, 135, 176–77.

101. Ibid., 173, 138. See also U.S. House Committee on Military Affairs, 74th Cong., 1st sess., *Taking the Profits out of War,* 1935, 582–84.

102. Williamson and Lindert, *American Inequality,* 81, 109.

103. Cuff, *War Industries Board*, 176, esp. note 77.

104. Ibid., 95n; Robert D. Cuff, "The Dollar-a-Year Men of the Great War," *Princeton University Library Chronicle* 30 (autumn 1968): 15.

105. James, "The Profiteer Hunt: The Other Side," 24.

106. Marquis James, "The Profiteer Hunt, IX. The Other Side of the Shield," *American Legion Weekly*, 8 June 1923, 26.

107. Ibid., 7, 22.

108. Ibid., 24–25; Marquis James, "The Profiteer Hunt, III. The Biggest Pay-off in the World's History," *American Legion Weekly*, 6 April 1923, 4.

109. Alfred D. Chandler Jr. and Stephen Salsbury, *Pierre S. du Pont and the Making of the American Corporation* (New York: Harper, 1971), 368–370, 402–3; U.S. Senate, Nye committee, *Hearings*, 163, 1021–26, 1137.

110. Construction of the Nitro works also prompted controversy. Plagued by chronic labor shortages, Thompson-Starrett was forced to employ workers who suffered from syphilis, amputation, mental retardation, and even delirium tremens. Wages for these workers were princely by prewar standards. Crozier, *Ordnance and the World War*, 249–69; U.S. House, Graham committee, *Ordnance*, 1:794, 4:4310, 4135–52; Chandler and Salsbury, *du Pont*, 410–27.

111. Chandler and Salsbury, *du Pont*, 426.

8. Grave Objections

1. Quoted in Lawrence Sullivan, "Work of the War Policies Commission," *Current History* 35 (November 1931): 240.

2. Associationalism is most commonly identified with the policies of Herbert Hoover during his service as secretary of commerce. Under this plan, the state would encourage business to serve the public interest voluntarily and through self-regulation. The state would attempt to organize voluntary cooperation rather than to use coercion. The instrument of state action would become the trade associations it would foster. The defense industry, with its technical sophistication and its necessarily close relation to the needs of the public, was an excellent location for associationalism to thrive. On associationalism in the defense industry, see Jacob Vander Meulen, *The Politics of Aircraft: Building an American Military Industry* (Lawrence: University Press of Kansas, 1991), 63–64, and Robert D. Cuff, "An Organizational Perspective on the Military-Industrial Complex," *Business History Review* 52 (summer 1978): 262–65. On Hoover, see Ellis Hawley, "Herbert Hoover, the Commerce Secretariat, and the Vision of an 'Associative State,' 1921–1928," *Journal of American History* 61 (June 1974): 116–40.

3. Marquis James, "Who Got the Money?" *American Legion Weekly*, 8 September 1922, 6.

4. U.S. House Select Committee on Expenditures in the War Department, *Hearings*, 66th Cong., 1st sess., 1919, 1:99–129, 266–71, 298–312, 360–75 (cited hereafter as Graham committee). The government supply was crucial because privately held copper inventories were also high: between 750 million and 1 billion pounds. Ibid., 1:388.

5. U.S. House, Graham committee, subcommittee 5, *Ordnance*, 1:153–67 (cited hereafter as Graham committee, *Ordnance*).

6. U.S. House, Graham committee, 1921, H. Rept. 1307, 3.

7. Ibid., 9.

8. Ibid., 9–10.

9. Ibid., 16, 19; Marquis James, "Who Got the Money? IV. Selling Out at Bargain Prices," *American Legion Weekly*, 29 September 1922, 6.

10. The search for extortionate prices could also prove profitable. When the Justice Department began investigating fraud in war contracts, one potential source offered to present the results of his personal investigation in return for a payment of $10,000. There is no

record that the informant, H. L. Scaife, received any compensation. See War Transactions Section Advisory Council, "Memo to A.G.," 15 November 1923, Records of the War Transactions Section, Department of Justice, RG 60, National Archives (cited hereafter as WTS.)

11. W. O. Watts, "Memorandum for the Attorney General," 1 April 1922, WTS, box 21; Marquis James, "The Profiteer Hunt, VIII. The President Hears about It," *American Legion Weekly*, 25 May 1923, 7; James, "Who Got the Money? IV. Selling Out at Bargain Prices," 28–29; "The War-Contract Situation," *American Legion Weekly*, 28 March 1924, 8; William Pencak, *For God and Country: The American Legion, 1919–1941* (Boston: Northeastern University Press, 1989), chap. 5; Wayne S. Cole, *Senator Gerald P. Nye and American Foreign Relations* (Minneapolis: University of Minnesota Press, 1962), 122. The Veterans of Foreign Wars (founded 1899) held views similar to the American Legion, but the VFW was less terse in expressing them. See U.S. House, *War Policies Commission*, 71st Cong., 2d sess., 1931, 7.

12. Marion C. Early to Attorney General Harry Daugherty, 7 August 1923, box 17; John A. Parker to Daugherty, 26 January 1924, box 28, and Frank J. Hogan to Daugherty, 24 January 1924, box 28, WTS; U.S. House, *Universal Mobilization for Military Service*, 178. Cf. "The War-Contract Situation," *American Legion Weekly*, 28 March 1924, 8. On the weakness of associationalism, see Robert D. Cuff, "The Dilemmas of Voluntarism: Hoover and the Pork-Packing Agreement of 1917–1919," *Agricultural History* 53 (October 1979): 731, 747.

13. U.S. House, Graham committee, subcommittee 4, *Quartermaster Corps*, 11–12, subcommittee 2, *Camps*, 2:2931–33; "Memorandum of Facts, *U.S. vs. J. L. Philips et al.*," WTS; Marquis James, "Who Got the Money? IV. Selling Out at Bargain Prices," *American Legion Weekly*, 29 September 1922, 5.

14. Marquis James, "Who Got the Money? V. Meat, Sugar, and Mosquito Bars," *American Legion Weekly*, 6 October 1922, 5–6.

15. U.S. House, Graham committee, subcommittee 1, *Aviation*, 1:458.

16. In the 1920s, the War Department continued the campaign against Du Pont described in the previous chapter. To assure that there had been no profiteering in the construction of the Old Hickory plant, the War Department conducted a lengthy and thorough audit of the Du Pont contract. Completed in 1924 at a cost of $249,042.33, the contract contained no fraud. Major General C. C. Williams, "Resume of the construction of the Old Hickory powder plant," 11 January 1924, WTS, box 49. Cf. Marquis James, "Who Got the Money? IV. Selling Out at Bargain Prices," 3–4.

17. U.S. House, Graham committee, *Quartermaster Corps*, 915, 1072; U.S. House, Graham committee, 1921, H. Rept. 1408, *Return of Salmon to Canners*, passim; Marquis James, "Who Got the Money?" *American Legion Weekly*, 8 September 1922, 4.

18. U.S. House, Graham committee, *Ordnance*, 1:1172, 1191. Cf. "The Unprofitable Side of Our Great Munitions Contracts," *Literary Digest*, 10 March 1917, 679–80.

19. U.S. House, Graham committee, *Hearings*, 1:1981; Graham committee, *Ordnance*, 1:1211–13; Marquis James, "The Profiteer Hunt, IV. The Air Service Starts a House-Cleaning," *American Legion Weekly*, 13 April 1923, 5–6, 24–25.

20. James, "The Profiteer Hunt, IV. The Air Service Starts a House-Cleaning," 5–6; Marquis James, "Who Got the Money? III. Camps, Powder, Shells, and Guns," *American Legion Weekly*, 22 September 1922, 6.

21. U.S. House, Graham committee, *Ordnance*, 2:1440; James, "The Profiteer Hunt: IV. The Air Service," 24–25.

22. "Memorandum on Behalf of Packard Motor Car Company re Liberty Motors" [1922?], WTS, box 29; Marquis James, "The Profiteer Hunt, VI. The Case of the Packard Motor Car Company," *American Legion Weekly*, 27 April 1923, 27–28.

23. Marquis James, "The Profiteer Hunt, V. Sought from Two Contractors, $3,500,000," *American Legion Weekly*, 20 April 1923, 5–6, 28; James, "The Profiteer Hunt, VII. The Government Looks into the Affairs of Five Prosperous Companies," *American Legion Weekly*, 11 May 1923, 9–10; James, "The Profiteer Hunt, VIII. The President Hears about It," *American Legion Weekly*, 5 May 1923, 8, 17.

24. U.S. House, Graham committee, 1921, H. Rept. 1400.

25. *Inaugural Addresses of the Presidents of the United States*, 87th Cong., 1st sess., 1961, H. Doc. 218, 209–11.

26. U.S. House, *Universal Mobilization for War Purposes*, 180–82; "Drafting the Dollar," *American Legion Weekly*, 13 July 1923, 8.

27. Robert K. Murray, *The Harding Era: Warren G. Harding and His Administration* (Minneapolis: University of Minnesota Press, 1969), 298; "Bad News for War Grafters," *Literary Digest*, 27 May 1922, 14; Marquis James, "The Profiteer Hunt: The Quest Begins and the Trail Grows Hot," 6; Daugherty to James A. Frear, 12 June and 12 July 1922, Frear Papers, State Historical Society of Wisconsin, Madison, box 1.

28. Report of the War Transactions Section, 1 December 1923, WTS, box 21; James A. Frear, weekly letter to constituents, 2 January, 2 February, 12 May, 25 May 1922, clippings in Frear Papers, box 5; *Congressional Record*, 1 April 1930, 6314, 6316; "Bad News for War Grafters," 14–15; Marquis James, "Who Got the Money?" *American Legion Weekly*, 8 September 1922, 28; Murray, *Harding Era*, 297–98; "Congratulations, Mr. Daugherty," *American Legion Weekly*, 29 June 1923, 8; "A Clean Bill for 'War Grafters,'" *Literary Digest*, 3 July 1926, 11; Pencak, *For God and Country*, 173–74.

29. Advisory Council to Attorney General, 2 October 1922, 25 October 1923, Report of the War Transactions Section, 1 December 1923, box 21, and F. D. Enfield, "Report from the Committee on Transactions," 11 January 1924, box 24, WTS.

30. In the other cases, no action was specified. See Jerome Mitchell, director, War Transactions Section, to Major General John A. Hull, 23 July 1925, box 49, and "Classification of Cases as of August 15, 1925," box 31, WTS.

31. In a strange reverse of fortune, F. G. Palmbeck, the general agent of the United Brotherhood of Carpenters at Camp Sherman, was indicted for witness tampering because he told union members not to cooperate with the Justice Department. Press release, 3 September 1922, memoranda of Charles Kerr, 20 November 1922, T. M. Bigger, 28 November 1922, and "Subject: Contract for Construction of Camp Sherman at Chillicothe, Ohio," 21 November 1922, WTS, box 21.

32. "Clean Bill for 'War Grafters,'" 11.

33. Charles L. Mee Jr., *The Ohio Gang: The World of Warren G. Harding* (New York: Evans, 1981), 136–37; Oswald Garrison Villard, "Mr. Weeks and the War Frauds," *Nation*, 8 November 1922, quoted in *Oswald Garrison Villard: The Dilemmas of the Absolute Pacifist in Two World Wars*, ed. Anthony Gronowicz (New York: Garland, 1983), 207–8.

34. Mee, *Ohio Gang*, 156–57.

35. James E. Darst, "That the Country May Know," *American Legion Weekly*, 9 April 1920, passim; James C. Olson, *Historical Dictionary of the 1920s: From World War I to the New Deal, 1919–1935* (Westport, Conn.: Greenwood Press, 1988), 4; Marquis James, "The Profiteer Hunt: The Quest Begins and the Trail Grows Hot," *American Legion Weekly*, 23 March 1923, 5, 30.

36. "A New Thrust at Compensation," *American Legion Weekly*, 27 January 1922, 13; "In Reply to the Chamber: The Legion Demonstrates the Fallacy of Anti-Compensation Arguments," *American Legion Weekly*, 3 February 1922, 13–14; "The Figures Are Mr. Mellon's," *American Legion Weekly*, 30 November 1923, 10; Marquis James, "Again the Mellon Touch," *American Legion Weekly*, 22 February 1924, 7; James, "Zero Hour in the Fight for Compensation," *American Legion Weekly*, 28 March 1924, 7; "What the Legion Fights," *American Legion Weekly*, 4 April 1924, 4. Cf. James A. Frear, "Weekly Letter," 26 February 1928, Frear Papers, box 5. In 1917–18 the U.S. Chamber of Commerce had opposed excess profits. See U.S. House, *War Policies Commission*, 287.

37. In Thomas Boyd's early and typically angry war novel, there is no mention of war profiteering. The main theme is a vivid description of the horrors of trench warfare. Thomas Boyd, *Through the Wheat* (1923; reprint, Carbondale: Southern Illinois University Press, 1978).

38. Laurence Stallings, *Plumes* (New York: Harcourt, 1924), 3, 10, 62, 342; Alfred S. Shivers, *The Life of Maxwell Anderson* (New York: Stein, 1983), 87. Stallings later mellowed

332 I NOTES TO PAGES 190-193

and softened his views considerably. See Stallings, *The Doughboys: The Story of the AEF, 1917–1918* (New York: Harper, 1963), 1–7.

39. Maxwell Anderson and Laurence Stallings, *Three American Plays* (New York: Harcourt, 1926), 75.

40. Alfred Harding, "What Price Censorship?" *American Legion Weekly,* 28 November 1924, 16; Shivers, *Maxwell Anderson,* 100–101.

41. U.S. House Committee on Military Affairs, 68th Cong., 1st sess., *Universal Mobilization for War Purposes,* 1924; Vander Meulen, *Politics of Aircraft,* 79.

42. During the war, President Wilson forbad Hoover to commandeer supplies unless Bernard Baruch gave his consent. Hoover agreed to cooperate, but may have chafed at the restriction. See Wilson to Hoover, 3 September 1918, and Hoover to Wilson, 6 September 1918, Hoover Papers, Hoover Institution Archives, Stanford University, Stanford, Calif. See also U.S. House, *Universal Mobilization for War Purposes,* 193, 196.

43. Ibid., 237; *Congressional Record,* 1 April 1930, 6314, 6321; U.S. House, *War Policies Commission,* 89, 112–13, 353.

44. Marquis James, "The Story of a Five-Billion-Dollar Box," *American Legion Weekly,* 21 September 1923, 28.

45. Thomas Amory Lee to Hoover, 29 May 1910, Hoover Papers, box 10; U.S. House, *Universal Mobilization for War Purposes,* 40, 67–69, 73, 75, 184, 217; "Should Wealth Be Conscripted?" *Nation,* 24 April 1935, 469.

46. U.S. House, *Universal Mobilization for War Purposes,* 199–211.

47. Peter A. Soderbergh, "*Aux Armes!:* The Rise of the Hollywood War Film, 1916–1930," *South Atlantic Quarterly* 65 (autumn 1966): 514; *Congressional Record,* 1 April 1930, 6321; U.S. House, *Universal Mobilization for War Purposes,* 76–81.

48. Soderbergh, "*Aux Armes!,*" 519.

49. Larry Wayne Ward, *The U.S. Government Goes to War: The U.S. Government Film Effort during World War I* (Ann Arbor: UMI Research Press, 1985), 53–58. On the early war themes see Charles V. Genthe, *American War Narratives, 1917–1918* (New York: Lewis, 1969), 98.

50. Michael T. Isenberg, *War on Film: The American Cinema and World War I, 1914–1941* (Rutherford, N.J.: Fairleigh Dickinson University Press, 1981), 109.

51. The best of the antiwar American novels was perhaps Ernest Hemingway's *A Farewell to Arms* (1929). It was joined in 1929 by an outpouring of angry English and European war literature. Prominent examples included Robert Graves's *Goodbye to All That,* a moving memoir of trench warfare, Edmund Blunden's *Undertones of War,* Robert Sherriff's evocative play *Journey's End* (which had extraordinary success in New York and Chicago as well as in London), and Erich Maria Remarque's incomparable *All Quiet on the Western Front.* The following year saw the publication of Siegfried Sassoon's *Memoirs of an Infantry Officer.* All were combat veterans of the Great War. Charles Edmonds, a lieutenant who did not share his comrades' disenchantment, wrote an account of his war experience in 1919, but it could not be published until a decade later. He acidly recalled "when the reaction against the war had reached hysterical proportions." See Arthur Marwick, *War and Social Change in the Twentieth Century: A Comparative Study of Britain, France, Germany, Russia, and the United States* (London: Macmillan, 1974), 83–84; Burns Mantle, ed., *The Best Plays of 1928–29* (New York: Dodd, Mead, 1929), 10, 12, 348; Charles Edmund Carrington [Charles Edmonds], *A Subaltern's War* (1930; New York: Arno, 1972), preface, 8. See also William Manchester, *The Last Lion, Winston Spencer Churchill: Alone, 1932–1940* (Boston: Little, Brown, 1988), 47.

52. Upton Sinclair, *Jimmie Higgins: A Story* (Lexington: University Press of Kentucky, 1970), 93. Cf. page 46 for the war fought in the defense of American millionaires. John Dos Passos's novel *1919,* an antiwar *cri de coeur,* includes antiprofiteering as a minor theme. See Dos Passos, *1919* (New York: Harcourt, 1932), 230, 337–40. By 1962 Dos Passos had moderated his judgments while keeping much of his bitterness. See Dos Passos, *Mr. Wilson's War* (Garden City, N.Y.: Doubleday, 1962), passim.

53. William Edward March Campbell [William March, pseud.], *Company K* (New York: Smith, 1933), 30.

54. Dalton Trumbo, *Johnny Got His Gun* (New York: Lippincott, 1939), 148.

55. "Little Orphan Annie," *Cartoonist Profiles* 1 (November 1970): 71; Thomas Craven, ed., *Cartoon Cavalcade* (Chicago: Consolidated, 1945), 230.

56. It is instructive in considering the mood of the 1920s and early 1930s to compare Gray's popularity with the plight of a predecessor, Art Young, the cartoonist of the socialist journal *The Masses*. During the war, Young drew a cartoon called "Having Their Fling," which depicted an editor, a politician, a minister, and a capitalist in a mad war dance in which they cavort among coins and cash. Young was prosecuted for conspiring to interfere with enlistment. After two divided juries, the Justice Department dropped the case, but Young was nearly ruined financially and socially. A decade later, such skepticism of the war's purpose was de rigueur. William Murrell, *A History of American Graphic Humor, 1865–1938* (New York: Macmillan, 1938), 2:70, 194–97.

57. U.S. House, *Universal Mobilization for War Purposes,* 195.

58. "Profits in Blood," *Nation,* 29 June 1932, 713–14; "Armament Makers," 33.

59. Two American firms, the Colt's Patent Fire Arms Company and the Winchester Repeating Arms Company, did, however, participate in an international cartel of arms firms that sought to prevent the implementation of the League of Nations restrictions. An International Congress of Gun Makers met in Paris on 16–17 February 1925 for this purpose. See U.S. Senate Special Committee Investigating the Munitions Industry, *Hearings,* 73d Cong., 2d sess., 1934, 2131–32, 2135, 2137 (cited hereafter as Nye committee).

60. In his role as secretary of commerce, Herbert Hoover had advised the arms makers to form an industrial association to advance their interests, thus offering another example of the government encouraging cooperation among arms makers. See "Profits in Blood," 713; John Gunther, "Slaughter for Sale," *Harper's,* May 1934, 658; Murray Stedman, *Exporting Arms: The Federal Arms Exports Administration, 1935–1945* (Morningside Heights, N.Y.: King's Crown, 1947), 8–11; Paul Hutchinson, "The Arms Inquiry," *Christian Century,* 8 May 1935, 652; Fenner Brockway and Frederic Mullaly, *Death Pays a Dividend* (London: Gollancz, 1945), 37.

61. League of Nations, *Report of the Temporary Mixed Commission for the Reduction of Armaments* (Geneva, 1924), 4–5, 7, 21. In 1934 Herbert Hoover issued a public statement declaring that as secretary of commerce in 1925 he had called a conference that resulted in a modification of the League of Nations Arms Traffic Convention that would make it suitable to American gun manufacturers. See *Washington Times,* 6 December 1934.

62. Benjamin F. Cooling, Peter Karsten, Daniel Schirmer, and Paul Koistinen date the appearance of the military-industrial complex in the late nineteenth and early twentieth century. The more persuasive case is made by those historians who do not find it in existence before World War II, such as James Abrahamson, Jacob Vander Meulen, Robert D. Cuff, Dean Allard, Stephen Skowronek, and Terrence Gough. See Benjamin F. Cooling, *Gray Steel and Blue Water Navy: The Formative Years of America's Military-Industrial Complex, 1881–1917* (Hamden, Conn.: Archon, 1979), 55, 81, 109; Daniel B. Schirmer, *Republic or Empire: America Resistance to the Philippine War* (Cambridge: Schenkman, 1972), 133; Paul A. C. Koistinen, *The Military-Industrial Complex: A Historical Perspective* (New York: Praeger, 1980); James L. Abrahamson, *America Arms for a New Century: The Making of a Great Military Power* (New York: Free Press, 1981), 128–29, 137–39, 143–44; Cuff, "An Organizational Perspective on the Military-Industrial Complex," 265; Stephen Skowronek, *Building a New American State: The Expansion of National Administrative Capacities, 1877–1920* (Cambridge: Cambridge University Press, 1981), 236; Terrence J. Gough, "Soldiers, Businessmen, and U.S. Industrial Mobilisation Planning between the World Wars," *War and Society* 9 (May 1991): 69.

63. U.S. House, Graham committee, *Aviation,* 1:658–59, 3:2924–25, 3480–81, 3630–31, especially the testimony of Col. Edgar T. Gorrell.

64. Vander Meulen, *Politics of Aircraft,* 43, 78, 197.

65. Koistinen, *Military-Industrial Complex*, 11; Marquis James, "The Story of a Five-Billion-Dollar Box," *American Legion Weekly*, 21 September 1923, 6; Trotter, "Merchants of Death Theory," 167–69; Gough, "Soldiers, Businessmen, and U.S. Industrial Mobilisation Planning," 69; U.S. House, *War Policies Commission*, 188–89, 260; U.S. Senate, Nye committee, *Hearings*, 1131, 448, 12404.

66. Frear to constituents, 2 February 1928, clipping in Frear Papers, box 5. Cf. weekly letters of 28 March 1928, box 5, and 10 January 1929, box 4.

67. Jonathon Mitchell, "The Armaments Scandal, II: Sowing Death Abroad," *New Republic*, 23 May 1934, 37; Hutchinson, "The Arms Inquiry," 656; U.S. Senate, Nye committee, *Hearings*, 780, 5066; Francis Delaisi, "Corruption in Armaments," *Living Age* 341 (September 1931): 51.

68. U.S. Senate, Nye committee, *Hearings*, 6687–89, 6700.

69. J. H. Kitchens Jr., "The Shearer Scandal and Its Origins: Big Navy Politics and Diplomacy of the 1920s" (Ph.D. diss., University of Georgia, 1968), 76, 177–79, 253; David Burner, *Herbert Hoover: A Public Life* (New York: Knopf, 1979), 291; "Statement of F. P. Palen" and C. L. Bardo to W. M. Flook, 1 January 1929, Records of the Special Committee Investigating the Munitions Industry, box 146, RG 154, National Archives (cited hereafter as NCR). Cf. William Shearer, *The Cloak of Benedict Arnold* (privately published, 1928), NCR, box 145.

70. Kitchens, "Shearer Scandal," 155; "Shearer," memorandum to Sen. Clark, 12 February 1935, NCR, box 46.

71. Shearer claimed that he had accomplished ten years' work in a single year. Since he claimed that his salary was $25,000 per year, he felt he deserved to receive $250,000. This compensation, however, was for lobbying in favor of the naval construction appropriation for 1929, not for poisoning the Geneva conference. After his dismissal by C. L. Bardo, Shearer received a retainer of $2,000 per month from William Randolph Hearst to lobby against the World Court and the League of Nations. Kitchens, "Shearer Scandal," 155, 215, 217; Shearer to S. W. Wakeman, 30 January 1928, and F. W. La Rouche, unpublished notes [1935?], NCR, box 146; Charles A. Beard, *The Navy: Defense or Portent?* (New York: Harper, 1932), 118.

72. C. L. Bardo to Shearer, 20 February 1928, and statement of F. P. Palen, NCR, box 146. Bardo denied that he wanted the Geneva Conference to fail; the shipbuilders wanted stability in ship construction. Stephen G. Rockwell, "Memorandum on Shearer Investigation," in Rockwell to La Rouche, 6 December 1934, NCR, box 141.

73. The money came from Newport News Shipbuilding and Drydock, New York Shipbuilding, and Bethlehem Shipbuilding. The settlement allowed the Big 3 to disclaim the validity of Shearer's charges. See E. A. Adams, secretary-treasurer, Newport News Shipbuilding, to Melbourne Bergman, attorney for Shearer, 8 June 1930, NCR, box 141. On 30 January 1929, contrary to his other claims, Shearer declared that he had been paid to conduct a campaign of naval preparedness. Shearer to Wakeman, NCR, box 142. See also Kitchens, "Shearer," 225–30; *New York Times*, 12 January 1930, 1, 28; "An Under-Cover Lobby to Defeat Disarmament?" *Literary Digest*, 21 September 1929, 7–9; "The Shearer Show," *Literary Digest*, 12 October 1929, 5–7. As recently as 1971, one scholar accepted the claim that Shearer had been paid to wreck the disarmament conference. See Charles Chatfield, *For Peace and Justice: Pacifism in America, 1914–1941* (Knoxville: University of Tennessee Press, 1971), 166.

74. "An Under-Cover Lobby to Defeat Disarmament," 11; *New York Times*, 6 January 1930, 5; Kitchens, "Shearer Scandal," 211–13, 224–25; U.S. Senate, 71st Cong., 1st sess., Committee on Naval Affairs, *Alleged Activities at the Geneva Conference*, 1930, passim; U.S. Senate, Nye committee, *Hearings*, 5945.

75. The consensus of scholars is that the 1927 Geneva conference failed for technical reasons, specifically an inability of Britain and the United States to agree to limit the strength of their cruiser fleets. See Armin Rappaport, *The Navy League of the United States* (Detroit: Wayne State University Press, 1962), 109; Merze Tate, *The United States and*

Armaments (Cambridge: Harvard University Press, 1949), 156–57; F. P. Walters, *A History of the League of Nations* (London: Oxford University Press, 1960), 366–68; "American, British, and Japanese Proposals at the Geneva Conference and Their Bearing on Competitive Naval Bidding," *Iron Age,* 15 September 1927, 691; and "The Failure of the Geneva Conference," *World's Work* 54 (September 1927): 469.

76. Kitchens, "Shearer Scandal," 225–30, 259; *Chicago Tribune,* 3 April 1932; Helmuth Engelbrecht and Frank G. Hanighen, *Merchants of Death: A Study of the Industrial Armament Industry* (New York: Dodd, Mead, 1934), 208–10; George Seldes, *Iron, Blood, and Profits: An Exposure of the World-Wide Munitions Racket* (New York: Harper, 1934), 155; Merle Curti, *Peace or War: The American Struggle, 1636–1936* (New York: Norton, 1936), 265–66; U.S. House, *War Profits Commission,* 493; U.S. Senate, Nye committee, *Hearings,* 5840–42, 5941–85, 6063–64, 6072–75.

9. Profits or Peace?

1. C. Hartley Grattan, *Why We Fought* (1929; reprint, Indianapolis: Bobbs-Merrill, 1969). Foreword by Keith L. Nelson; afterword by the author. For the book's thesis, see 132ff, esp. 140–44, and the afterword, xxix. For Grattan's background, see C. Hartley Grattan, *The Deadly Parallel* (New York: Stackpole, 1939), 175, Nelson's foreword, x–xi, and Grattan's prelude, xiii–xiv. Warren I. Cohen argues in *The American Revisionists: The Lessons of Intervention in World War I* (Chicago: University of Chicago Press, 1967), 207, that revisionist ideas were colored by the ethnic heritage of the writers. In Grattan's case, this seems not to be true. There is evidence, however, that Grattan was deeply affected by the Civil War. See Grattan, *Bitter Bierce: A Mystery of American Letters* (1929; New York: Cooper Square, 1966), 14, 17, and *Why We Fought,* xxiii.

2. Grattan later changed his assessment and directed his attack toward the pro-British biases of Woodrow Wilson and Ambassador Walter Hines Page. See Grattan, *How America Was Forced into the World War: The Walter Hines Page Legend* (Girard, Kans.: Haldeman-Julius, n.d.), 32. Other writers, such as Harry Elmer Barnes, maintained a belief in the munitions makers hypothesis. Congressman John M. Nelson, progressive Republican of Wisconsin, cited Barnes's *World Politics in Modern Civilization* six times in a single speech. See untitled speech MSS [1932?], Nelson Papers, State Historical Society of Wisconsin, Madison, box 12. The Women's International League of Peace and Freedom included sessions on "The Economic Menace to Peace" and "Armament Industries as War Breeders" at its 1932 convention. Dorothy Detzer recommended that each local chapter of the WILPF purchase copies of George Seldes's *Iron, Blood, and Profits* and Helmuth Engelbrecht and Frank Hanighen's *Merchants of Death.* See Detzer to branch and legislative chairmen, 2 May 1934, Women's International League of Peace and Freedom Records, State Historical Society of Wisconsin, Madison, box 1.

3. Quoted in Wayne S. Cole, *Senator Gerald P. Nye and American Foreign Relations* (Minneapolis: University of Minnesota Press, 1962), 65.

4. Michael S. Sherry, *The Rise of American Air Power: The Creation of Armageddon* (New Haven: Yale University Press, 1987), 58–60; George Seldes, *Iron, Blood, and Profits: An Exposure of the World-Wide Munitions Racket* (New York: Harper, 1934), 112–13; Detzer, "Memorandum on the Findings of the War Policies Commission," Madison Chapter, WILPF Papers, box 1, State Historical Society of Wisconsin, Madison.

5. The New Deal recovery program also included $15 million for the manufacture of military aircraft. Jacob Vander Meulen, *The Politics of Aircraft: Building an American Industry* (Lawrence: University Press of Kansas, 1991), 118; U.S. Senate Special Committee Investigating the Munitions Industry, *Hearings,* 74th Cong., 2d sess., 1936 (cited hereafter as Nye committee), 5720, 5727; Mabel Vernon to friend, 27 July 1933, and William T. Stone, "International Ramifications of the U.S. Naval Program," memorandum [1934?], Madison Chapter, WILPF Papers, box 1; *Washington Times,* 22 February 1935; Seldes, *Iron, Blood, and*

Profits, 310–11; Robert Dallek, *Franklin D. Roosevelt and Foreign Policy, 1932–1945* (New York: Oxford University Press, 1979), 75.

6. U.S. Senate, Nye committee, *Hearings,* 5785–86. Cf. p. 6186 for similar views expressed by John Flynn, the most vigorous advocate of controlling profiteering of the entire decade.

7. The isolationism of the 1930s was much more than simple agrarian radicalism, but the isolationists found their greatest support in rural America. On the composition of isolationism, see Manfred Jonas, *Isolationism in America, 1935–1941* (Ithaca: Cornell University Press, 1966), 15–16, 17n, 22–23.

8. Cole, *Nye,* 62; Cole, *Roosevelt and the Isolationists, 1932–45* (Lincoln: University of Nebraska Press, 1983), 37–38; J. C. Vinson, "War Debts and Peace Legislation: The Johnson Act of 1934," *Mid-America* 50 (July 1968): 214, 217; Arthur Capper to James A. Frear, 13 January 1934, Frear Papers, State Historical Society of Wisconsin, Madison.

9. Quoted in Fred Greene, "The Military View of American National Policy," *American Historical Review* 66 (January 1961): 358–59; Sherry, *American Air Power,* 59.

10. Anthony Cave Brown, *"C": The Secret Life of Sir Stewart Graham Menzies, Spymaster to Winston Churchill* (New York: Macmillan, 1987), 123; Thomas C. Kennedy, "Beard vs. FDR on National Defense and Rearmament," *Mid-America* 50 (January 1968): 23n, 26.

11. Kennedy, "Beard vs. FDR," 31.

12. Charles A. Beard, *The Navy: Defense or Portent?* (New York: Harper, 1932), 170–72; Selig Adler, *The Isolationist Impulse: Its Twentieth-Century Reaction* (London: Abelard, 1957), 259; Kennedy, "Beard vs. FDR," 23n, 26; Cole, *Roosevelt and the Isolationists,* 43; Scott Nearing, *War: Organized Destruction and Mass Murder by Civilized Nations* (New York: Vanguard, 1931), 38–43, 65, 180. On Hoover, see Charles Chatfield, *For Peace and Justice: Pacifism in America* (Knoxville: University of Tennessee Press, 1971), 160; Robert H. Levine, *The Politics of American Naval Rearmament, 1930–1938* (New York: Garland, 1988), 47–48; Armin Rappaport, *The Navy League of the United States* (Detroit: Wayne State University Press, 1962), 142, 152; Maj. Gen. George Van Horn Moseley, "One Soldier's Journey," typescript, Moseley Papers, Hoover Institution, Stanford University, Stanford, Calif., 2:153; and Herbert Hoover, *The Memoirs of Herbert Hoover,* vol. 2, *The Cabinet and the Presidency* (New York: Macmillan, 1952), 338.

13. Kennedy, "Beard vs. FDR," 23n, 26.

14. Sherry, *American Air Power,* 59; Merze Tate, *The United States and Armaments* (Cambridge: Harvard University Press, 1948), 85.

15. Levine, *Politics of Naval Rearmament,* 229; Dallek, *Roosevelt and Foreign Policy,* 7–9, 19, 75, 90.

16. Kennedy, "Beard vs. FDR," 31, 37.

17. Moseley, "One Soldier's Story," 2:114–16, 118 (quote), 135. For an account of the origins of Eisenhower's use of the term *military-industrial complex,* see *Washington Post,* 31 March 1969.

18. Moseley, "One Soldier's Journey," 2:135; U.S. House, *War Policies Commission,* 71st Cong., 2d sess., 1931, 38, 188–89, 260, 355; Lawrence Sullivan, "Work of the War Policies Commission," *Current History* 35 (November 1931): 243; Seymour Waldman, "Breeding War," *New Republic,* 13 May 1931, 358.

19. U.S. House, *War Policies Commission,* quote 355, and 362–66, 408–70. See also "The Production of Munitions: A Statement of War Department Policy," *Army Ordnance* 15 (January/February 1935): 205 and "What about War Losses? An Editorial," *Army Ordnance* 15 (March/April 1935): 299.

20. Bishop Francis McDonnell of the Federal Council of Churches opposed war planning for this reason. See *Amarillo (Tex.) News,* 4 June 1931, clipping in Records of the War Policies Commission, box 177, RG 107, National Archives (cited hereafter as WPC). Cf. Seymour Waldman, "War Policies and Peace," *New Republic,* 17 June 1931, 112–13.

21. *Congressional Record,* 1 April 1930; *New York Times,* 14 and 29 January 1931; Robert H. Montgomery, "Summarized Analysis of Testimony," 10 March 1932, WPC, box 175; U.S.

Senate, Nye committee, *Hearings*, 5988–89; Paul A. C. Koistinen, *The Military-Industrial Complex: A Historical Perspective* (New York: Praeger, 1980), 54. Koistinen overlooks the 1924 investigations, which may be reviewed in U.S. House Committee on Military Affairs, *Universal Mobilization for War Purposes*, 68th Cong., 1st sess., 1924. See also William Pencak, *For God and Country: The American Legion, 1919–1941* (Boston: Northeastern University Press, 1989), chap. 5.

22. Memorandum of Robert H. Montgomery, 25 November 1931, WPC, box 178.

23. Col. James D. Fife to Gen. Van Horn Moseley, 24 June 1930, WPC, box 176. Fife listed the probable members of the WPC and categorized their views on military matters as "sound" or "unsound." On the determination of the army to control wartime mobilization, see Terrence J. Gough, "Soldiers, Businessmen, and U.S. Industrial Mobilisation Planning Between the World Wars," *War and Society* 9 (May 1991): 80–81. See *Amarillo (Tex.) News*, 4 June 1931, clipping in WPC, box 177, for opposition to all war planning by the Federal Council of Churches.

24. Stephen E. Ambrose, *Eisenhower*, vol. 1, *Soldier, General of the Army, President-Elect, 1890–1952* (New York: Simon and Schuster, 1983), 91–92; *Congressional Record*, 1 April 1930, p. 6310.

25. By 1935, there was also considerable business opposition to the concept of a universal draft. A leading business journal argued that if there were a universal draft, a war party would gain "absolute control" over the U.S. economy. This would create an incentive to declare war. See "How Not to Prevent War," *Business Week*, 13 April 1935, 40.

26. *New York Times*, 2 April 1930, 2; Seymour Waldman, "A Seven Per Cent War," *World Tomorrow* 14 (July 1931): 216.

27. Frank H. Simonds, "The Collapse of the Peace Movement," *Annals of the American Academy* 174 (July 1934): 116; John Gunther, "Slaughter for Sale," *Harper's*, May 1934, 655; "War Pays," *Living Age* 345 (November 1933): 199.

28. Arthur Marwick, *War and Social Change in the Twentieth Century: A Comparative Study of Britain, France, Germany, Russia, and the United States* (London: Macmillan, 1974), 3, 7.

29. Detzer, "Memorandum on the Findings of the War Policies Commission and Statement of the W.I.L.P.F. to the Commission," [May 1931], Madison Chapter, WILPF Papers, box 1; Seldes, *Iron, Blood, and Profits* (New York: Harper, 1934), 92, 327; Beverly Nichols, *Cry Havoc!* (London: Cape, 1933), 67–68; "Our Treaty Navy," *Army Ordnance* 14 (September/October 1933): 112.

30. Detzer, "Memorandum on the Findings of the War Policies Commission," May 1931, Madison Chapter, WILPF Papers, box 1.

31. L. F. Haber, *The Poisonous Cloud: Chemical Warfare in the First World War* (New York: Oxford University Press, 1986), 230–35; Chatfield, *Peace and Justice*, 147, 160; Seldes, *Iron, Blood, and Profits*, 285–87, 318; "Our Treaty Navy," *Army Ordnance* 14 (1933): 112. For a differing view on the role of women in foreign policy, see Joan Hoff-Wilson, "Of Mice and Men," in *Women and Foreign Policy: Lobbyists, Critics, and Insiders*, ed. Edward P. Crapol (Westport, Conn.: Greenwood Press, 1987), 182, 185, 187n. Cf. John M. Craig, "Lucia True Ames Mead: American Publicist for Peace and Internationalism," in *Women and Foreign Policy*, 81. On Detzer see her article "What Neutrality Means," *Nation* 161 (4 December 1935): 642–43, "Memorandum on the Findings of the War Policies Commission," and Jonas, *Isolationism in America*, 107. A feminist's argument that a female perspective would reform the world may be found in Amy Woods, "Economics for Peace and Freedom," unpublished memorandum, May 1933, Madison Chapter, WILPF Papers, box 1, p. 3.

32. Dallek, *Roosevelt and Foreign Policy*, 85. See also Nicholas John Cull, *Selling War: The British Propaganda Campaign against American "Neutrality" in World War II* (New York: Oxford University Press, 1995), 10.

33. "Pacifism in the Colleges," *World Tomorrow* 16 (June 1933): 415.

34. "Now for a Repeal of Twaddle!" *Army Ordnance* 14 (November/December 1933): 101, 171; Chatfield, *Peace and Justice*, 271–72; Eric W. Rise, "Red Menace and Drinking

Buddies: Student Activism at the University of Florida, 1936–1939," *Historian* 48 (August 1986): 561; Ralph Brax, *The First Student Movement* (Port Washington, N.Y.: Kennikat, 1981), 52–53; "Investigating Armaments," *Nation*, 4 April 1934, 389; Merle Curti, *Peace or War: The American Struggle, 1636–1936* (New York: Norton, 1936), 300; "Philadelphia Pickets," *Army Ordnance* 25 (March/April 1935): 295.

35. I am indebted to James Baughman, the biographer of Henry Luce, for identifying the author. See "Arms and the Men," *Fortune*, March 1934, 52–56, 113–26 (quote on 86); Dwight MacDonald, "'Fortune' Magazine," *Nation*, 8 May 1937, 529; Cole, *Nye*, 66; Seldes, *Iron, Blood, and Profits*, 320; Paul Hutchinson, "The Arms Inquiry," *Christian Century*, 8 May 1935, 643; "Fortune's Arms and the Man," *Army Ordnance* 15 (September/October 1934): 105–6.

36. For the best summary of judgments of the Nye committee, see John E. Wiltz, *In Search of Peace: The Senate Munitions Inquiry, 1934–36* (Baton Rouge: Louisiana State University Press, 1963), chap. 10. See also Earl K. Molander, "Historical Antecedents of Military-Industrial Criticism," in *War, Business, and Society*, ed. Benjamin F. Cooling (Port Washington, N.Y.: Kennikat, 1977), 184; Charles DeBenedetti, *The Peace Reform in American History* (Bloomington: Indiana University Press, 1980), 126, Koistinen, *Military-Industrial Complex*, 116, Robert James Leonard, "The Nye Committee: Legislating against War," *North Dakota History* 41 (fall 1974): 26–27, and Robert H. Ferrell, "The Merchants of Death, Then and Now," *Journal of International Affairs* 26, no. 1 (1972): 32–33.

37. Detzer to branch and legislative chairmen, 2 April 1934, Madison Chapter, WILPF Records, box 1.

38. Cole, *Nye*, 10, 41, 65, 68; "Who Are the War Profiteers?" *Literary Digest*, 29 September 1917, 9–10.

39. Nye had, however, supported intervention in 1917. Cole, *Nye*, 18, 22, 41, 65, 68; Darrel Leroy Ashby, "Progressivism against Itself: The Senate Western Bloc in the 1920s," *Mid-America* 50 (October 1968): 292–93, 304; Robert A. Divine, *The Illusion of Neutrality* (Chicago: University of Chicago Press, 1962), 64; Walter Johnson, *The Battle against Isolation* (Chicago: University of Chicago Press, 1944), 14–15.

40. Vandenberg also maintained a link to Luce and *Fortune*. C. David Tompkins, *Senator Arthur H. Vandenberg* (East Lansing: Michigan State University Press, 1970), 1, 124–28; MacDonald, "'Fortune' Magazine," 529.

41. Kennedy, "Beard vs. FDR," 25; Robert H. Levine, *The Politics of American Naval Rearmament* (New York: Garland, 1988), 233.

42. Marie Louise Degen, *History of the Women's Peace Party* (1939; reprint, New York: Franklin, 1974), 245; Anthony Sampson, *The Arms Bazaar: From Lebanon to Lockheed* (New York: Viking, 1977), 76; Divine, *Illusion of Neutrality*, 57–58; Detzer to branch chairmen and secretaries, 7 February 1934, Madison Chapter, WILPF Papers, box 1; Seldes, *Iron, Blood, and Profits*, 325. In 1964 Gerald P. Nye told a historian that he had allowed the story that Dorothy Detzer had originated the investigation to persist because he liked her and she enjoyed believing the story. Agnes Anne Trotter, "The Development of the Merchants of Death Theory of American Intervention in World War I, 1914–1939" (Ph.D. diss., Duke University, 1966), 229n.

43. Dorothy Detzer, *Appointment*, 3, 10, 12.

44. Detzer to branch and legislative chairmen, 19 March, 13 April 1934, Madison Chapter, WILPF Papers, box 1; Cole, *Nye*, 69; Detzer, *Appointment*, 156; Cole, *FDR and the Isolationists*, 78.

45. In 1927 Nye aspired unsuccessfully to organize a voting bloc in the Senate that would express western principles. See Ashby, "Progressivism against Itself," 292–93, 304. Cf. Adler, *Isolationist Impulse*, 170–71.

46. Bone was angry about stock profits gained by William Boeing, a major Seattle aircraft manufacturer. See Vander Meulen, *Politics of Aircraft*, 142.

47. Hiss's chief accuser, Whittaker Chambers, wrote that Hiss was a communist when he served on the Nye committee. Hiss's appointment came because of the committee's

shortage of funds. Two members, Senators Homer T. Bone and James P. Pope, served on the agriculture committee, and Hiss was a member of its staff. They arranged to have Hiss temporarily detached and assigned to the Nye committee. Whittaker Chambers, *Witness*, (New York: Random House, 1952), 339; John Chabot Smith, *Alger Hiss: The True Story* (New York: Holt, Rinehart, 1976), 82, 86.

48. Cole, *FDR and the Isolationists*, 148; Chatfield, *Peace and Justice*, 179; Detzer, *Appointment*, 165; Alger Hiss, *Recollections of a Life* (New York: Holt, 1988), 76–85; Stephen Raushenbush, "Confidential Memorandum," 14 January 1936, Records of the Special Committee Investigating the Munitions Industry (Nye committee), box 156, RG 154, National Archives (cited hereafter as NCR); Jonas, *Isolationism in America*, 146; Stephen and Joan Raushenbush, *The Final Choice: America between Europe and Asia* (New York: Reynal, 1937).

49. Dallek, *Roosevelt and Foreign Policy*, 7–9, 19; Paul Hutchinson, "The Arms Inquiry," *Christian Century*, 8 May 1935, 652.

50. "Our War Profiteers, the Food Speculators," *Literary Digest*, 26 May 1917, 1584; U.S. Senate, 74th Cong., 2d sess., 1936, *Special Committee Investigating the Munitions Industry*, 4944. In turn, in 1918 the future president was charged by a clerk in the Bureau of Construction and Repair with having accepted improperly a contract at inflated prices. See E. David Cronon, ed., *The Cabinet Diaries of Josephus Daniels* (Lincoln: University of Nebraska Press, 1963), 304, 315; Franklin D. Roosevelt, *Complete Presidential Press Conferences of Franklin D. Roosevelt* (New York: Da Capo, 1972), 4, no. 164 (12 December 1934): 269–73, 9, no. 358 (6 April 1937): 250; Samuel I. Rosenman, ed., *The Public Papers and Addresses of Franklin D. Roosevelt* (New York: Harper, 1950), 4:67, 182–92, 7:67.

51. Franklin D. Roosevelt, *The Papers and Addresses of Franklin D. Roosevelt*, vol. 3, *The Advance of Recovery and Reform, 1934* (13 vols.; New York: Random House, 1938–1950), 239–40.

52. Divine, *Illusion of Neutrality*, 55; Levine, *Politics of Naval Rearmament*, 229; Rappaport, *Navy League*, 142.

53. The great flaw in the arms control plan was that each nation had to authorize international inspection and licensing of their armaments industries. Since Germany, Britain, Italy, and Japan would not accept inspection, disarmament failed. *Congressional Record* 78, pt. 1 (1934), p. 9095, quoted in Trotter, "Merchants of Death Theory," 239; John Eppstein, "Traffic in Arms: The Task of the Royal Commission, *Contemporary Review* 147 (June 1935): 417–22, reprinted in Thomas A. Rousse, *Nationalization of Munitions*, Bulletin no. 3638 (Austin: University of Texas, 1936), 187; Anne Hartwell Johnstone and Elizabeth Armstrong Hawes, *Control of the Arms Traffic* (Washington, D.C.: National League of Women Voters, 1935), reprinted in Rousse, *Nationalization of Munitions*, 72. On the relationship between munitions and narcotics, see André Giraud, "Who Should Make War Munitions?" *Rotarian* 45 (August 1934): 15, reprinted in Rousse, *Nationalization of Munitions*, 184, and Laura Puffer Morgan, "A Possible Technique of Arms Control: Lessons from the League of Nations Experience in Drug Control," *Geneva Studies* 11 (November 1940): 47–50, 78–79; Chatfield, *Peace and Justice*, 167; and Murray Stedman, *Exporting Arms: The Federal Arms Exports Administration, 1935–1945* (Morningside Heights, N.Y.: Kings Crown, 1947), 8–11.

54. Curti, *Peace or War*, 277; Raymond Leslie Buell, "American Neutrality and Collective Security," *Geneva Special Studies* 6, no. 6 (1935): 8.

55. Noel-Baker became the leading opponent of the private manufacture of arms in the 1930s. See his book *The Private Manufacture of Armaments* (New York: Oxford University Press, 1937), esp. 64–72, 88–92. For the specification of the charges, see League of Nations, *Report of the Temporary Mixed Commission on Armaments* (Geneva, 15 September 1921), 11. On the controversy, see Otto Lehmann-Russbuldt, *War for Profits*, trans. Pierre Loving (New York: King, 1930), 158; Tate, *The United States and Armaments*, 66; Walters, *League of Nations*, 48, 58–59, 218, 706; Sampson, *Arms Bazaar*, 69; Pertinax [pseud.], "Who Should Make War Munitions? The Private Interests, Regulated," *Rotarian* 45 (August 1934): 15; J. D. Scott, *Vickers: A History* (London: Weidenfeld, 1962), 239–40; Fenner Brockway and Frederic Mullaly, *Death Pays a Dividend* (London: Gollancz, 1945), 11–12.

56. "'Facts' about Munitions Makers, II," *Army Ordnance* 15 (July/August 1934): 41.

57. Flynn to Raushenbush, [January 1935] and 4 February 1935, and Raushenbush to Flynn, 8 March 1935, NCR, box 156. Flynn was widely known as the author of several muckraking books. He also wrote a daily newspaper column, a weekly magazine column, and many articles for national magazines. He was the financial columnist for the *New Republic*. After the war, he became a militant anticommunist critic of Roosevelt. See Flynn, *The Roosevelt Myth* (New York: Devin-Adair, 1948, 1956), xii, and passim.

58. William T. Stone, "The Munitions Industry: Analysis of the United States Senate Investigation," *Geneva Special Studies* 5, no. 9 (1934): 83; Bruce Winton Knight, *How to Run a War* (1936; reprint, New York: Arno, 1972), 200–219; Rousse, *Nationalization of Munitions*, 14–39.

59. Rappaport, *Navy League*, 137–38, 176; Chatfield, *Peace and Justice*, 109; Kennedy, "Beard vs. FDR," 27. The Navy League was formed at the New York Yacht Club in 1902. Many of its early members were wealthy yachtsmen but not principally shipbuilders. Seldes, *Iron, Blood, and Profits*, 270–74.

60. Kennedy, "Beard vs. FDR," 27; U.S. Senate, Nye committee, *Hearings*, 287–96, 2131–32, 2135, 2137.

61. "The War Cry," unpublished MSS, NCR, box 158; "Confidential Memorandum," 17 June 1935, NCR, box 154. For Nye's continued insistence on this charge, see *Washington Times*, 3 April 1935.

62. "Facts about 'Munition Makers': An Editorial," *Army Ordnance* 14 (May/June 1934): 361; Helmuth C. Engelbrecht and F. C. Hanighen, "The War on the Arms Industry," *Commonweal*, 16 February 1934, 427; Engelbrecht, "The Arms Industry—An Appraisal," *World Tomorrow*, 7 December 1933, 661; Stone, "Munitions Industry," 2; League of Nations, *Statistical Information on the Trade in Arms, Ammunition, and Matériel of War* (Geneva, 1924), 65.

63. U.S. Senate, Nye committee, *Hearings*, 1934, 2274.

64. Armin Rappaport, *Henry L. Stimson and Japan, 1931–33* (Chicago: University of Chicago Press, 1962), 89; Chatfield, *Peace and Justice*, 225.

65. Divine, *Illusion of Neutrality*, 26–30, 46; Jonathan Mitchell, "The Armaments Scandal, II: Sowing Death Abroad," *New Republic*, 23 May 1934, 37. As early as 1928, Fish had argued against foreign arms sales on the grounds that they would drag the nation into war. See Seldes, *Iron, Blood, and Profits*, 167, 332.

66. Walters, *League of Nations*, 528, 533–34; Noel-Baker, *Private Manufacture of Armaments*, 48.

67. "United States Naval Mission to Peru," unpublished MSS, NCR, box 158; "The Munitions Revelations," *New Republic*, 19 September 1934, 144–45.

68. As described in chap. 8, New York Shipbuilding had asked the navy to send a cruiser to Brazil to demonstrate its quality to the Brazilian navy. The Curtiss-Wright aircraft firm had solicited the endorsement of Captain Ernest J. King, helping sales of its "Falcon" warplane to the Dominican Republic. The Nye committee also discovered that in past years a U.S. ambassador had helped sell submarines to Spain, the navy had helped sell warships to both Turkey and Colombia, and the War Department had helped sell anti-aircraft guns to Poland. Nye admitted to having been "astonished" by these revelations.

69. Stone, *The Munitions Industry*, 9; "Senate Munitions Investigation," *Commercial and Financial Chronicle*, 15 September 1934, 1637.

70. U.S. Senate, Nye committee, *Hearings*, 640, 780, 1157, 1185, 1684, 12471.

71. "United States Naval Mission to Brazil," unpublished typescript, NCR, box 158. Anthony Sampson has argued, without proof, that bribery was more common in the arms trade than in other areas of commerce. He reasons that since negotiations were usually secret and the decision of one or two individuals could prove decisive, there was a greater incentive to bribery in the arms business. See Anthony Sampson, *The Arms Bazaar: From Lebanon to Lockheed* (New York: Viking, 1977), 52.

72. U.S. Senate, Nye committee, *Hearings*, 2092–2102, 2220; Raushenbush to Vandenberg, 8 August 1934, NCR, box 155.

73. The evidence of collusion on cruiser contracts may be found in U.S. Senate, Nye committee, *Hearings*, 5840–42, 6687–89, 6700, 12010–12. Cf. Robert Wohlforth to Donald Y. Wimple, 19 February 1935, NCR, box 141. For other extensive attempts to establish collusion on shipbuilding contracts for 1927–34, see Nye committee, *Hearings*, 4743–48, 5158.

74. U.S. Senate, Nye committee, *Hearings*, 11388, 11868; S. G. Rockwell to F. W. LaRouche, 6 December 1934, box 140, and N. P. Alifas to Nye, NCR, box 141; George J. Earl to S. W. Wakeman, 3 November 1927, NCR, box 142. One officer testified to the Committee on Naval Affairs that government shipyards included interest on capital investment, depreciation, repairs, administrative costs, and insurance costs in their accounting, but did not use these costs when bidding for a contract. See statement of Brigadier General W. S. Peirce to Committee on Naval Affairs, NCR, box 142.

75. Stephen Raushenbush to William T. Stone, 11 July 1935, Stone to Raushenbush, 11 July 1935, NCR, box 154.

76. Robert D. Cuff, "We Band of Brothers—Woodrow Wilson's War Managers," *Canadian Review of American Studies* 5 (fall 1974): 137.

77. Newton D. Baker, *Why We Went to War* (New York: Council on Foreign Relations, 1936), 118–22; Trotter, "Merchants of Death Theory," 271. Cf. *New York Times*, 2 and 3 January 1935.

78. "Confidential Memorandum," 27 June 1935, NCR, box 154. In 1934 Charles Warren, a leading international lawyer who had been assistant attorney general during the Wilson administration, argued that it was nonmunitions contraband that had caused the main friction before World War I. The Nye committee ignored this argument. See Divine, *Illusion of Neutrality*, 71.

79. This line of argument, championed by Senator Clark, may be followed in U.S. Senate, Nye committee, *Hearings*, 8487–8504, 8575–79. It is summarized in the *Washington Times*, 2 March 1935.

80. Nye believed that the Morgan firm was blocking construction of the St. Lawrence Seaway (which would be of major benefit to North Dakota wheat farmers) because the construction of its hydroelectric plants would be competitors of Morgan-controlled electric utilities. See Cole, *Nye*, 64. For examples of fierce attacks on the House of Morgan, see the writings of Oswald Garrison Villard, esp. "The War and the Pacifists," *Nation*, 23 October 1935, and "Neutrality and the House of Morgan," *Nation*, 13 November 1935, quoted in *Oswald Garrison Villard: The Dilemmas of the Absolute Pacifist in Two World Wars*, ed. Anthony Gronowicz (New York: Garland, 1983), 448, 450.

81. Lansing to Wilson, 6 September 1915, in "Confidential Memorandum," 27 June 1935, NCR, box 154. Cf. U.S. Senate, Nye committee, *Hearings*, 7883, 7915–16, 7921.

82. Grattan, *Why We Fought*, 131; Robert Lansing, *War Memoirs of Robert Lansing* (Indianapolis: Bobbs-Merrill, 1935), 22. Stephen and Joan Raushenbush dismiss those who accepted this argument as "casual commentators" in their memoir of the Nye committee, *Final Choice*, 244.

83. Stephen Raushenbush, "Confidential Memorandum," 14 January 1936, NCR, box 156.

84. Cole, *Nye*, 85; Trotter, "Merchants of Death Theory," 264; *Washington Times*, 20 December 1934.

85. U.S. Senate, Nye committee, *Hearings*, 8512–13; Wiltz, *In Search of Peace*, 63–64, 202–3; Trotter, "Merchants of Death Theory," 268–69; Jonas, *Isolationism in America*, 148.

86. Robert H. Ferrell, "Woodrow Wilson and Open Diplomacy," in *Issues and Conflicts: Studies in Twentieth-Century American Diplomacy*, ed. George L. Anderson (Lawrence: University Press of Kansas, 1959), 200–204; Trotter, "Merchants of Death Theory," 291.

87. "The Uses of Woodrow Wilson," *Nation*, 29 January 1936, 118; Wiltz, *In Search of Peace*, 202–208; Jonas, *Isolationism in America*, 137–40. For Hull's defense of Wilson's "scrupulous honesty," see Hull, *The Memoirs of Cordell Hull* (New York: Macmillan, 1948), 404. Some liberals had broken with Wilson over the Red Scare of 1919–20, but this memory had dimmed by the 1930s. See Adler, *Isolationist Impulse*, 70.

88. Wiltz, *In Search of Peace,* 141.

89. Flynn to Raushenbush [January 1935], 4 and 25 February 1935, and Raushenbush to Flynn, 20 February and 8 March 1935, NCR, box 156; Adler, *Isolationist Impulse,* 302; Chatfield, *Peace and Justice,* 319n; Cole, *Nye,* 86; U.S. Senate, Nye committee, *Hearings,* 6186, 6193; Trotter, "Merchants of Death Theory," 293–94. The quotation is from "Abolishing War Profits," *Nation,* 3 April 1935, 377. Flynn's views on war profits are outlined in his article "Other People's Money: Net Results of the Morgan Inquiry," *New Republic,* 19 February 1936, 46.

90. "Abolishing War Profits," *Nation,* 3 April 1935, 377; Trotter, "Merchants of Death Theory," 277–78; Detzer to branch and legislative chairmen, 13 April 1934, Madison Chapter, WILPF Records, box 1.

91. The historian of the Nye inquiry, John E. Wiltz, argues that its importance to neutrality legislation was limited, as this would have passed anyway, but this issue is not appropriate to this study. See Wiltz, *In Search of Peace,* 227–30. Cf. Sampson, *Arms Bazaar,* 80; Selig Adler, *The Uncertain Giant: American Foreign Policy between the Wars* (New York: Macmillan, 1965), 163–66; Chatfield, *Peace and Justice,* 236; Thomas H. Guinsburg, *The Pursuit of Isolationism in the United States Senate from Versailles to Pearl Harbor* (New York: Garland, 1982), 177. The Roosevelt administration interpreted the term *implements of war* narrowly. It permitted sales of oil, copper, trucks, and scrap iron, all of which had military value. See "Have the War Profiteers No Conscience?" *Christian Century,* 27 November 1935, 1507.

92. In May 1934, before the Nye inquiry was well under way, the Congress also approved a joint resolution declaring an embargo on sales of arms to the Chaco belligerents. Stone, *Munitions Industry,* 4, 6; "The United States and World Organization during 1935," *Geneva Special Studies* 6, no. 10 (1935): 12–13; Laura Puffer Morgan, "A Possible Technique of Disarmament Control: Lessons from the League of Nations Experience in Drug Control," *Geneva Studies* 11 (November 1940): 68.

93. Nye maintained his belief in the Merchants of Death theory until at least 1939. Arthur H. Vandenberg wavered sooner. See Robert H. Ferrell, "The Merchants of Death, Then and Now," *Journal of International Affairs* 26, no. 1 (1972): 32, and C. David Tompkins, *Senator Arthur H. Vandenberg: The Evolution of a Modern Republican, 1884–1945* (East Lansing: Michigan State University Press, 1970), 124–26. Cf. Dallek, *Roosevelt and Foreign Policy,* 101; Helmuth C. Engelbrecht, *Revolt against War* (New York: Dodd, Mead, 1937), 11; Curti, *Peace or War,* 277; Keith L. Nelson, "The Warfare State: History of a Concept," *Pacific Historical Review* 40, no. 2 (1971), reprinted in *The Military-Industrial Complex,* ed. Carroll W. Pursell Jr. (New York: Harper, 1972), 16.

94. Warren I. Cohen, *The American Revisionists: The Lessons of Intervention in World War I* (Chicago: University of Chicago Press, 1967), 174, 180; Cole, *Nye,* 137; Dallek, *Roosevelt and Foreign Policy,* 129; Jonas, *Isolationism in America,* 149; Seldes, *Iron, Blood, and Profits,* 318–20; Divine, *Illusion of Neutrality,* 163; Adler, *Isolationist Impulse,* 169. Editors of the *World Tomorrow* included such notables in the antiprofiteering campaign as Helmuth Engelbrecht, Dorothy Detzer, Reinhold Niebuhr, and Devere Allen. See *World Tomorrow,* 23 November 1933, 626, and "War Resistance Leaps Ahead," *World Tomorrow* 16 (June 1933): 415. Even the military journal *Our Navy* endorsed the Flynn plan for controlling war profits. See U.S. Senate, Nye committee, *Hearings,* 7397.

95. Charles Callan Tansill, *America Goes to War* (Boston: Little, Brown, 1938), 52–55, 657; Linley Gordon, "War and the Munitions Racket," *Review of Reviews* (February 1935), reprinted in Rousse, *Nationalization,* 201; Gronowicz, *Villard,* 449; "How Not to Prevent War," *Business Week,* 13 April 1935, 40. See also *Interdependence,* League of Nations Society of Canada, 14th annual conference, 13, nos. 3–4 (1936): 218–31.

96. Pierre du Pont, "Do Business Men Want War?" *Nation's Business* 22 (August 1934): 34–38, reprinted in Rousse, *Nationalization,* 156–59. Du Pont's ideas are also explicated in an interview with the *Washington Times,* 8 December 1934. Cf. Helmuth C. Engelbrecht and Frank C. Hanighen, "The War on the Arms Industry," *Commonweal,* 16 February 1934, 428.

Angell's arguments are summarized in *The Unseen Assassins* (New York: Harper, 1932), 35–36, 44–45, 138, 148, 185. Jonathon Mitchell presents a pre-Nye critique of the theory in *Goose Steps to Peace* (Boston: Little, Brown, 1931), 270–71, 284.

97. Jonas, *Isolationism in America,* 145–47; Trotter, "Merchants of Death Theory," 310.

98. Sampson, *Arms Bazaar,* 85–86. In 1935 a British referendum known as the Peace Ballot resulted in eleven million Britons voting for the abolition of private arms sales, while only 780,000 supported their continuation. See Noel-Baker, *Private Manufacture of Armaments,* 83. The Royal Commission also concluded that without United States arms sales during the Great War, Britain would have lost. Harold J. Laski heaps scorn on the commission in "The British Arms Inquiry," *Nation,* 4 March 1936, 272.

99. *Washington Times,* 18 December 1934.

100. Leonard, "Nye Committee," 23; Gronowicz, *Villard,* 484. Canada had established government-owned defense corporations, which finessed the profiteering issue. See David Carnegie, *History of Munitions Supply in Canada, 1914–1918* (London: Longman, 1925), 265.

101. Baruch, *Taking the Profit out of War,* 119–25; Baruch, "Recapture of War-time Profits: Price Control and Taxation the Only Equitable Methods," *Army Ordnance* 15 (September/October 1934): 75–77; *New York Times,* 24 June 1934, 20C; Cole, *Nye,* 85.

102. Brent Dow Allinson, "Senator Nye Sums Up," *Christian Century,* 16 January 1935, 80–81; Guinsburg, *Pursuit of Isolationism,* 244.

103. U.S. Senate, Nye committee, *Hearings,* 12408–9; Nye, "Profiting from Experiences," text of speech at Carnegie Hall, New York, 27 May 1935, NCR, box 154.

104. U.S. Senate, Nye committee, *Hearings,* 517, 1712.

105. *Washington Times,* 6 December 1934.

106. Bone quoted in Raymond Leslie Buell, "American Neutrality and Collective Security," *Geneva Special Studies* 6, no. 6 (1935): 8; U.S. Senate, Nye committee, *Hearings,* 6009, 6204. Cf. Vander Meulen, *Politics of Aircraft,* 142.

107. George H. Gallup, *The Gallup Poll: Public Opinion, 1945–1971* (New York: Random House, 1972), 1:54, 189–93; *Public Opinion Quarterly* 2 (July 1938): 387–88, cited in Trotter, "Merchants of Death Theory," 316.

108. Detzer, *Appointment,* 171; Cole, *Nye,* 146; Wiltz, *In Search of Peace,* 22; Charles DeBenedetti, *The Peace Reform in American History* (Bloomington: Indiana University Press, 1980), 126; Trotter, "Merchants of Death Theory," 94, 270; Koistinen, *Military-Industrial Complex,* 11, 116; Sampson, *Arms Bazaar,* 77–80; Harold C. Syrett, "The Business Press and American Neutrality, 1914–1917," *Mississippi Valley Historical Review* 32 (September 1945): 215; George Thayer, *The War Business: The International Trade in Armaments* (New York: Simon and Schuster, 1969), 17–34. Leonard argues forcefully but unconvincingly that Nye was unlike McCarthy. He believes that the press was equally guilty for spreading Nye's distortions. See Leonard, "The Nye Committee," 24, 26–27. Hoff-Wilson, "Of Mice and Men," 181.

109. Quoted in Earl F. Molander, "Historical Antecedents of Military-Industrial Criticism," *Military Affairs* 40 (April 1976): 184.

10. Penning the Warhog

1. Its sponsors were Congressman Carl Vinson of South Carolina, chairman of the House Naval Affairs Committee, and Sen. Park Trammell of Florida, who chaired the Senate Naval Affairs Committee. See Jacob Vander Meulen, *The Politics of Aircraft: Building an American Military Industry* (Lawrence: University Press of Kansas, 1992), 141; Robert Dallek, *Franklin D. Roosevelt and Foreign Policy, 1932–1945* (New York: Oxford University Press, 1979), 75; and Wayne S. Cole, *Roosevelt and the Isolationists, 1932–45* (Lincoln: University of Nebraska Press, 1983), 264.

2. E. H. Foley Jr., general counsel, Treasury Department, to Treasury Secretary Henry Morgenthau Jr., 8 March and 1 July 1940, Henry Morgenthau Jr., Diary, vol. 246, p.

350, and vol. 278, p. 181, and transcript, meeting re taxes, 1 July 1940, Morgenthau Diary, vol. 278, p. 124, Franklin D. Roosevelt Library, Hyde Park, N.Y. (Morgenthau diaries are cited hereafter as MD); Mark H. Leff, *The Limits of Symbolic Reform: The New Deal and Taxation, 1933–1939* (Cambridge: Cambridge University Press, 1984), 127; Cole, *Roosevelt and the Isolationists*, 424–26; John Morton Blum, *V Was for Victory: Politics and American Culture during World War II* (New York: Harcourt Brace, 1976), 128.

3. The nickname was coined by Ray Stannard Baker, a Progressive journalist, and the faction included George Norris, Gerald Nye, Arthur Vandenberg, Bennett Champ Clark, and Robert La Follette Jr. See Vander Meulen, *The Politics of Aircraft*, 131.

4. Assistant Secretary of the Treasury John L. Sullivan to Morgenthau, 15, 16, and 17 April 1940, MD, vol. 254, pp. 222, 322, vol. 255, p. 54; Foley to Morgenthau, 1 July 1940, MD, vol. 278, pp. 181–83.

5. Some of these planes were even named with grisly but unintentional irony: the Douglas *Devastator* and the Brewster *Buffalo*.

6. The Treasury Department was well aware of the poor profitability in shipbuilding and aircraft. See Sullivan to Morgenthau, 5 August 1940, vol. 289, p. 78, and George J. Mead, director, aeronautical division, National Defense Advisory Council, to Morgenthau, 4 September 1940, MD, vol. 305, pp. 75–77; Vander Meulen, *Politics of Aircraft*, esp. 196–98, 203, and Blum, *V Was for Victory*, 119.

7. U.S. House Committee on Military Affairs, *Universal Mobilization for War Purposes*, 68th Cong., 1st sess., 1924; U.S. House Committee on Military Affairs, *Taking the Profits out of War*, 74th Cong., 1st sess., 1935; U.S. Senate Committee on Military Affairs, *To Prevent Profiteering in War*, 74th Cong., 1st sess., 1935; U.S. House Committee on Military Affairs, *Taking the Profits out of War*, 75th Cong., 1st sess., 1937; U.S. Senate Committee on Military Affairs, *To Prevent Profiteering in Time of War*, 75th Cong., 1st sess., 1937; U.S. Senate Committee on Military Affairs, *To Draft the Use of Money in Time of War*, 75th Cong., 3d sess., 1938.

8. U.S. House, *Taking the Profits out of War*, 1935, 522–23; U.S. Senate, *To Prevent Profiteering in War*, 1935, 21, 45; U.S. Senate, *To Prevent Profiteering in War*, 1937, 4–5; U.S. House, *Taking the Profits out of War*, 1937, 108–9.

9. Cole, *Roosevelt and the Isolationists*, 265–66; Dallek, *Roosevelt and American Foreign Policy*, 172–73.

10. In 1917 Roosevelt wrote to the House Naval Affairs Committee to denounce yacht owners who demanded prices "away beyond reason" to convert their vessels to minesweepers. He also deplored prices demanded by owners of tugs and fishing boats as "outrageous," and he disallowed several claims presented by the Fore River Shipbuilding Corporation for payment of officers' bonuses. See "Our War Profiteers, the Food Speculators," *Literary Digest*, 26 May 1917, 1584, and U.S. Senate, Special Committee Investigating the Munitions Industry, *Hearings*, 74th Cong., 2d sess., 1936, 4944.

11. Frank Freidel, *Franklin D. Roosevelt: The Apprenticeship* (Boston: Little, Brown, 1952), 211–19, quote on 296; E. David Cronon, ed., *The Cabinet Diaries of Josephus Daniels* (Lincoln: University of Nebraska Press, 1963), 304, 315; Carroll Kilpatrick, ed., *Roosevelt and Daniels: A Friendship in Politics* (Chapel Hill: University of North Carolina Press, 1952), 62–63; Rexford G. Tugwell, *In Search of Roosevelt* (Cambridge: Harvard University Press, 1972), 266.

12. Memorandum for the secretary's diary, 8 July 1940, MD, vol. 280, p. 334; Rexford Guy Tugwell, "The Compromising Roosevelt," *Western Political Quarterly* 6 (June 1953): 328.

13. Franklin D. Roosevelt, *The Papers and Addresses of Franklin D. Roosevelt*, vol. 3: *The Advance of Recovery and Reform, 1934* (13 vols.; New York: Random House, 1938–50), 239–40.

14. Franklin D. Roosevelt, *Complete Presidential Press Conferences of Franklin D. Roosevelt* (New York: Da Capo, 1972) 4, no. 164 (12 December 1934): 269–73.

15. Roosevelt and Secretary of State Cordell Hull, at the instigation of the British gov-

ernment, may have attempted unsuccessfully to dissuade Nye from probing the files of the J. P. Morgan Company. Thus Roosevelt may have had a specific reason for distrusting Nye's efforts. See Raymond Gram Swing, "The Morgan Nerve Begins to Jump," *Nation*, 1 May 1935, 504. The quotation is from the entry of 15 May 1942, Morgenthau presidential diary, vol. 5, p. 1093, Franklin D. Roosevelt Library, Hyde Park, N.Y. (cited hereafter as MPD). It is cited in Warren F. Kimball, *The Juggler: Franklin Roosevelt as Wartime Statesman* (Princeton: Princeton University Press, 1991), 7.

16. Roosevelt, *Presidential Press Conferences* 3, no. 122 (16 May 1934): 352–53, 5, no. 192 (20 March 1935): 169–70; "Taking the Profits out of War," *Christian Century*, 1 May 1935, 566; *Washington Times*, 12 December 1934; John Edward Wiltz, *In Search of Peace: The Senate Munitions Committee, 1934–36* (Baton Rouge: Louisiana State University Press, 1963), 34, 119–22, 140.

17. On 6 March 1935, Roosevelt said that his war profits committee had made a verbal report and placed some material in the State Department for presidential use. Roosevelt, *Presidential Press Conferences*, vol. 5, no. 188 (6 March 1935), 149; vol. 5, no. 192 (20 March 1935), 169.

18. Roosevelt, *Papers and Addresses of Franklin D. Roosevelt*, vol. 4, *The Court Disapproves, 1935*, 191; Davis R. B. Ross, *Preparing for Ulysses: Politics and Veterans during World War II* (New York: Columbia University Press, 1969), 19.

19. Roosevelt, *Papers and Addresses of Franklin D. Roosevelt*, 4:193.

20. Franklin Roosevelt may also have had personal reasons that led him to oppose war profiteering. Roosevelt was a careful student of his family history, and the charge of war profiteering blemished his family's reputation. Roosevelt was a very wealthy man, deriving most of his fortune from a huge estate assembled by his maternal grandfather. Although ruined financially by the economic depression of 1857, Grandfather Warren Delano recouped his losses after the outbreak of the Civil War. He did so by selling opium—the most widely prescribed analgesic of the war—from his post in China. Delano's profits amounted to about $1 million at a time when a person who earned $1,000 in a year could consider himself quite prosperous and when a private in the Union army earned $11 a month. By contrast, FDR's great-uncle, William Henry Aspinwall, who supplied arms to the Union cause, voluntarily returned a check for $25,000 to the U.S. government as his share of the profits on a major sale. During World War II, the muckraking journalist Westbrook Pegler disclosed the source of Grandfather Delano's wealth. Although this revelation bothered Eleanor Roosevelt and FDR's uncle, there is no evidence that the aspersion troubled the president, but he was well aware of it. See Daniel R. Fusfeld, *The Economic Thought of Franklin D. Roosevelt and the Origins of the New Deal* (New York: Columbia University Press, 1956), 9; Freidel, *Roosevelt: The Apprenticeship*, 13; Kenneth S. Davis, *FDR: The Beckoning of Destiny, 1882–1928* (New York: Putnam, 1972), 42; Geoffrey C. Ward, *Before the Trumpet: Young Franklin D. Roosevelt* (New York: Harper, 1985), 88n, 91n; James G. Randall and David Donald, *The Civil War and Reconstruction* (2d ed.; Boston: Heath, 1961), 487. On Roosevelt's knowledge of his grandfather's involvement in the China trade, see William D. Hassett, *Off the Record with FDR, 1942–1945* (New Brunswick, N.J.: Rutgers University Press, 1958), 101–2. Warren Delano's reputation is defended in Daniel W. Delano Jr., *Franklin Roosevelt and the Delano Influence* (Pittsburgh: Nudi, 1946), 163. There is evidence that FDR's great-grandfather, Isaac Roosevelt, prospered greatly during the American Revolution, but there is no evidence that FDR was aware of or bothered by this circumstance. See Davis, *FDR: Beckoning of Destiny*, 19, and Bernard Mason, "Entrepreneurial Activity in New York during the American Revolution," *Business History Review* 40 (summer 1966): 190n.

21. Samuel I. Rosenman, ed., *The Public Papers and Addresses of Franklin D. Roosevelt* (New York: Harper, 1950), 7:67; Roosevelt, *Presidential Press Conferences* 15, no. 645 (21 May 1940): 356.

22. Rosenman, *Public Papers and Addresses of Roosevelt*, 9:236–38, 276, 299, 351–52, 419, 660, 671; 10:67, 99, 144–46, 177, 192, 285, 307; 11:13–17, 28–30, 219, 221, 367, 372–74, 399,

402; 12:19–20, 28, 67, 93, 159–60, 209–11; 13:28–28, 37, 473; Roosevelt, *Presidential Press Conferences* 15, no. 645-A (23 May 1940): 370, 17, no. 707 (7 January 1941): 40–41; Hassett, *Off the Record*, 26.

23. By January 1941, the proportion of Americans who thought that the Great War had not been a mistake had risen to 43 percent. George H. Gallup, *The Gallup Poll: Public Opinion, 1945–1971* (New York: Random House, 1972), 192–93, 253, 272.

24. Entries of 25 September and 3 October 1939, MD, vol. 2, pp. 309–10, 318–19; Roosevelt, *Presidential Press Conferences* 14, no. 585 (3 October 1939): 212–13; entries of 13 November 1938, 16 May 1939, and 12 March 1940, MPD, vol. 1, pp. 55, 99, vol. 2, p. 439.

25. Roland Stromberg, "American Business and the Approach of War," *Journal of Economic History* 13 (winter 1953): 73–75; Gerald T. White, *Billions for Defense: Government Financing by the Defense Plant Corporation during World War II* (University: University of Alabama Press, 1980), 45; Leff, *Limits of Symbolic Reform*, 158–60.

26. Transcript, meeting re closing agreements, 29 November 1939, MD, vol. 225, pp. 16–19, 27.

27. Morgenthau to Secretary of the Navy Charles Edison, 29 November 1939, MD, vol. 225, pp. 31–32, and Foley to Morgenthau, MD, vol. 246, p. 349.

28. Minutes, Treasury Department conference, 4 December 1939, MD, vol. 226, p. 182.

29. Although the sections of the Vinson-Trammell Act limiting profits on defense contracts did not apply to foreign orders, the provision that limited the rate of tax amortization was in effect for foreign orders. See transcripts, meetings of 29 November 1939, MD, vol. 225, pp. 16–17, 30 January 1940, MD, vol. 238, pp. 293–94, and 4 February 1940, MD, vol. 240, pp. 20–21, and Morgenthau to Roosevelt, 1 December 1939, MD, vol. 226, p. 113.

30. Morgenthau to Roosevelt, 19 April 1940, MD, vol. 255, p. 263, and transcripts, meeting of Morgenthau, Arthur Purvis et al., 12 April 1940, MD, vol. 254, p. 20, and meeting re Allied Purchasing, 26 May 1940, MD, vol. 276, p. 38; Kenneth S. Davis, *FDR: Into the Storm, 1937–1940* (New York: Random House, 1993), 405–6.

31. White, *Billions for Defense*, 7–8.

32. Samuel P. Huntington, *The Soldier and the State* (Cambridge: Harvard University Press, 1957), 338–39; Albert A. Blum, "Birth and Death of the M-Day Plan," in *American Civil-Military Relations*, ed. Harold Stein (University: University of Alabama Press, 1963), 66–70, 76–78; R. Elberton Smith, *The Army and Economic Mobilization* (Washington: U.S. Army, 1959), 98–105; Davis, *FDR: Into the Storm*, 404.

33. Blum, "Birth and Death of the M-Day Plan," 79–83; minutes, Treasury Department conference, 4 September 1939, MD, vol. 209, pp. 157, 162–63.

34. Diary entry, 30 December 1942, MPD, vol. 4, p. 1060; Blum, *V Was for Victory*, 118.

35. Blum, "Birth and Death of the M-Day Plan," 86–87; Dallek, *Roosevelt and Foreign Policy*, 255; Davis, *FDR: Into the Storm*, 552–53, 604.

36. Transcripts, group meeting, 22 May 1940, MD, vol. 265, p. 42, first meeting re excess profits, 1 July 1940, MD, vol. 278, pp. 98 (quote), 111–16, second meeting re excess profits tax, 1 July 1940, MD, vol. 278, p. 142, meeting re excess profits, 9 July 1940, MD, vol. 281, pp. 1–5, and memorandum for secretary's diary, 8 July 1940, MD, vol. 280, p. 333; entry of 14 August 1940, MPD, vol. 3, p. 634. Harold G. Vatter in his *U.S. Economy in World War II* (New York: Columbia University Press, 1985), 35, correctly compliments the NDAC on its many accomplishments, but somewhat overstates its achievements in controlling defense profits.

37. Roswell Magill, "Notes on the Excess Profits Tax," [1939], MD, vol. 213, pp. 268–73.

38. These included Jacob Viner of the University of Chicago, Roswell Magill of Columbia, Harold Groves of Wisconsin, and ablest of them all, Randolph Paul of Yale. In 1920 Viner had published a meticulous study of the economic management of the Great War aptly titled "Who Paid for the War?" and together his colleagues had published more than a dozen books on the theory of public finance. See Viner, "Who Paid for the War?" *Journal of Political Economy* 28 (January 1920): 46–76.

39. Transcript, meeting re excess profits tax, 8 July 1940, 10:15 A.M., MD, vol. 280, pp. 268–89, Morgenthau to Roosevelt, 31 July 1941, and Roosevelt to Rep. Robert L. Doughton,

31 July 1941, MD, vol. 426, pp. 141–45; Charles Gilbert, *American Financing of World War I* (Westport, Conn.: Greenwood Press, 1970), 93.

40. Transcripts, meeting re excess profits tax, 10:15 A.M., 8 July 1940, MD, vol. 280, p. 278, and group meeting re taxes, 7 May 1941, MD, vol. 395, p. 296; Gilbert, *American Financing of World War I,* 96.

41. Transcript, meeting re excess profits tax, 8 July 1940, MD, vol. 280, p. 272, Roosevelt to Morgenthau, 9 October 1940, MD, vol. 320, p. 305, and Dave H. Morris Jr. to Morgenthau, 31 October 1941, MD, vol. 456, pp. 202–7.

42. Transcript, meeting re taxes, 9 May 1941, MD, vol. 396, p. 130.

43. The quotation is by John L. Sullivan, assistant secretary of the treasury. Randolph Paul, the leading tax adviser, added, "I see no basis in my present thinking for differentiating between war profits arising out of contracts with the Government and any other profits that are attributable to the emergency." See transcript, meeting re excess profits, 8 July 1940, MD, vol. 280, p. 282.

44. W. Elliot Brownlee Jr., "Wilson and the Modern State: The Revenue Act of 1916," *Proceedings of the American Philosophical Society* 129, no. 2 (1985): 198–99, 204; Gilbert, *American Financing of World War I,* 96.

45. Transcripts, meetings re excess profits tax, 16 July 1940, 10 A.M., MD, vol. 282, p. 20, and 7 August 1940, 10:15 A.M., MD, vol. 290, p. 181; transcript, meeting re social security, 31 October 1941, MD, vol. 456, p. 128. Mark H. Leff, in *The Limits of Symbolic Reform,* 225, 263–75, argues that the Treasury staff was conservative and that in 1939 Morgenthau made the agency "a mouthpiece for the conservative position." Morgenthau's words and actions at other points serve to counter Leff's position.

46. Sullivan, "Comparison of Treasury Plan and Stam Plan for Excess Profits Tax," memorandum, 5 August 1940, MD, vol. 289, p. 76.

47. Transcript, telephone conversation, Roosevelt and Morgenthau, 4 June 1940, MPD, vol. 3, p. 569; Sullivan to Morgenthau, 25 July 1940, MD, vol. 287, pp. 89–90, transcript, group meeting, 5 August 1940, MD, vol. 289, p. 55.

48. Transcript, telephone conversation, Sullivan and Morgenthau, 26 August 1940, MD, vol. 296, pp. 13–14; D. W. Bell, acting secretary of the treasury, report on cabinet meeting, 23 August 1940, MD, vol. 295, p. 192, transcript, second meeting re taxes, 9 May 1941, MD, vol. 396, p. 227; Leff, *Limits of Symbolic Reform,* 206.

49. Sullivan to Morgenthau, 30 July 1941, MD, vol. 426, p. 18.

50. Roosevelt to Doughton, 31 July 1941, MD, vol. 426, p. 141.

51. Standard Statistics Company, *Excess Profits Tax* (1940), MD, vol. 321, p. 12; transcript, meeting re taxes, 20 February 1942, MD, vol. 499, p. 12; Leff, *Limits of Symbolic Reform,* 37.

52. Presidential speech writer Samuel I. Rosenman noted Roosevelt's great tolerance for persons who disagreed with him, but also observed that this tolerance had a limit. "When he [Roosevelt] felt that he was right in a political or economic view, and that the opposition was coming from people who were actuated by selfish economic or political considerations, he was bitter, vindictive, and unforgiving, and became personally unfriendly toward those who led the opposition." Rosenman named Walter George as an opponent whom Roosevelt despised. See Samuel I. Rosenman, *Working with Roosevelt* (New York: Harper, 1952), 468; Davis, *FDR: Into the Storm,* 278–80, 293; Rhoda D. Edwards, "The Seventy-eighth Congress on the Home Front: Domestic Economic Legislation, 1943–44" (Ph.D. diss., Duke University, 1966), 294.

53. U.S. House, *Universal Mobilization for War Purposes,* 188; "The Legion's Campaign for a Universal Draft Law," *American Legion Weekly,* 12 December 1924, 10; Wiltz, *In Search of Peace,* 46; transcript, meeting re taxes, 14 September 1942, MD, vol. 569, pp. 17, 20; entry of 16 May 1945, MPD, vol. 7, p. 1602; Edwards, "Seventy-eighth Congress," 294; Walter George, "Your $100 Billion Tax Bill," *American Magazine* 135, no. 3 (1943): 144.

54. In fact, the savings would have been considerably more. These figures are based on the assumption of a ceiling of 12 percent of invested capital, whereas the Treasury's re-

quest was for a ceiling of 10 percent. See Morris to Morgenthau, 31 October 1941, MD, vol. 456, p. 207, and Standard Statistics Co., *Excess Profits Taxes*, 6.

55. Transcript, meetings re excess profits tax, 11 September 1940, MD, vol. 305, p. 91, and 20 September 1940, MD, vol. 307, p. 235.

56. Mark H. Leff argues forcefully but not entirely convincingly that the redistributionalist policies of Franklin Roosevelt and the New Deal were never seriously intended to be effective. See *The Limits of Symbolic Reform*, passim, esp. 164–68.

57. See Tugwell, "The Compromising Roosevelt," reprinted in his *In Search of Roosevelt*, esp. 267; Perkins, *The Roosevelt I Knew* (New York: Viking, 1946), 380; Blum, *V Was for Victory*, 117–22; Kimball, *The Juggler*, 186–87.

58. Entries of 8 July, 25 September 1939, MPD, vol. 1, p. 170, vol. 2, pp. 309–10; minutes, Treasury Department conferences, 4 September and 29 November 1939, vol. 209, pp. 157, 162–63, vol. 225, p. 11; meeting of Morgenthau and Harry Dexter White, 27 September 1939, MD, vol. 214, p. 21. In 1938 congressional conservatives had won an important victory. They had succeeded in repealing the undivided profits tax, which was one of the New Deal's key weapons against economic royalism. See James T. Patterson, *Congressional Conservatism and the New Deal: The Growth of the Conservative Coalition in Congress, 1933–1939* (Lexington: University Press of Kentucky, 1967), 229–33.

59. Roosevelt to Morgenthau, 9 October 1940, MD, vol. 320, p. 305.

11. A Prescription for Profiteering

1. Comparative studies of the wartime economies of the United States, Great Britain, Germany, and Japan are Joe R. Feagin and Kelly Riddell, "The State, Capitalism, and World War II: The U.S. Case," *Armed Forces and Society* 17 (fall 1990): 53–79, and Geoffrey Mills and Hugh Rockoff, "Compliance with Price Controls in the United States and the United Kingdom during World War II," *Journal of Economic History* 47 (March 1987): 197–213.

2. Prominent officials of the Great War who were available to offer counsel included Bernard Baruch, chairman of the War Industries Board, other board members such as Hugh Johnson, General Robert E. Wood, the quartermaster general of the army, Walter S. Gifford, executive director of the Council of National Defense, and Navy Secretary Josephus Daniels. See Albert A. Blum, "Birth and Death of the M-Day Plan," in *American Civil-Military Relations*, ed. Harold Stein (University: University of Alabama Press, 1963), 75.

3. Before World War II, there was some trading with Japan, which was a *potential* enemy. In 1938–39 Seversky Aircraft created a dummy corporation to allow it to sell aircraft, sheet aluminum, and machine guns to the Japanese. See Jacob Vander Meulen, *The Politics of Aircraft: Building an American Military Industry* (Lawrence: University Press of Kansas, 1991), 189.

4. Kenneth S. Davis, *FDR: Into the Storm, 1937–40* (New York: Random House, 1993), 533, 552; Feagin and Riddell, "The State, Capitalism, and World War II," 66–68; Blum, "Birth and Death of the M-Day Plan," 85, 89; Nelson Lichtenstein, *Labor's War at Home: The CIO in World War II* (Cambridge: Cambridge University Press, 1982), 93–94; William D. Emerson, "Franklin D. Roosevelt as Commander-in-Chief in World War II," *Military Affairs* 22 (winter 1958/59): 184; John Braeman, "The New Deal and the `Broker State': A Review of Recent Scholarly Literature," *Business History Review* 46 (winter 1972): 427; James McGregor Burns, *Roosevelt: The Soldier of Freedom* (New York: Harcourt Brace, 1972), 256; Paul A. C. Koistinen, "Warfare and Power Relations in America: Mobilizing the World War II Economy," in *The Home Front and War in the Twentieth Century: The American Experience in Historical Perspective*, ed. James Titus (Washington: GPO, 1984), 99–101, 104, 242n. Roosevelt's wartime administration is defended in Harold G. Vatter, *The U.S. Economy in World War II* (New York: Columbia University Press, 1985), 80–83, Allen M. Winkler, *Home Front U.S.A.: America during World War II* (Arlington Heights, Ill.: Harlan Davidson, 1986),

90, Richard M. Ketchum, *The Borrowed Years, 1938–1941* (New York: Random House, 1989), 620–22, and in Richard Polenberg, *War and Society: The United States, 1941–1945* (Philadelphia: Lippincott, 1972), 7.

5. Transcript, telephone conversation, Treasury Secretary Henry Morgenthau Jr. and General George C. Marshall, 18 May 1940, Morgenthau Diaries, vol. 263, p. 321, Franklin D. Roosevelt Library, Hyde Park, N.Y. (Morganthau Diaries are cited hereafter as MD); John Morton Blum, *Roosevelt and Morgenthau* (Boston: Houghton Mifflin, 1970), 448.

6. He truly was going fishing. Roosevelt left that evening for a ten-day fishing trip on the north shore of Lake Huron in Canada. See entry of 30 July 1943, Morgenthau presidential diary, vol. 5, pp. 1249, 1275, Franklin D. Roosevelt Library, Hyde Park, N.Y. (cited hereafter as MPD); William D. Hassett, *Off the Record with FDR, 1942-1945* (New Brunswick, N.J.: Rutgers University Press, 1958), 194–95; Blum, *Roosevelt and Morgenthau,* 442–44; Warren F. Kimball, *The Juggler: Franklin Roosevelt as Wartime Statesman* (Princeton: Princeton University Press, 1991), 8; Koistinen, "Warfare and Power Relations," 242n.

7. Although Roosevelt had majored in economics as an undergraduate, when Rexford Tugwell joined the Roosevelt campaign in 1932, he found the future president very weak in economic understanding. Tugwell began a crash program to educate the candidate. John Maynard Keynes recorded a similar impression when he conferred with FDR in 1936. See Daniel R. Fusfeld, *The Economic Thought of Franklin D. Roosevelt and the Origins of the New Deal* (New York: Columbia University Press, 1956), 4, 33; Frank Freidel, *Franklin D. Roosevelt: The Triumph* (Boston: Little, Brown, 1956), 265. Cf. Davis, *FDR: Into the Storm,* 537; James T. Patterson, *Congressional Conservatism and the New Deal: The Growth of the Conservative Coalition in Congress* (Lexington: University Press of Kentucky, 1967), 58–59, 74; Mark H. Leff, *The Limits of Symbolic Reform: The New Deal and Taxation, 1933–1939* (Cambridge: Cambridge University Press, 1984), 220; Wilson D. Miscamble, "Thurman Arnold Goes to Washington: A Look at Antitrust Policy in the Later New Deal," *Business History Review* 56 (spring 1982): 15.

8. Entry of 13 November 1938, MPD, vol. 1, p. 55.

9. Entry of 16 December 1942, MD, vol. 597, p. 27.

10. Transcript, meeting re excess profits, 14 July 1941, MD, vol. 420, p. 326. For other examples, see transcripts, meeting re taxes, 15 May 1941, vol. 398, p. 127, meeting re excess profits, 14 July 1941, MD, vol. 420, p. 241; entry of 12 December 1944, MPD, vol. 6, p. 1469, and Blum, *Roosevelt and Morgenthau,* 442.

11. Leff, *Limits of Symbolic Reform,* 221.

12. Samuel I. Rosenman, *Working with Roosevelt* (New York: Harper, 1952), 468; entry of 25 September 1939, MPD, vol. 2, pp. 309–10; minutes, meeting of Morgenthau and Harry Dexter White, 27 September 1939, MD, vol. 214, p. 21, and meeting re tax statement, 7 August 1941, MD, vol. 429, p. 233; John Morton Blum, *From the Morgenthau Diaries: Years of Urgency, 1938–1941* (Boston: Houghton Mifflin, 1965), 310.

13. See chap. 10. Patterson, *Congressional Conservatism,* 58–59; Randolph Paul to FDR, 13 November 1939, MD, vol. 222, pp. 192–95, transcript, second meeting re excess profits, 1 July 1940, MD, vol. 278, p. 142, John Hanes, memorandum to the files, 9 November 1939, MD, vol. 22, p. 94, transcript, group meeting, 28 May 1940, 4:00 P.M., MD, vol. 277, p. 125.

14. Quoted in MPD, vol. 5, p. 1155. On Hopkins's role see Kimball, *The Juggler,* 9. During the spring of 1944, FDR told his personal aide, William D. Hassett, that he had been working only four hours a day and intended to continue. See Hassett, *Off the Record,* 241–44. Cf. Leff, *Limits of Symbolic Reform,* 221; Robert Dallek, "Franklin Roosevelt as World Leader," *American Historical Review* 76 (December 1971): 1506. Emerson, "Roosevelt as Commander-in-Chief," 192, shows that FDR's control of the military was incomplete. Writers who have claimed that Roosevelt was preoccupied with the war include Mark A. Stoler, "U.S. Civil-Military Relations in World War II," *Parameters* 21 (August 1991): 64–67; Donald H. Riddle, *The Truman Committee: A Study in Congressional Responsibility* (New Brunswick, N.J.: Rutgers University Press, 1964), 28; Frances Perkins, *The Roosevelt I Knew* (New York: Viking, 1946), 376; and Ketchum, *The Borrowed Years,* 621.

15. Davis, *FDR: Into the Storm*, 363; Winkler, *Home Front U.S.A.*, 78–80.

16. Rhoda D. Edwards, "The Seventy-eighth Congress on the Home Front: Domestic Economic Legislation" (Ph.D. diss., Duke University, 1966), 4, 11; Robert E. Ficken, "Political Leadership in Wartime: Franklin D. Roosevelt and the Elections of 1942," *Mid-America* 57 (January 1975): 33–34; Patterson, *Congressional Conservatism*, 214; Wayne S. Cole, *Roosevelt and the Isolationists, 1938–1945* (Lincoln: University of Nebraska Press, 1983), 293–94, 404, 538–39; Leff, *Limits of Symbolic Reform*, 265; Blum, *Roosevelt and Morgenthau*, 451–52; Blum, *V Was for Victory*, 231–32; Rosenman, *Working with Roosevelt*, 427; Mary Hedge Hinchey, "The Frustration of the New Deal Revival, 1944–46" (Ph.D. diss., University of Missouri, 1965), 25; Winkler, *Home Front U.S.A.*, 79–85.

17. Blum, *V Was for Victory*, 228–30, 241–43; Geoffrey Mills and Hugh Rockoff, "Compliance with Price Controls in the United States and the United Kingdom during World War II," *Journal of Economic History* 47 (March 1987): 201; Polenberg, *War and Society*, 28. Congress also took the lead in accelerating the rate of surplus property disposal, which proved to be a windfall for business. Davis R. B. Ross, *Preparing for Ulysses: Politics and Veterans during World War II* (New York: Columbia University Press, 1969), 91, 201–2.

18. Leff, *Limits of Symbolic Reform*, esp. 292; U.S. House Committee on Military Affairs, 74th Cong., 1st sess., 1935, *To Prevent Profiteering in War*, 56–57.

19. *New York Times*, 4, 8, and 9 September 1942; "Notes on Cabinet Meeting," 22 May 1942, MPD, vol. 5, pp. 1128–29; "Statement of Secretary Morgenthau before the Joint Committee on Internal Revenue Taxation," 28 May 1942, MD, vol. 533, p. 160, Morgenthau to Philip Murray, 21 March 1942, MD, vol. 510, p. 98, transcript, meeting of 2 June 1942, MD, vol. 530, pp. 130–32; Lichtenstein, *Labor's War at Home*, 99; Hassett, *Off the Record*, 66.

20. John L. Sullivan to Morgenthau, 5 November 1942, MD, vol. 581, p. 215.

21. Entry of 3 December 1942, MPD, vol. 5, p. 1202; Paul to Morgenthau, 11 January 1943, MD, vol. 603, p. 9; *New York Times*, 7 and 8 September 1942; Leff, *Limits of Symbolic Reform*, 291; Edwards, "Seventy-eighth Congress," 38, 106.

22. F. W. Paish, "Economic Incentive in Wartime," *Economics* 8 (August 1941): 239–48; transcript, meeting of 4 September 1940, re British Purchasing Program, MD, vol. 302, p. 289, and vol. 482, p. 75.

23. Transcript, meeting re inflation, 13 April 1942, MD, vol. 516, p. 11.

24. White to Morgenthau, 13 October 1941, MD, vol. 456, pp. 27–31.

25. Transcript, meeting re inflation, 10 April 1942, MD, vol. 515, p. 90.

26. R. Elberton Smith, *The Army and Economic Mobilization* (Washington: U.S. Army, 1959), 220–25, 300; Morgenthau to Roosevelt, 23 May 1945, MPD, vol. 7, p. 1563; telephone conversation, Morgenthau and Donald Nelson, 30 June 1942, MD, vol. 544, pp. 198–99.

27. Hugh Rockoff, *Drastic Measures: A History of Wage and Price Controls in the United States* (Cambridge: Cambridge University Press, 1984), 164–65; Harvey C. Mansfield et al., *A Short History of OPA* (Washington: Office of Price Administration, 1947), 271; Mills and Rockoff, "Compliance with Price Controls," 210.

28. Riddle, *Truman Committee*, 137–38.

29. Robert J. Benes, *Studies in Industrial Price Control*, pt. 1: *Iron and Steel Scrap*, ed. Harvey C. Mansfield (Washington: Office of Price Administration, 1947), passim.

30. Polenberg, *War and Society*, 32.

31. Mills and Rockoff, "Compliance with Price Controls," 201ff.; Winkler, *Home Front U.S.A.*, 40, 149; Paul M. O'Leary, "Wartime Rationing and Governmental Organization," *American Political Science Review* 39 (December 1945): 1103; Richard R. Lingeman, *Don't You Know There's a War On? The American Home Front, 1941–1945* (New York: Putnam, 1970), 243.

32. Virgil B. Zimmermann, *Problems in Price Control: National Office Organization and Management* (Washington: Office of Price Administration, 1947), 123.

33. Judith Russell and Renee Fantin, *Studies in Food Rationing* (Washington: Office of Price Controls, 1947), 330; Rockoff, *Drastic Measures*, 149.

34. O'Leary, "Wartime Rationing," 1104; James A. Maxwell and Margaret N. Balcon, "Gasoline Rationing in the United States," *Quarterly Journal of Economics* 56 (November

1946): 134, 154. Cf. Roy Hoopes, *Americans Remember the Home Front* (New York: Hawthorn, 1973), 89, and Lingeman, *Don't You Know There's a War On?* 241–43.

35. Emmette S. Radford, *Field Administration of Wartime Rationing* (Washington: Office of Price Administration, 1947), 75–82; Mansfield et al., *Short History of OPA,* 271–72.

36. E. H. Foley Jr. to Morgenthau, 6 October 1941, MD, vol. 448, pp. 65–68.

37. Perkins, *Roosevelt I Knew,* 374–75; Smith, *Army and Economic Mobilization,* 69.

38. Smith, *Army and Economic Mobilization,* 437–38.

39. Ibid., 474. See also Gerald T. White, *Billions for Defense: Government Financing by the Defense Plant Corporation during World War II* (University: University of Alabama Press, 1980), 9.

40. Transcripts, meeting re excess profits tax, group meeting, 6 September 1940, MD, vol. 303, pp. 153–83, 226–31.

41. White, *Billions for Defense,* 31–33; Smith, *Army and Economic Mobilization,* 476–502; White, "Financing Industrial Expansion for War," 173.

42. White, *Billions for Defense,* 80, 90; White, "Financing Industrial Expansion for War," 158; Frederick J. Dobney, "The Evolution of a Reconversion Policy: World War II and Surplus Property Disposal," *Historian* 36 (May 1974): 506–7; Rosenman, *Working with Roosevelt,* 393. A recent study asserts that the federal government owned approximately 40 percent of the nation's capital assets, but this figure is based on an older report and is probably erroneous. See Gregory Hooks and Leonard E. Bloomquist, "The Legacy of World War II for Regional Growth and Decline: The Cumulative Effects of Wartime Investments on U.S. Manufacturing, 1947–1972," *Social Forces* 71 (December 1992): 305.

43. The argument that World War II destroyed an opportunity to move toward socialism continues to find scholarly support. See Gregory Hooks, "The United States of America: The Second World War and the Retreat from New Deal Era Corporatism," in *Organising Business for War: Corporatist Economic Organisation and the Second World War,* ed. Wyn Grant, Jan Nekkers, and Frans van Waarden (New York: Berg, 1991), 93–99. See also White, *Billions for Defense,* 91; Max Lerner, *Public Journal: Marginal Notes on Wartime America* (New York: Viking, 1945), 366–69; transcript, group meeting, 8 May 1940, MD, vol. 260, pp. 207–8, and Sullivan to Morgenthau, 12 December 1942, MD, vol. 597, p. 43.

44. White, *Billions for Defense,* 91–92; Barton J. Bernstein, "The Removal of War Production Board Controls on Business, 1944–46," *Business History Review* 39 (summer 1965): 259–60. Certain business executives preferred a slower rate of reconversion, as did some military officers. See Feagin and Riddell, "The State, Capitalism, and World War II," 69; Lichtenstein, *Labor's War at Home,* 205; and Polenberg, *War and Society,* 229–30.

45. White, *Billions for Defense,* 123; George W. Steinmeyer, "Disposal of Surplus War Property: An Administrative History" (Ph.D. diss., University of Oklahoma, 1969), 136, 153, 175–76, 197, 204.

46. Transcript, conference re closing agreements, 1 March 1940, MD, vol. 244, pp. 17–18. Kenneth E. Boulding argues that the reconversion program was not too rapid and that in general it was most successful. See Boulding, "The Deadly Industry: War and the International System," in *Peace and the War Industry* (Chicago: Aldine, 1970), 3–4.

47. James W. Fesler et al., *Industrial Mobilization for War: History of the War Production Board and Predecessor Agencies, 1940–1945* (Washington: War Production Board, 1947), 656–57; White, *Billions for Defense,* 130–31.

48. White, *Billions for Defense,* 131; U.S. Office of Price Administration, *Corporate Profits, 1936–1946: Wartime vs. Peacetime Earnings, Part II,* War Profits Study no. 17 (Washington, 1947), 17. See also a similar analysis in Feagin and Riddell, "The State, Capitalism, and World War II," 74–75.

49. U.S. Office of Price Administration, *Wartime vs. Peacetime Earnings,* 72–76; Smith, *Army and Economic Mobilization,* 392; Robert E. Kline, "Renegotiation of War Contracts," lecture to the War Law Institute, National University School of Law, Washington, D.C., 19 July 1943, in Leon Henderson Papers, box 33, National Archives.

50. U.S. Office of Price Administration, *Wartime vs. Peacetime Earnings*, 12, 24, 32, 36, 48, 52.

51. Vatter concludes that "the corporate pecuniary sacrifice was modest." Since the after-tax increase from 1940 to 1943 was 77 percent, it would appear that he somewhat understates the case. Vatter, *U.S. Economy in World War II*, 56.

52. U.S. Office of Price Administration, *Wartime vs. Peacetime Earnings*, 48–51.

53. Transcript, meeting re taxes, 28 May 1942, MD, vol. 533, pp. 3–21, "Statement of Secretary Morgenthau before the Joint Committee on Internal Revenue Taxation," 28 May 1942, MD, vol. 533, pp. 158–61.

54. Wisconsin was selected as the source of the sample because it was heavily engaged in war production and because corporate income tax records retained in the Wisconsin state archives include key data about the economic well-being of corporate executives. The incompleteness of some of the tax returns restricted the sample somewhat, but for the companies selected the data were nearly complete.

55. The Wisconsin Department of Revenue does not allow the publication of data in any way that would identify specific individuals or corporations. The records specify the salary of corporate officers before personal income taxes were assessed. They do not state the income of executives after taxes. See Hoopes, *Americans Remember the Home Front*, 83–85, 87, and Lingeman, *Don't You Know There's a War On?* 107–31, for the representativeness of the various industries.

56. A recent study of Wisconsin corporations during World War II did not support the contention that smaller businesses were unpatriotic and tried to circumvent controls in an effort to increase short run profits. The study considered only two firms, however. See Richard H. Keene and Gene Smiley, "Small Business Reaction to World War II Government Controls," in *Essays in Economic and Business History* (East Lansing: Michigan State University Press, 1990), 312.

12. War Profits and Cold War Culture

1. George H. Gallup, *The Gallup Poll: Public Opinion, 1945–1971* (New York: Random House, 1972), 189, 679

2. Charles Gilbert, *American Financing of World War I* (Westport, Conn.: Greenwood Press, 1970), 226–30.

3. Gallup, *Gallup Poll*, 331, 374, 495.

4. Ibid., 270–71.

5. Jeffrey G. Williamson and Peter H. Lindert, *American Inequality: A Macroeconomic History* (New York: Academic Press, 1980), 84, 315–16; Carole Shammas, "A New Look at Long-Term Trends in Wealth Inequality in the United States," *American Historical Review* 98 (April 1993): 425–27; Neil A. Wynn, "War and Social Change: The Black American in Two World Wars," in *War and Society: A Yearbook of Military History*, ed. Brian Bond and Ian Roy (London: Croom Helm, 1977), 2:54; Allan M. Carter, "Income Shares of Upper Income Groups in Great Britain and the United States," *American Economic Review* 44 (December 1954): 877, 883; Selma F. Goldsmith, "Changes in the Size Distribution of Income," *American Economic Review* 47 (May 1957): 507, 509–10, 517–18; Barry Chiswick and Jacob Mincer, "Time Series Changes in Personal Income Inequality in the United States from 1939, with Projections to 1985," *Journal of Political Economy* 80 (May/June 1972): 534. A dissenting view is in Paul A. C. Koistinen, "Warfare and Power Relations in America: Mobilizing the World War II Economy," in *The Home Front and War in the Twentieth Century: The American Experience in Historical Perspective*, ed. James Titus (Washington: GPO, 1984), 234–35.

6. Robert Higgs, "Wartime Prosperity? A Reassessment of the U.S. Economy in the 1940s," *Journal of Economic History* 52 (March 1992): 41ff.; Paul Fussell, *Wartime: Understanding and Behavior in the Second World War* (New York: Oxford University Press, 1989), 195–207; Gallup, *Gallup Poll*, 488.

7. George A. Christy, "Risk and Opportunity in the Stock Market: A Quasi-Current Appraisal," *Southwestern Social Science Quarterly* 47 (December 1966): 284.

8. Many doughboys were paid in French francs. They deposited the money in French banks, which refused to allow the funds to be removed from the country. In 1919 France devalued the franc, and the soldiers lost their wages. See Walter Rundell Jr., *Black Market Money: The Collapse of U.S. Military Currency Control in World War II* (Baton Rouge: Louisiana State University Press, 1964), 60.

9. Lee Kennett, *G.I.: The American Soldier in World War II* (New York: Scribner, 1987), 95–99.

10. Davis R. B. Ross, *Preparing for Ulysses: Politics and Veterans during World War II* (New York: Columbia University Press, 1969), 124, 203, 215.

11. Fussell, *Wartime,* 171–78, quote 174; Studs Terkel, *"The Good War": An Oral History of World War Two* (New York: Pantheon, 1984). Terkel's phrase has been recast and revised by Michael C. C. Adams in *The Best War Ever: America and World War II* (Baltimore: Johns Hopkins University Press, 1994), 2–4, 139–55. See also Paul D. Casdorph, *Let the Good Times Roll: Life at Home in America during World War II* (New York: Paragon, 1989).

12. Published in New York by Harcourt, Brace (1948). Catton's breezy book assailed the predominance of businessmen in war mobilization. Other histories of the home front during World War II have tended to be brief, e.g., Richard Polenberg, *War and Society: The United States, 1941–1945* (Philadelphia: Lippincott, 1972); Allen M. Winkler, *Home Front U.S.A.: America during World War II* (Arlington Heights, Ill.: Harlan Davidson, 1986); impressionistic, e.g., Richard R. Lingeman, *Don't You Know There's a War On? The American Home Front, 1941–1945* (New York: Putnam, 1970), John Morton Blum, *V Was for Victory: Politics and American Culture during World War II* (New York: Harcourt Brace, 1976); Fussell, *Wartime;* or written from the governmental perspective, e.g., R. Elberton Smith, *The Army and Economic Mobilization* (Washington: U.S. Army, 1959).

13. The film was based on MacKinlay Kantor's minor novel, *Glory for Me* (New York: Coward, McCann, 1945).

14. Lillian Hellman, *The Collected Plays* (Boston: Little, Brown, 1972), 402–7.

15. Arthur Miller, *All My Sons* (New York: Dramatists Play Service, 1947), quote 67.

16. Joseph Heller, *Catch-22* (New York: Simon and Schuster, 1955), 249–51.

17. Kurt Vonnegut Jr., *God Bless You, Mr. Rosewater; or, Pearls before Swine* (New York: Holt, Rinehart, 1965), 19–21; Peter Aichinger, *The American Soldier in Fiction, 1880–1963: A History of Attitudes toward Warfare and the Military Establishment* (Ames: Iowa State University Press, 1975), 81, 84; Emily Stipes Watts, *The Businessman in American Literature* (Athens: University of Georgia Press, 1982), 129.

18. Henry Morford, *The Days of Shoddy: A Novel of the Great Rebellion of 1861* (Philadelphia: Peterson, 1863); Wayne Charles Miller, *An Armed America, Its Face in Fiction: A History of the American Military Novel* (New York: New York University Press, 1970), 85.

19. Robert Endicott Osgood, *Ideals and Self-Interest in America's Foreign Relations: The Great Transformation of the Twentieth Century* (Chicago: University of Chicago Press, 1953), esp. 383–87; Huntington, *Soldier and the State,* 458–59.

20. Warren F. Kimball, *The Juggler: Franklin Roosevelt as Wartime Statesman* (Princeton: Princeton University Press, 1991), esp. 13, offers the views of a New Realist.

21. *Henry V,* 3.3.22. Paul A. Jorgensen, *Shakespeare's Military World* (Berkeley: University of California Press, 1956, 1973), 152.

22. George J. Stigler and Claire Friedland, "Profits of Defense Contractors," *American Economic Review* 56 (September 1971): 692–94; Douglas R. Bohi, "Profit Performance in the Defense Industry," *Journal of Political Economy* 81 (May/June 1973): 728.

23. Kenneth D. Alford, *The Spoils of World War II: The American Military Role in the Stealing of Europe's Treasures* (New York: Carol, 1994), esp. 278.

Bibliographical Note

The sources used in this study include the records of numerous governmental investigations, both published and in manuscript, the personal papers of critics of war profits, and an extensive selection of contemporary and recent books, articles, and pamphlets. Each source used in this study is cited fully in the first reference in each chapter in which it is used.

Books

No previous book that has come to my attention deals expressly with the topic presently considered. Nevertheless, the research and insights of three generations of scholars who have examined various aspects of the subject have made substantial contributions to this work.

War profits emerged as a scholarly issue in America when the muckraking journalist Gustavus Myers published his seminal *History of the Great American Fortunes* (1909). Myers intended to smudge the luster of the Civil War, and he accomplished this by attributing the origin of J. P. Morgan's immense fortune to profits gained during the Civil War. The Myers thesis was shattered by R. Gordon Wasson in *The Hall Carbine Affair: A Study in Contemporary Folklore* (1941, 1948, 1971).

It was generally not until the issue of war causation arose after World War I, however, that war profits inspired the earnest attention of professional historians. The most comprehensive early treatments of what became known as "profiteering" were by European scholars, Richard Lewinsohn's *The Profits of War through the Ages* (1936) and Philip Noel-Baker's *The Private Manufacture of Armaments* (1936). A variety of texts generally seeking to link profit seeking with warmongering and other abuses found a ready market in the 1930s. A selection of titles in this lurid genre includes Seymour Waldman, *Death and Profits* (1932); Fenner Brockway, *The Bloody Traffic* (1933); Helmuth C. Engelbrecht and Frank C. Hanighen, *merchants-of-death* (1934); George Seldes, *Iron, Blood, and Profits* (1934); and Stephen and Joan Raushenbush, *War Madness* (1937).

The merchants-of-death literature of the early 1930s was vintage Progressive history. Although most authors were well-trained scholars, the books were alarmist, carried Marxist overtones, and sought to proselytize. After World War II, this scholarly disposition evolved into a more careful study of civil-military relations that focused on a search for the origins and characteristics of what became commonly known as "the military-industrial complex" (often abbreviated as MIC). Unlike the nefarious arms traders, the military-industrial complex was not explicitly condemned as a bringer of war, but was indicted on lesser charges as a bloated wastrel and a disturber of the peace. The concept originated with C. Wright Mills's influential and controversial book, *The Power Elite* (1956), and it gained great respectability when Dwight D. Eisenhower, a professional soldier with impeccable credentials, included it in his 1961 presidential valedictory address.

The concept of a military-industrial complex, like the merchants-of-death theory, was heavily freighted with ideology. In some degree, most of the writing on war profits during the cold war had the flavor of an exposé. This is evident from some of the titles of books devoted to the subject, which rivaled their ancestors in grisliness. The genre includes Fred J. Cook, *The Warfare State* (1964), Richard J. Barnet, *The Economy of Death* (1969), Richard F. Kaufman, *The War Profiteers* (1972), and Ivan Melada, *Guns for Sale* (1983), to cite a few. These books seem intended more to disturb than to inform, and a reader can become numb to the message.

Scholars who have written about the military-industrial complex have been considerably more restrained than were the writers of the merchants-of-death school, but an echo of the earlier outrage can still be heard. Benjamin F. Cooling, who searched diligently for the origin of the military-industrial complex in *Gray Steel and Blue Water Navy* (1979), reported that "the process was certainly not complete by 1893, but the first tentacles of a modern military-industrial complex were there." Paul A. C. Koistinen, one of the most resolute scholars who have probed the topic, asserted in *The Military-Industrial Complex* (1980) that "this MIC has helped to perpetuate Cold War tensions." Much of the military-industrial complex literature discloses that progressive history, although toned down, has not entirely disappeared. A more recent book by Gregory Hooks, *Forging the Military-Industrial Complex* (1991), while thoroughly researched and clothed in scholarly objectivity, still blamed the MIC for "postwar conservatism" and "the New Deal's failure to culminate in a social democratic breakthrough."

There is a third school of thought. In the 1930s the merchants-of-death theory was broadly accepted, but not unanimously. Its foremost critic was Norman Angell, a prolific British pacifist who argued in *The Unseen Assassins* (1932) that nationalist impulses were far more important as war causatives than profit seeking. In a classic study, *The Soldier and the State* (1957), Samuel P. Huntington pointed to the strong tradition of business pacifism and noted the irony of portraying businessmen as warmongers. Huntington regarded the alleged dangers of military-business cooperation as exaggerated and un-proven, and in any case he expected this cooperation to be both limited and temporary.

By 1972 the idea of a military-industrial complex had attracted numerous other critics. Robert H. Ferrell wrote in the *Journal of International Affairs* (1972) that the concept was so vague that it "could not stand by itself," and Carroll W. Pursell Jr. complained in *The Military-Industrial Complex* (1972) that it has no "theoretical framework."

The present study has benefited vastly from the work of those scholars who have questioned the utility of the notion of a military-industrial complex. Besides Huntington, Ferrell, and Pursell, the civil-military relations school—a variant of the "New Military History"—includes James L. Abrahamson (*America Arms for a New Century* [1981]), Dean Allard ("The Influence of the United States Navy upon the American Steel Industry" [1959]), Robert D. Cuff ("An Organizational Perspective on the Military-Industrial Complex" [1984]), Terrence J. Gough ("Soldiers, Businessmen, and U.S. Industrial Mobilisation Planning between the World Wars" [1991]), Stephen Skowronek (*Building a New American State* [1981]), Jacob Goodwin (*Brotherhood of Arms* [1985]), and Jacob Vander Meulen (*The Politics of Aircraft* [1991]). Rather than seeking to discredit the MIC, the New Military History concentrates less opprobriously on understanding civil-military relations.

Governmental Records

While this study is thus indebted to the insights of three bodies of historical interpretation—the merchants-of-death school, the military-industrial com-plex school, and the civil-military relations school—its toil was greatly eased by the existence of a wide variety of published sources and manuscript collec-tions relating to the subject. For the early period, there is now available a rich selection of published records of the various colonial governments. The colo-nial records of Pennsylvania, Connecticut, Massachusetts Bay, and Rhode Island were particularly helpful, as was Great Britain's *Calendar of State Papers, Colonial Series*. For the Revolution, the papers of George Washington, Nathanael Greene, Robert Morris, and John Sullivan were indispensable.

For the early national and Jacksonian periods there is much to be gained from careful inspection of *Niles' Weekly Register* and *Niles' National Register*. This is also the period when very useful governmental reports and congres-sional investigations become available. In particular, these included the reports of the commissary general of subsistence (1847), of the quartermaster general (1847-48), and of the secretary of war (1848). Investigations by both Congress and state legislatures proliferated during the Civil War and were indispensable to this study. The most useful were those by New York and Ohio, by the House committee on government contracts, and by Senate committees on ordnance contracts, heavy ordnance, traffic with rebels, naval supplies, and the Banks expedition. The records of the joint committee on the conduct of the war were also of considerable help.

The post–Civil War controversy can be traced in the Senate's *Sale of Arms by Ordnance Department* (1872) and the House's *Sale of Arms to France* (1871). The naval rearmament controversy of the 1890s is discussed usefully in Robert Hessen, *Steel Titan: The Life of Charles M. Schwab* (1975) and in Cooling, *Gray Steel and Blue Water Navy* (1979). Pacifist views of the late nineteenth and early twentieth centuries are best developed in the papers of David Starr Jordan and the Women's International League of Peace and Freedom.

Opposition to preparedness is well described in various issues of *La Follette's Weekly Magazine* (1912-14) and in Anne Trotter's articles and dissertation. The controversy over supplies for World War I is best addressed by examination of congressional investigations by the Thomas, Graham, and Nye committees and by the House Committee on Military Affairs. The published records of the Nye committee should be supplemented by inspection of its archives. Also invaluable are records of the war transactions section of the Justice Department (1922-24) and of the War Policies Commission (1931). The scholarship of Robert D. Cuff is particularly rich and insightful.

The hearings of the War Policies Commission and of the Nye committee are important for the 1930s, but they should be supplemented by the opposing views expressed in the journal *Army Ordnance.* Liberal journalists were very active in this period, and their views are best examined in such journals as the *Nation, Literary Digest, Christian Century, New Republic,* and *World Tomorrow.*

For World War II, the diaries of Henry Morgenthau Jr. are both voluminous and indispensable for understanding the attempt to control wartime profits. Reports of the Office of Price Administration are also thorough and pertinent to the same problem. The records of the Wisconsin Department of Revenue supplied valuable inside data, but unfortunately they have now been closed to researchers. John Morton Blum's *V Was for Victory* (1976) remains the most useful treatment of the World War II home front.

INDEX